Dairy and Food Product Technology

The Editors

Birendra Kumar Mishra is working as Assistant Professor in the Department of Rural Development and Agricultural Production at North-Eastern Hill University, Tura campus, Meghalaya. He received a Ph.D. degree in Animal husbandry and Dairying (Dairy Technology) in 2003 from Institute of Agricultural sciences, BHU, Varanasi. He completed his Master Degree in Dairy Science from C.S.A. Agriculture University, Kanpur in1995 with gold medal for obtaining first class and position. He obtained his B.Sc. (Ag) in 1991 with first class and first position from Gorakhpur University, Gorakhpur. Dr. Mishra has published more than two dozen research papers and articles in referred journals, twenty book chapters and also published seven books titled-Livestock Production and Rural Development in India, Sustainability and Economic Development in Hill Agriculture, Livestock Production and Management, Advances in Livestock Production and Management, Dairy and Food Processing Industry-Recent Trends in two volumes and Dairy Product Technology. He served private dairy industry in various capacities as Quality control Manager and Technical officer etc. during 1995 to 1999.He is also a member and life member of various professional bodies and societies at national and international level. He is the recipients of Young Scientist Award 2010 from BIOVED research society, Allahabad, U.P, Founder Fellow Award 2011, from Indian Society of Hill Agriculture for his outstanding contribution in the field of Rural Development and Livestock Production and BIOVED Fellowship award 2012 from BIOVED research society, Allahabad, U.P. He is also guiding the research scholars at Ph.D. level. Dr. Mishra is handling externally funded research projects in the field of food and nutrition.

Subrota Hati is working as an Assistant Professor in Department of Dairy Microbiology, SMC College of Dairy Science, Anand Agricultural University, Anand, Gujarat. He graduated from West Bengal University of Animal and Fishery Sciences in Dairy Technology in 2006. He did his M.Sc in Dairy Microbiology from National Dairy Research Institute, Karnal. He awarded Doctoral degree in Dairy Microbiology from National Dairy Research Institute, Karnal in 2012. He is presently working on Isolation and characterization of milk derived bioactive peptides and its biofunctional properties. He also served Dairy Industry as Quality Assurance Executive in Mother Dairy, New Delhi. He is also handling externally funded projects by ICAR, DST and DBT as PI or Co-PI. He was the recipient of various awards: URMILABALA GOLD MEDAL by Indian Dairy Association, New Delhi; PROF. SUKUMER DE GOLD MEDAL as a Topper by WBUAFS, Kolkata; Two Best Paper awards by IDA, New Delhi; 03 Best Poster awards; Silver Medal by All Bengal Teachers' Association, Govt. West Bengal, Young Scientist Award by BIOVED, Allahabad; Best Young Scientist award by SASNET-FF, AAU, Anand and Lund University, Sweden and Best Oral Presentation Award by IIFANS, New Delhi etc. He is the recipient of Junior and Senior Research fellowship during his Master and Doctoral programmes in National Dairy Research Institute, Karnal, Haryana. Dr.Subrota is a member of various Scientific Societies: Life member of Dairy Technology Society of India, Karnal; Life membership of SASNET-Fermented Foods, Anand; Indian Dairy Associations, New Delhi. He was also acting as Editorial Board member of American Journal of food and Nutrition, Journal of Dairy and Food Technology and also reviewers of National and International Journals. He was recently awarded the Young scientist Project by DST, Govt. India, New Delhi on ACE inhibitory Bioactive Peptides derived from fermented milks. He has published 18 research papers, 05 Review Articles, 22 Technical articles in various National and International peer reviewed Journals and also published 16 Book Chapters and presented 35 abstracts in various National and International Seminars/Conferences.

Dairy and Food Product Technology

— Editors —

Birendra Kumar Mishra

Department of Rural Development and Agricultural Production
North-Eastern Hill University
Tura Campus-794002
Meghalaya (India)

Subrota Hati

Dairy Microbiology Department
SMC College of Dairy Science
Anand Agricultural University, Anand
Gujarat (India)

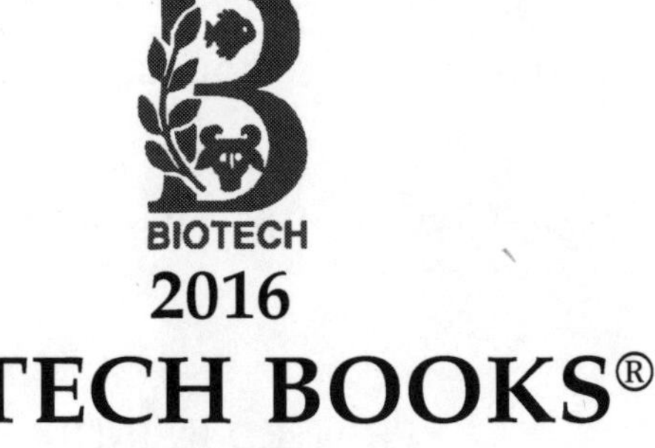

2016

BIOTECH BOOKS®

ISBN 978-81-7622-363-8

Published by: **BIOTECH BOOKS**®
4762-63/23, Ansari Road, Darya Ganj,
New Delhi - 110 002
Phone: +91-011-23262132
E-mail: biotechbooks@yahoo.co.in

Digitally Printed at : **Replika Press Pvt. Ltd.**

PRINTED IN INDIA

Foreword

It gives me great pleasure to write a Foreword to this book entitled ***Dairy and Food Product Technology***. The present publication is a collection articles on dairy and food product technology.

With the global populations set to rise from the present 7 billion to 8 billion by 2030 and 9 billion by 2050, there are considerable opportunities for Indian food producers to respond to this growing market. FAO has analyzed global dairy trends as far ahead as 2050. Their analysis predicts that as incomes increase people generally prefer to spend a higher share of their food budget on animal protein-dairy and meat consumption tends to grow faster than that of food crops. Expectedly, the past three decades have seen buoyant growth in the consumption of livestock products, especially in newly industrializing countries and emerging markets.

Rising population coupled with increasing urbanization, education and general awareness have stimulated a trend wherein consumers are not only conscious about taste but also about health, safety, attractiveness and convenience of foods. In future there will be increased demand for higher value added and product performance creating greater challenges for new product development. On domestic front, affluent middle class is growing considerably. This will be the target group for most of the dairy and food products, especially for higher value added products. Post WTO scenario provides India with an opportunity to market their products in other parts of the word, particularly to more than 150 million non-resident Indians settled all over the globe. Tremendous export potential, therefore, exists for unique traditional dairy and food products. The potential for producing, healthy functional foods incorporating valuable dairy ingredients in existing and new product formulations will also have to be exploited. This calls for dairy and food scientists and entrepreneurs to adopt a holistic approach to product development encompassing new dimensions of value addition, unfolding newer processing know how, international quality and safety standards as also global environmental practice.

In recent years, there have been significant advances in production and processing of dairy and food products, probably the most important part of the human diet. Improvements in process technology have been accompanied by massive changes in the scale of many food product operations, and the manufacture of a wide range of dairy and other related food products. The authors, who are all specialists in these products, have been chosen from around the country. There is no doubt that the book will have a scientific recognition by dairy and food scientists, students, researchers and dairy operatives, and will become an important component of the dairy and food products sector.

I extend my appreciation to Dr. B.K. Mishra and Dr. Subrato Hati for their endeavor in bringing together researchers in this area of study culminating in this volume. I look forward to it receiving a wide circulation among the educationists and policy makers.

Eugene D. Thomas

Pro-Vice-Chancellor,
North-Eastern Hill University,
Tura Campus, Chasingre
Meghalaya

Preface

India is currently the largest producer of milk in the world and this status has maintained since the late nineties. Milk production is estimated at 139.67 mt in 2013-14.It is the biggest agricultural produce contributing 22 per cent to India's agricultural GDP in the financial year 2013-14.Overall dairy industry is estimated at USD70 billion in 2013; expected to double to USD 140 billion by 2020.Over the last decade, the Indian dairy sector has grown at an average of 4.04 per cent against the world average of 2.2 per cent.The per capita milk availability in India stands at 296 grams per day, higher than the world average. This has been largely achieved through the combination of favorable policies and an institutional network that has helped support millions of rural households in pursuing their livelihoods through small scale dairy farming. About one fifth of the milk produced is collected and processed by the organized sectors. The dairy sector in our country is going through major changes with the liberalization policies of the Government and restructuring of the economy. This has brought greater participation of the private sector. This is also consistent with the global trends, which could hopefully lead to greater integration of Indian dairying with the world market for milk and milk products. India today is the words largest and fastest growing market for the milk and milk products with an annual growth rate of about 6.1 per cent.

India is witnessing winds of change because of improved milk availability, a changeover to market economy, globalization and entry of private sector in this sector. The value addition and variety in the availability of milk products are on every consumer's agenda. There is an increasing demand for new products and processes.

In this dynamic dairy business environment, the driving trends of Indian dairy industry will be changing demographic, rise in disposable income, increasing urbanization, life style, aspiration, increasing nuclear family with more working women, demand for functional and healthy foods, concern over food safety and increased on the- go-consumption. As a result of this, the dairy industry has to equip itself to meet more expenditure of food and subsequently on milk, rising demand of value added products, nutritious and healthy foods in convenient packing, milk being a complete diet will see rising consumption, shift from loose to branded and rise in demand of small portion packs. This will certainly necessitate timely research and development initiatives in dairy product manufacturing and new product development continuously for value addition and cost reduction in processing of dairy products

India is the world's second largest producer of food next to China, and has the potential of being the biggest with the food and agricultural sector. The food processing industry is one of the largest industries in our country, ranked fifth in terms of production, consumption, export and expected growth. The food industry is on a high as Indians continue to have a feast. Fuelled by what can be termed as a perfect ingredient for any industry -large disposable incomes - the food sector has been witnessing a marked change in consumption patterns, especially in terms of food.

Increasing incomes are always accompanied by a change in the food basket. The proportionate expenditure on cereals, pulses, edible oil, sugar, salt and spices declines as households climb the expenditure classes in urban India while the opposite happens in the case of milk and milk products, meat, egg and fish, fruits and beverages. Accounting 32 per cent of the country's total food market, the food processing industry is one of the largest industries in India and is ranked fifth in terms of production, consumption, export and expected growth. The total food production in India is likely to double in the next 10 years with the country's domestic food market estimated to reach US$ 258 billion by 2015. The food processing industry forms an important segment of the Indian economy in terms of contribution to GDP, employment and investment, and is a major driver in the country's growth in the near future. This industry contributes as much as 9-10 per cent of GDP in agriculture and manufacturing sector.

The Confederation of Indian Industry (CII) has estimated that the foods processing sectors has the potential of attracting US$ 33 billion of investment in 10 years and generate employment of 9 million person-days (Ministry of Food Processing, Govt. of India Annual Report,2014-15).

Food processing is a large sector that covers activities such as agriculture, horticulture, plantation, animal husbandry and fisheries. It also includes other industries that use agriculture inputs for manufacturing of edible products. The Ministry of Food Processing, Government of India indicates the following segments within the Food Processing industry:

1. Dairy
2. Fruits and vegetable processing
3. Grain processing
4. Meat and poultry processing
5. Fisheries

Consumer foods including packaged foods, beverages and packaged drinking water. Though the industry is large in size, it is still at a nascent stage in terms of development of the country's total agriculture and food produce, only 2 per cent is processed.

The Indian food industry stood around US$ 39.03 billion in 2013 and is expected to grow at a rate of 11 per cent to touch US$ 64.31 billion by 2018.

Indian agricultural and processed food exports during April-May 2014 stood at US$ 3,813.63 million, according to data released by the Agricultural and Processed Food Products Export Development Authority (APEDA).

The various competitive advantages in the food processing sector in India have been analyzed under the frame work given below:

Given the size of the industry and the nascent development stage, the food processing sector may be the key focus area for the Government of India. The importance of the sector is further enhanced by the fact that over 70 per cent of the population depends upon agricultural activity for livelihood. The government has therefore been focusing on commercialization and value addition to agricultural produce, minimizing pre/post-harvest wastage, generating employment and export growth in this sector, through a number of regulatory and fiscal incentives. The industry is largely unorganized, with a small but growing organized sector.

The popularity of food and agro products is not surprising when the sector is now offering a growth of more than 150 per cent in sales. With such promise in the sector, a number of foreign companies have joined the fray. While US brands such as McDonald's, Pizza Hut and Kentucky Fried Chicken have become household names, more are on their way.

The new wave in the food industry is not only about foreign companies arriving here attracted by the prospective size of the market. It is also about the migration of the Made in India tag on food products traveling abroad. Indian food brands and fast moving consumer goods (FMCGs) are now increasingly finding prime shelf-space in the retail chains of the US and Europe. These include Cobra Beer, Bikanervala Foods, MTR Foods' ready-to-eat food stuff, ITC's Kitchen of India and Satnam Overseas' Basmati rice.

The Government has formulated and implemented several schemes to provide financial assistance for setting up and modernizing of food processing units, creation of infrastructure, support for research and development and human resource development in addition to other promotional measures to encourage the growth of the processed food sector.

The Central government has permitted under the Income Tax Act a deduction of 100 per cent of profit for five years and 25 per cent of profit in the next five years in case of new agro processing industries set up to package and preserve fruits and vegetables.

Excise Duty of 16 per cent on dairy machinery has been fully waived off and excise duty on meat, poultry, and fish products has been reduced from 16 per cent to 8 per cent.

Most of the processed food items have been exempted from the purview of licensing under the Industries (Development and regulation) Act, 1951, except items reserved for small-scale sector and alcoholic beverages

Infrastructure Development in Food Processing Sector

There is a lack of suitable infrastructure in the shape of cold chain, packaging centers, value added center, modernized abattoirs etc. Improvement in general infrastructure is also an aid for energizing of sector. Government attaches highest priority to development and expansion of physical infrastructure for facilitating prompt growth of industries. In order to address the problem of infrastructure in food processing sector, the Government has implemented the scheme for infrastructure development comprising the following components:

Challenges and Opportunities

The future of the Indian farmer depends on the success of the food industry as India's prosperity is predominantly linked to the growth of incomes in the agrarian sector of the economy. Increasing liberalization of the economy has tried to lift the protection that the food and agriculture sector once enjoyed in the country. This has exposed the sector both to the opportunities and challenges of the global food economy.

The market forces are compelling the Indian agriculture producers to increase the quality of their farm produce while continuing to maintain their cost competitiveness in order to be able to compete effectively in the global food market. Even in the domestic market, rising per capita incomes and changing demographic profile of the population has ensured the growing demand for processed and convenience foods. Increasing consumer awareness about health and hygiene has shifted the focus of the market to "safe" foods. The Indian food-processing sector is undergoing a veritable revolution - all the way from the plate to the plough.

Indian food processing industry has seen significant growth and changes over the past few years, driven by changing trends in markets, consumer segments and regulations. These trends, such as changing demographics, growing population and rapid urbanization are expected to continue in the future and, therefore, will shape the demand for value added products and thus for food processing industry in India. The Government of India's focus towards food processing industry as a priority sector is expected to ensure policies to support investment in this sector and attract more FDI. India, having access to vast pool of natural resources and growing technical knowledge base, has strong comparative advantages over other nations in this industry. The food processing sector in India is clearly an attractive sector for investment and offers significant growth potential to investors.

Challenges Faced by the Indian Food Industry

The most crucial challenge today that the Indian food processing industry is facing is the lack of suitable infrastructure in the shape of cold chain, packaging centers, value added center, modernized abattoirs etc.

Improvement in general infrastructure is also a must requirement for the industry to progress. Some other important initiatives that are needed are:

1. Promotion of appropriate crossbreeds while conserving indigenous breeds of livestock Establishment of livestock marketing system

2. Promotion of rural backyard poultry in a cooperative marketing setup Development of cooperative dairy firms
3. Enhancing livestock extension services
4. Encouraging private veterinary clinic
5. Institutionalizing a framework for utilizing synergy between restoration and creation of water bodies for water harvesting and fishery
6. Provision of an insurance package to avoid distress

Strengths and Opportunities that India Enjoys

It is the seventh largest country, with extensive administrative structure and independent judiciary, a sound financial and infrastructural network and above all a stable and thriving democracy. Due to its diverse agro-climatic conditions, it has a wide-ranging and large raw material base suitable for food processing industries. Presently a very small percentage of these are processed into value added products. It is one of the biggest emerging markets, with over 1 billion population and a 250 million strong middle class. Rapid urbanization, increased literacy and rising per capita income, have all caused rapid growth and changes in demand patterns, leading to tremendous new opportunities for exploiting the large latent market. An average Indian spends about 50 per cent of household expenditure on food items. However, demand for processed/convenience food is constantly on the rise in our country.

India's comparatively cheaper workforce can be effectively utilized to setup large low cost production bases for domestic and export markets. Liberalized overall policy regime, with specific incentives for high priority food processing sector, provides a very conducive environment for investments and exports in the sector. Very good investment opportunities exist in many areas of food processing industries, the important ones being : fruit and vegetable processing, meat, fish and poultry processing, packaged, convenience food and drinks, milk products etc.

In the specific field of dairy and food product technology numerous important problems need solution keeping in view this book has been written. This book containing 18 chapters and all the authors were requested to contribute with regards to particular aspects related to advances in dairy and food product technology.

I am highly indebted to prof. E.D. Thomas, Pro-Vice-Chancellor, North-Eastern Hill University, Tura Campus, Tura, Meghalaya for not only writing the Foreword, but also for encouraging me in my all academic jobs at all times.

We wish to record a word of appreciation to *Mr. Sunil Mittal* of *Biotech Books*, Publisher, New Delhi to accept this manuscript and bringing out this book in a highly graceful touch

B.K. Mishra

Subrota Hati

Contents

List of Contributors

AM Shendurse, Assistant Professor, Shri GN Patel Dairy Science and Food Technology College, Sardarkrushinagar Dantiwada Agricultural University, Sardarknagar – 385 506, District Banaskantha, Gujarat
E-mail: amshendurse@gmail.com .

Anuj Kumar, Ph.D. Scholar, Division of Dairy Technology, NDRI, Karnal, Haryana-132001, India

A.K. Chaurasiya, Assistant Professor, Department of Horticulture, North-Eastern Hill University, Tura Campus, Tura-794002, Meghalaya

AK Singh, M S Swaminathan School of Agriculture, CUTM, Gajapati, Odisha -761211, (India)

BK Mishra, Assistant Professor, Department of RDAP, North-Eastern Hill University, Tura Campus, Tura-794002, Meghalaya

Bharat Bhushan, Dairy Microbiology Division, National Dairy Research Institute, Karnal, Haryana-132001, India
E-mail: bpsingh03@gmail.com

Brij Pal Singh, Dairy Microbiology Division, National Dairy Research Institute, Karnal, Haryana-132001, India

B Paul, Head, Department of Zoology, Don Bosco College, Tura, Meghalaya.

Chaudhari A, Department of Dairy Engineering, S.M.C College of Dairy Science, A.A.U, Anand

D.C. Rai, Professor and Head, Department of animal Husbandry and Dairying, Banaras Hindu University, Varanasi-221 005
E-mail: dcrai.bhu@gmail.com

Deependra Singh, Dairy Microbiology Division, National Dairy Research Institute, Karnal, Haryana-132001, India

Dilip Kumar, Department of animal Husbandry and Dairying, Banaras Hindu University, Varanasi-221 005

Heena Sharma, Division of Livestock Products Technology, Indian Veterinary Research Institute, Izatnagar, Bareilly, (U. P.)-243122.

K.K. Datta, Principal Scientist, DESM Division, National Dairy Research Institute, Karnal-132001, Haryana

Kumaresh Halder, Assistant Professor, NIFTEM, Plot No. 97, Sector 56, HSIIDC Industrial Estate, Kundli, Sonipat – 131 028, Haryana
E-mail: kumar.halder@gmail.com.

M. Verma, Department of Home Science, Banaras Hindu University, Varanasi-221 005

Manju Gare, Ph.D. Scholar, Division of Dairy Microbiology, NDRI, Karnal, Haryana-132001, India

Manpreet Kaur, Division of Livestock Products Technology, Faculty of Veterinary Science and Animal Husbandry, Sher-e-Kashmir University of Agricultural Sciences and Technology, R.S. Pura, Jammu-181102, J and K

Meena Goswami, Assistant Professor, Department of Livestock Products Technology, College of Veterinary Sciences and Animal Husbandry, DUVASU, Mathura (U.P.)-281001
E-mailL dr.goswami2008@yahoo.co.in

Nalanda Balamurgan, Research Scholar, Department of RDAP, North-Eastern Hill University, Tura Campus, Tura-794002, Meghalaya

P. Narender Raju, Scientist, Dairy Technology Division, National Dairy Research Institute, Karnal
E-mail: pnr.ndri@gmail.com.

P. R. Patel, Department of animal Husbandry and Dairying, Banaras Hindu University, Varanasi-221 005

Parminder Singh, Department of Livestock Products Technology, College of Veterinary Science, Guru AngadDev Veterinary and Animal Sciences University, Ludhiana-141004, (Punjab),

Pavan Kumar, Assistant Professor, Department of Livestock Products Technology, College of Veterinary Science, Guru AngadDev Veterinary and Animal Sciences University, Ludhiana-141004, (Punjab)

Richa Badola, Ph.D. Scholar, Division of Dairy Technology, NDRI, Karnal, Haryana-132001, India

Rohit Charan, Department of Dairy and Food Business Management, SDAU, Banaskantha-385506, Gujarat, India

S. Bera, Faculty of Dairy Technology, WBUAFS, Mohanpur-741252, West Bengal
E-mail: msubarna@rediffmail.com.

S. Makhal, ITC Ltd., R and D Center, Bangalore

Sawale Pravin Digambar, Ph.D. Scholar, Division of Dairy Technology, NDRI, Karnal, Haryana-132001, India

Shaik Abdul Hussain, Scientist, Division of Dairy Technology, NDRI, Karnal, Haryana-132001, India
E-mail: abdulndri@gmail.com

Shilpa Vij, Principal Scientist, Dairy Microbiology Division, National Dairy Research Institute, Karnal, Haryana-132001, India

Shiv Raj Singh, Department of Dairy and Food Business Management, SDAU, Banaskantha-385506, Gujarat, India
E-mail: shivagritech2007@gmail.com

SK Roy, Dean, Faculty of Dairy Science and Food Technology, Sardarkrushinagar Dantiwada Agricultural University, Sardarknagar – 385 506, District Banaskantha, Gujarat

Subrota Hati, Assistant Professor, Dairy Microbiology Department, S.M.C. College of Dairy Science, Anand Agricultural University, Anand-388001, Gujarat, India

Sunil Kumar, Associate Professor and Head, Division of Livestock Products Technology, Faculty of Veterinary Science and Animal Husbandry, Sher-e-Kashmir University of Agricultural Sciences and Technology, R.S. Pura, Jammu-181102, J and K.

Tarunpal Singh, Department of Livestock Products Technology, College of Veterinary Science, Guru AngadDev Veterinary and Animal Sciences University, Ludhiana-141004, (Punjab)

Upadhyay J. B., Professor Dairy Engineering, Deptt of Dairy Engineering, S.M.C College of Dairy Science, A.A.U, Anand
E-mail: upadhyay2004@yahoo.co.in

2016, Dairy and Food Product Technology Pages 1–17
Editors: **Birendra Kumar Mishra and Subrota Hati**
Published by: **BIOTECH BOOKS, NEW DELHI**

Chapter 1

Supply Chain Management in Dairy Industry

***A.M. Shendurse*[1] *and S.K. Roy*[2]**

[1]*Assistant Professor,*
Shri GN Patel Dairy Science and Food Technology College,
[2]*Dean,*
Faculty of Dairy Science and Food Technology,
Sardarkrushinagar Dantiwada Agricultural University,
Sardarknagar – 385 506, Dist. Banaskantha, Gujarat

Introduction

Dairy industry in India has witnessed a remarkable journey in last few decades. From being a laggard and net importer of dairy products in 1950s and 1960s, India has covered a lot of ground. India has transformed from a country of acute milk shortage to the world's leading milk producer. India now is world's largest producer of milk with a production of 121.8 million tonnes in 2010-11. The credit of this phenomenal success is largely attributed to "Operation Flood", a co-operative led movement started in 1970s which took in its fold millions of small holding farmers who joined the three tier co-operative structure and increased India's milk output. The three tier structure constitutes village societies, district unions and state federations.

Apart from smallholdings and three tier co-operative structures, Indian dairy industry' suniqueness also emanates from imbalance in the collection and consumption pattern. Milk is predominantly collected from millions of smallholding farmers residing in villages and is sold to the urban centres through a complex distributionstructure. The processing sector is dominated by co-operatives, though the prominence of private dairy processors is gradually increasing. Apart from this, the unorganised sector (local sweet makers etc) processes a large portion of the total liquid milk output. In fact, the unorganised sector outnumbers the organised sector (both co-operatives and private players combined) in dairy processing.

The consumption pattern of dairy products in India is very different from many western countries, the primary reason being conventional dietary habits of Indian households. Approximately 55-60 per cent of the milk produced is consumed in liquid form. The rest of the milk products consumed are predominantly traditional Indian dairy products like ghee, paneer, chhana, dahi and other traditional sweets. Universal dairy products like cheese, table butter, ice-creams are consumed in moderate amounts, even though the growth rate of some of these products are healthy.

In spite of having largest milk production, India is a very minor player in the world dairy trade. Till the 1980s, India had adopted strategies like import-substitution, quantitative restrictions on imports and exports and canalisation. Competition within the organized sector was regulated through licensing provisions, which prohibited new entrants into the milk-processing sector. However, in early 1990s the Government of India introduced major trade policy reforms that favoured increasing privatization and liberalization of the economy. The dairy industry was further opened up after India became a member of World Trade Organization (WTO). In 2000, the Union Government allowed free import and export of most dairy products.

In recent years, India has maintained a positive trade balance for dairy products. The advantages India leverage are low farm gate prices (due to low milk production costs) and proximity to milk deficit markets of Asia and Middle East. However, India's export performance is not up to its potential. The key reasons attributed for its below par export performance are low quality and hygiene standards, lack of experience in marketing dairy products in international markets, and significant increase in consumption of milk and milk products in domestic market leading to limited surplus of exports. If India has to emerge as an exporting country, it is imperative that we should develop proper production, processing and marketing infrastructure, which is capable of meeting international quality requirements. A comprehensive strategy for producing quality and safe dairy products should be formulated with suitable legal backup (Karmakar and Banerjee, 2006).

The government of India has expressed strong interest in maintaining self sufficiency in dairy and other agricultural products. However, there has been a steady increase in import of dairy products to India after trade liberalization. The key items imported to India are butter oil, whey products, cheese, and milk powders. The key nations that export dairy products to India are Denmark, Nepal, USA, France, Netherlands and Italy. Many dairy products from other major milk producing nations face non-tariff barriers in India. In past few years, India has been able to achieve a fair degree of self sufficiency in all aspects of the dairy value chain. Indian dairy industry's uniqueness has also called for many distinctive requirements and ingenious solutions that one normally doesn't find in other large dairy producing nations.

Indian Dairy Value Chain: A Cooperative Movement

India is the largest producer of milk in the world with a total production of 121.8 million tonnes in 2010-11. It also has the world's largest livestock population, housing 57 per cent and 16 per cent of the globe's buffalo and cattle respectively. However, a majority of its primary producers involved in dairying activities are small and marginal farmers, with 80 per cent of the animals kept in small farms with an average of 2-8 animals (Hemme *et al.*, 2003). Dairying is an important economic activity in India as livestock distribution is more equitable than land distribution and thus it is considered an area that can be the focus of anti-poverty and equity-oriented programs (Staal *et al.*, 2008).

Ninety two percent of the primary producers are concentrated in rural areas and have limited access to marketing and infrastructural facilities. Only 20 per cent of milk produced is marketed through formal systems, thus limiting the scope of processing, value addition and better market penetration in the regional and international markets (Naik and Abraham, 2009). Most States (sub-nationals) in India have producer cooperative unions, where members sell their milk to be processed and marketed.

The birth of cooperative movements in India was in the dairy sector, which began as a humble initiative in 1947 in the village Anand in Gujarat, where milk producers with State support joined hands to have a stake in their produce and share the profitsgained by selling their produce. Such societies federated at the district level to form a cooperative union which helps to add value to the milk through processing and marketmilk products in distant markets. The success of Kaira District Milk Cooperative Union, known as Amul where Anand village society is a member, led to the adoption of the Kaira model as a blueprint in all milk cooperatives in the country under the three phases of "Operation Flood"(1970–1996). During operation flood, rural milk shed areas were linked to urban markets through the development of a network of village cooperatives for procuring and marketing milk. And milk production and productivity were enhanced by ensuring the availability of veterinary services, artificial insemination (AI), feed and farmer education. The investment paid off, promoting production gains of 4–5 percent per annum (Punjabi, 2009). Today almost all states in the India have their own dairy cooperatives supporting millions of small and marginal farmers.

During its initial years, the focus of these cooperatives was primarily on liquid milk. Overtime they progressively moved onto higher value milk based products. Quality and price competitiveness have helped these cooperatives remain competitive in the wake o femerging competition from multinational companies like Unilever, Nestle and Britannia. A key source of competitive advantage has been their ability to implement best practices across all levels of the network: the federation, the unions, the village societies and the distribution channel.

Due to its high perishability, milk and milk products have one of the most stringent codes of standards. Conforming to standards depends much on how milk is handled throughout the supply chain, through technology and good supply chain management, particularly in the context of a cooperative setup, where producers are many.

Technological intervention at the grassroots through the adoption of Bulk Milk Coolers (BMC) at the village society level have helped increase the scope for the improving milk quality in the cooperatives. In their efforts to diversify to higher value products, this intervention can help reduce sour milk content and spoilage during transportation and storage. However, in order for these benefits to materialize, there is a need to bring about parallel changes in operational practices. Therefore, improvement in the quality and competitiveness of milk and milk products depends on the synergy of technological and organisational improvements coupled with operational changes at all levels of the value chain (Naik and Abraham, 2009).

Opportunities in Indian Dairy Value Chain

Dairy has a lot of potential to improve rural incomes, nutrition and women empowerment, and hence is a very critical area for investment. A well-developed industry will enable millions of farmers to capitalize on the emerging opportunities and make a significant impact on rural incomes. On the flip side, weak efforts towards dairy development also can have a significant but negative impact on the dairy industry. The growth rate has been sluggish over the past few years (Punjabi, 2009). With an increase in demand on one hand and sluggish supply on the other, there is a likely shortfall in demand in the coming years.

Major areas of intervention in the dairy sector have been highlighted as below-

Growth and Market Opportunity

In the past two decades per capita milk consumption in India increased from 150 to 250 gram per day, with a predicted consumption of 370 gram daily in 2020. Current growth rate in milk production is only 3.8 per cent compared to 4.5 per cent in the 1990s and a projected growth in demand in the coming decade of 4 to 5 per cent. A demand–supply gap has existed in the past years resulting in price increases of 21 per cent on a year to year basis. This is severely affecting the ability of rural and urban poor to buy milk. In India, where over 50 per cent of the population is vegetarian, milk is a key source of dietary protein and other essential nutrients (ILRI, 2011). Thus this demand–supply gap has severe consequences for millions of poor in a country with an already chronic level of malnutrition.

Pro-poor Potential

With a vegetarian population of over 50 per cent, milk and milk products are a crucial source of protein and other nutrients in India. Despite wide-spread malnutrition amongst young children in India, it has not been shown that increasing sales of milk further weakens their nutritional situation. Rather, the increased regular income in relatively small amounts offers households the opportunity of improving their access to nutritious food. Female household members are generally responsible for feeding and management of livestock within the homestead (ILRI, 2011). Increased milk production along with improved training on basic nutrition, especially nutrition of women and children, and food safety could have a significant impact on food and nutrition security.

Dairy development has often been identified as having an especially large potential for providing pathways out of poverty for poor rural livestock keepers, because livestock already form a considerable share of their assets and family labour is often not their greatest constraint (Adisu *et al.*, 2012). Increasing the productivity of the existing assets has often been regarded as the most sustainable approach to alleviate poverty (Bachmann 2004). Millions of small and marginal farmers in dairying who own two to three animals and produce an average of 5 litres comprise a critical portion of India's dairy industry (Punjabi, 2009). Livestock development in general and dairy development activities in particular are key components of pro-poor development strategies because livestock distribution is much more equitable than land distribution. Thus, changes in the dairying environment have important implications for the smallholder farmers and for poverty reduction.

Researchable Supply Constraints

In India, out of the 180 million bovines, only about 12 per cent are cross bred with average milk yields (corrected to 365 day lactations) of 6.44 kg per day. Forty-one per cent are local cattle with average milk yields of only 1.97 kg per day. Forty-seven of milch animal are buffaloes yielding on average 4.40 kg of milk daily. The average milk yield across all bovine animals is 3.6 kg per day (ILRI, 2011).

Feed is the major financial input into dairy accounting for an estimated 70 per cent of the production cost and is a major constraint to increasing production. In India, reduced access to grazing and rising opportunity costs for producing fodder crops has led to considerable increases in feed prices. Thus, in many parts of the country, the price of cereal residues, *i.e.* straws and stovers, currently the major feed resource, accounting for almost half of all livestock feed, is now half the price of grain by weight (ILRI, 2011). Concentrate availability is limited because of the priority in most of the country of allocating land to food crops and by-product concentrates such as brans and oil cakes are exported in significant quantities. Improving feed supply through green fodder and forage production has largely failed because of severe constraints in the availability of arable land and irrigation water. Even in areas where land and water are available, attempts to increase fodder supply have generally failed because access to quality forage seeds is a major impediment. On the other hand, various feed resources remain underutilized and few opportunities for improving feed rations through supplementation or processing are implemented. Knowledge and extension on feeding remains inadequate.

Poor genetics of breeding animals is another constraint to improving dairy productivity (as stated above, only 12 per cent of dairy cattle are cross bred). Even where artificial insemination (AI) is being used, less than 15 per cent of the breeding bulls have been tested. Furthermore, conception rates after AI are only about 40 per cent when the service is supplied by government agencies although conception rates are higher when supplied by Non-Governmental Organizations (NGOs). Private AI services are in their infancy and often impaired by government policies. The introduction of exotic semen for cross-breeding is hardly regulated, leading to low fertility and survival rates of higher grade dairy animals. Breeding programs for buffaloes have not been very effective (ILRI, 2011).

Animal health services are very variable regionally within India with a large number of producers having little or no access to veterinary services. Even where veterinary services exist they often have inadequate facilities and a lack of operational funds. Only in limited areas have private suppliers been able to successfully establish the delivery of animal health services. Similarly livestock extension services are usually nonexistent or very limited. Animal diseases such as foot and mouth disease, brucellosis and haemorrhagic septicaemia cause large economic losses (ILRI, 2011).

In general the key services necessary to improve dairy animal productivity and management are found to be fragmented, uncoordinated and non-integrated. Services are rarely tailored to the need of smallholders. Their high transaction cost is a further element discouraging participation of poor producers in dairy value chains. Economically and institutionally viable models for integrated service delivery in the dairy sector are virtually absent or operating at an insufficient scale.

With the exception of the large urban centres, almost 80 per cent of the milk marketed is traded through the informal sector. Government policies hardly consider the importance of the informal trade. Most milk is consumed as liquid milk—only about 20 per cent of the milk produced is processed. Cooling facilities are limited and restricted to the organized sector, leading to considerable food safety issues (ILRI, 2011).Growing awareness in the Indian consumer about the quality aspects of milk and milk products has brought the organized dairy sector into the foreground. There is also an increasing demand for packaged traditional Indian dairy products, the production of which is still dominated by the unorganized sector. Organized dairy industry has only just begun to realize its vast potential in this area (with the introduction of packaged *dahi, kheer,* and *lassi* in tetra pack, etc.) (Chaudhary *et al.*, 2010).

Enabling Environment

The high number of smallholders and landless people engaged in dairy together with the significantly increasing demand for milk and other dairy products and the existing supply gaps has put dairy high on the development agenda in India. An ambitious 15-year National Dairy Development plan has been prepared by the National Dairy Development Board, which recognizes the potential role of the private sector (feed, AI, processing etc.) and marketing structures other than dairy cooperatives. The Government of India (GoI) has asked for Consultative Group on International Agricultural Research (CGIAR) involvement in this new program. Therefore, the probability of this Program influencing a major development program and leveraging significant development funds is very high. In addition, both private sector and cooperative dairies are showing renewed interest in investing in improved collection, processing and distribution systems after many years of stagnation (ILRI, 2011).

Existing Momentum

The choice of the dairy value chain in India was based on the above described importance of milk in food security, the importance to poor consumers and poor rural producers and ILRI's (International Livestock Research Institute) comparative advantage in research infrastructure, staffing and wide ranging established partnerships in India. ILRI's institutional and technical research experience and projects with its wide range of public and private partners in India and beyond will be harnessed and focus onto the dairy value chains. This comprises well developed research on key constraints in input and services through innovation platforms and hub structures, well tested approaches to mitigating key technical constraints in animal feeding, health and genetics and knowledge management strategies that bring those aspects together and that can deliver to target beneficiaries and other pertinent actors. The ILRI has already carried out research on aspects of the dairy value chain in the state of Assam, supported by the World Bank, which had led to changes in policy on the informal market sector and has developed several large projects that address technical constraints to dairy development in the feed, health and breeding sectors. These projects are being implemented in close collaboration with other CGIAR centres (*i.e.* International Crops Research Institute for the Semi Arid Tropics (ICRISAT), International Maize and Wheat Improvement Center (CIMMYT), International Rice Research Institute (IRRI)), the national agricultural research system (Indian Council of Agricultural Research) and several State Agricultural Universities), private enterprises (feed manufacturers, fodder traders, seed industry) and NGO's (Sir Ratan Tata Trust, BAIF, BASIX, RDT) (ILRI, 2011).

The following are the potential opportunity areas of trade/business across India's dairy value chain.

Breeding

- ☆ High quality imported bovine semen (particularly for Holstein-Friesian and Jersey varieties).

Feed and Nutrition

- ☆ Low cost supplementary feed and concentrates that can justify the economics of milk production by small farmers.
- ☆ Low cost feed supplements like protein feed and urea molasses block.
- ☆ Feed block making machines.

Healthcare and Herd Management

- ☆ Low cost dairy extension services (feed and nutrition, healthcare and vaccination, artificial insemination, total healthcare management, record keeping etc) at farmer's doorsteps.
- ☆ Sophisticated diagnostic kits for diseases like Brucellosis, Tuberculosis, Paratuberculosis, and Elisa test etc.
- ☆ Preventive health products teeth dips, dry cow formulations, milk replacers
- ☆ Machines for manure and fodder management

Milk Production, Storage and Transport

- ☆ Low cost hygienic milk collection equipment to be used by small holding dairy farmers. Clean-in-place (CIP) systems, small automatic milk collection machines can be purchased for a community of farmers and used collectively. This will address one of the biggest issues in Indian dairy industry, *i.e.* unhygienic milk collection practice.

Processing

- ✰ Packaging equipment for products like butter, cheese, UHT (Ultra High Temperature) milk, aseptic filling.
- ✰ Most of the capital intensive dairy processing equipment like self cleaning cream separators, homogenizers, continuous butter and cheese making machine, pasteurizer plates, large scale ice cream freezers, ammonia compressors etc
- ✰ Resource efficient processing equipment to suit Indian conditions
- ✰ Imported second hand dairy processing equipment (The outlook for second hand capital intensive processing equipment is improving. Many of private players have small to medium capacity processing plants and are willing to purchase imported second hand equipment to reduce capital investment and stay competitive)
- ✰ Industrial production of traditional Indian dairy products and sweets

Marketing, Distribution and Retail

- ✰ Low cost packaging technology to suit Indian conditions
- ✰ Sensors and automation equipment for fleet and cold storage management

Challenges in Indian Dairy Value Chain

India is home to one third of the world's undernourished children. This persistent undernutrition has devastating effect on human development and economic growth in the country. With over 50 per cent of the population being vegetarian, milk is a key source of dietary protein and other essential nutrients and plays a key role in the mitigation of undernutrition. Unfortunately a significant gap exists between demand and supply of milk despite milk production contributing about 18 per cent to agricultural GDP and being–by value–the single most important agricultural commodity. About 70 per cent of the milk is produced by small, marginal and landless farmers keeping up to 3 adult dairy animals. Even households supplying private dairies have an average herd-size of only about 10 animals. For 70 million rural households, 40 per cent of whom are landless, milk production is an important part of their livelihoods. About 70 per cent of labour in dairying is provided by women and engagement in dairying has been shown to provide pathways out of poverty (ILRI, 2011). Improving the dairy sector in India will therefore benefit producers by providing livelihoods and consumers by providing milk at affordable prices.

Milk and milk products are consumed by most households, both rich and poor, and consumption is increasing; milk sales are an important livelihood strategy for rural households most of which are poor. Milk processing and trading provide a livelihood for large numbers of vendors and processors. Hence, enhancement of the traditional milk value chain can simultaneously improve the welfare of large numbers of poor farmers, intermediaries in the milk chain and consumers.

The major challenge of dairy value chain is to create both demand for locally-produced, quality milk and capacity to supply it. This could be achieved through out-scaling of capacity-building methods and tools, and through up-scaling of policy processes of proven success. A study carried out by ILRI found that most urban consumers are not satisfied with milk quality and safety, while most producers and milk chain intermediaries are unaware of hygienic milk handling. The high costs of informality and poor linkages between actors further constrain dairy development. Moreover, given increased consumer demand for safer milk combined with potential regulations in the horizon aimed at mandating safety standards, informal sector participants' livelihoods in the dairy sector are under increased threat (ILRI, 2013).

It is anticipated that enhancing dairy quality and capability of supplying milk, while simultaneously improving linkages and decreasing transaction costs, will lead to a demand-driven production system which will further stimulate investment in production technologies (*e.g.* artificial insemination, feeding, credit and veterinary services) that are available but under-utilized. The knowledge, policy and institutional barriers could be overcome by improving the traditional milk marketing through two inter-linked strategies clustered around knowledge management (out-scaling) and policy engagement (up-scaling).The improvement is expected to result in significant livelihood benefits to those engaged in the value chain as more consumers purchase larger quantities of safer and higher quality milk, and farmers and traders see both their markets increase, costs decrease and losses from milk spoilage and wastage go down.

Following are the challenges for dairy value chain-

- Small herd size - it's a challenge to support millions of small herd sizes.
- Weak and uncoordinated input services and supplies.
- Low yield of milk per animals - 3.6 kg/day; local cattle 1.97 kg/day; buffalo 4.4 kg/day.
- Poor genetics *i.e.* only 12 per cent crossbred.
- No systematic performance recording in breeding programmes.
- Feed shortage.
- Variable animal health services.
- Very weak extension services.
- Informal trading of marketed milk.
- Poor infrastructure.
- Government weeded to co-operative model.
- Poor policy environment for private sector input suppliers.
- Poor co-ordination among agencies (including research agencies).
- Environmental impact of dairy sector.

These challenges across India's dairy value chain emanate from the critical issues and priorities witnessed by Indian dairy industry and its stake holders. The critical issues of Indian dairy chain are mentioned in Table 1.1, whereas Table 1.2

Table 1.1: Critical Issues in the Dairy Chain

Stage	*Priority*	*Issues*
Policy environment	1. Developing livestock policy 2. Breed development	1. Lack of a coherent livestock development policy 2. Ineffective implementation of policy and projects due to lack of clarity in roles of different agencies 3. Lack of resources 4. Lack of clarity between roles of different departments 5. Lack of regulation for quality of feed and medicines
Services	1. Disease control/health/breeding/extension services 2. Support to dairy farmer organizations/women's self-help groups	1. Inadequate coverage of veterinarian and breeding services 2. Non-existent extension services 3. Scope to enhance activities of NGOs in these areas 4. Lack of private sector involvement in dairy development services and activities
Inputs	1. Feed supply 2. Fodder 3. Medicines/vaccine supply	1. Quality/cost of feed 2. Ineffective approach for management of common property resources 3. Quality of medicines
	1. Formal credit for animal purchase	1. Very poor access to formal credit at the farm level
	1. Informal loans for animal purchase or other dairy needs	1. Very high rate of interest; farmer has to sell milk at low price to the trader if he/she has borrowed money from the trader
Production	1. Dairy farming 2. Selling milk to cooperatives/traders/private dairy agents	1. Poor management and feeding practices because of lack of information in the absence of extension activities. 2. Low productivity because of poor genetic potential, poor feeding and management practices, poor access to health and breeding services, lack of good-quality animals 3. Availability of milk per household very low 4. Low profitability from dairy enterprise

Contd...

Table 1.1–Contd...

Stage	Priority	Issues
Marketing/processing	1. Collection of milk from farmers through village society, processing and marketing of milk in cities and urban areas	1. Lack of coverage of villages 2. Lack of transparency in milk testing and pricing 3. Lack of democracy in village societies 4. Marketing only in peri-urban/urban areas 5. Maintaining quality of milk/infrastructure 6. Milk prices declared by cooperatives kept low and used as a benchmark price by other players
	1. Purchase milk from farmers and selling milk and processed products to consumers	1. No transparency in milk pricing 2. Adulteration and quality of milk and milk products 3. Unhygienic conditions for milk processing
	1. Purchase of milk from farmers through village agents, processing and selling milk	1. No transparency in pricing of milk 2. Quality of milk

Table 1.2: Priority Objectives of different Stakeholders along the Dairy Value Chain

Consumers	Processor	Trader	Producer
Safe dairy products	Better milk quality	Reduced cost of transport Protection of the raw milk market	Access to quality services (either publically or privately provided)
Product value for money	Increased access to markets (domestic and export)	Minimize competition with imports	Fair pricing for quality products
Variety of products	Input into policy/advocacy	Input into policy/advocacy	Stable farm income/and reduced income risks
Nutritional products	Optimizing plant capacity	Expand and diversify operations	Access to milk for home consumption
Readily available	Stable milk supplies	Stable milk supplies, regular access	Increased productivity of animals
Properly labelling and packaged	Assured access to quality inputs	More formalized market role	Availability to quality inputs

depicts the priority objectives of different stakeholders along the dairy value chain (Punjabi, 2009).

Factors Affecting Dairy Value Chain in India

Quality of Animals

The quality of animals is critical in determining its milk productivity and hence overall production. Currently, low productivity per animal hinders development of the dairy sector. Despite being the world's largest milk producer, India's productivity per animal is very low, at 987 kg per lactation, compared with the global average of 2038 kg per lactation(Punjabi, 2009).

Low Productivity

The low productivity is a result of ineffective cattle and buffalo breeding programmes, limited extension and management on dairy enterprise development, traditional feeding practices that are not based on scientific feeding methods, and limited availability and affordability of quality feed and fodder. In addition, the limited supply of quality animals is exacerbated by policies limiting interstate movement of animals. Indigenous cattle and buffalo make up 45 percent of the country's total milch population, in contrast to the cross-bred cows at 10 percent(Punjabi, 2009).

Animal Health and Breeding Services

Animal health and breeding services provision, veterinary infrastructure development and vaccinations are the responsibility of the state government. These services have traditionally been provided for free or at a very subsidized rate. In the past few years, there has been increasing awareness that the state pays heavily to offer these services, which are easily available to farmers. Consequently, many states have instituted partial or full-cost recovery fees for providing the services.

In addition to the State Department of Animal Husbandry, Dairying and Fisheries, the milk cooperatives and NGOs (BAIF, JK Trust) provide services in many states. So do trained private sector AI technicians, although for a fee. As well, state livestock development agencies are being set up as autonomous bodies to offer services in animal breeding in the form of procurement, production and distribution of breeding inputs (such as semen and liquid nitrogen), training and promotional activities.

Despite these initiatives, the availability of services remains limited. Cattle and buffalo breeding programmes have been initiated but have not had the desired impact because of a lack of coordination between the different state departments. And extension activities in dairy management are woefully lacking. Farmers have not been able to take advantage of the potential of their animals because they lack information on feeding and management practices. Extension, especially for women involved in livestock rearing, would enhance dairy production considerably.

Availability of Feed

Crop residues are the single largest bulk feed material available to farmers for feeding livestock, specifically ruminants. They include coarse straws, fine straws, leguminous straws, pulses straws and sugarcane tops. Fodder from common property resources is another major source of feed for animals. But lack of efficient management of common property resources is a major constraint in availability of these resources for fodder. The area under cultivated fodder production is limited only to 5 percent of the total cultivable land. In the states of Haryana, Punjab, Gujarat and some parts of Rajasthan, land use for green fodder production is estimated at 10 percent or more(Punjabi, 2009). There is a need for restructuring the land use strategy to elevate the overall proportion of cultivable lands for fodder production.

Concentrates used for fodder include coarse grains, such as maize, sorghum, bajra and other millets, and other cereal by-products, such as rice bran/ polish and various oil meals, including groundnut cake, mustard cake, coconut cake, soybean meal, cotton seed meal and sesame cake. The escalating price of feed ingredients is a major cause for concern. In many states, cooperatives are involved in producing feed concentrate and selling to farmers at subsidized rates.

Scarcity of Fodder Resources

Scarcity of fodder resources is likely to be a major constraint in the development of the dairy sector unless adequate measures are undertaken to augment them. Another important issue regarding feed is the lack of regulations to ensure quality. In the absence of a coherent policy, all kinds of substandard feeds are available in the market.

Formal/Informal Credit

Lack of access to credit to expand the herd is a critical problem for farmers. There is little access to formal credit through the cooperatives. Informal credit is available from private traders and agents of private companies, but the interest rate is very high. And these loans may or may not be linked to dairy activity. When taking a loan from a trader, the farmer is then tied to selling the milk to that trader, often at a low rate. The commercial banks are not favourably disposed to providing credit to livestock farmers and the cooperative credit system is very weak, resulting in excessive dependence of livestock farmers on informal sources and usually at exorbitant interest rates. Efforts should be put on correcting these distortions and ensure timely availability of inputs and services, including credit to livestock.

Vaccines/Medicines

The Government and the private sector are involved in producing medicines and vaccines. However, quality control is a critical issue. An important policy question is whether the government should be involved in the manufacturing and production of vaccines or should it instead take on a regulatory role to ensure quality and availability at a reasonable price.

SWOT Analysis of Indian Dairy Value Chain

Within the framework of the competitiveness drivers and issues, the smallholder dairy sector's strengths, weaknesses, opportunities and threats have been assessed. The strengths and weaknesses are factors that are directly controllable, while

Table 1.3: SWOT Analysis of Indian Dairy Value Chain

Strengths	*How to Build on Them*
● Large number of small and marginal farmers involved in dairying ● An effective marketing channel helps to meet the demands of the urban consumer ● Very large number of animals and huge scope to enhance productivity ● Self-sufficiency in medicine production and do not have to rely on exports	● Strengthen economic viability of dairy farms by interventions on the input side as well as ensuring more fair farmer prices ● Increase the link between rural production areas and urban markets ● Focus on strengthening the indigenous breed to help significantly enhance productivity ● Ensure availability of quality medicines by strengthening regulatory framework for quality
Weaknesses	*How to Correct Them*
● Large share of milk (70–85 per cent) of marketable surplus goes through informal channel where quality is a big concern ● Sometimes quality is an issue in the formal channel as well ● Very little competition to cooperatives because private sector was not allowed to participate in until recently ● Farmers do not share in the benefits of high demand because of poor governance of cooperatives ● Milk production is scattered over a large number of farmers producing miniscule quantities ● Milk distribution is limited to urban and peri-urban areas ● Low milk prices because of lower prices declared by cooperatives, which results in low prices of milk paid by all players ● Ad hoc export policies ● Quality of milk and milk products are a barrier to entry to the export market, especially the EU and the USA ● Lack of policy focus on strengthening indigenous breeds ● Non-existent extension facilities ● Because of low access to credit and risk-taking ability, farmers cannot increase their herd size	● Focus on quality issues even in the informal channel by training traders and by enforcing food quality regulations ● Develop infrastructure and training for clean milk production ● Support a fair playing field for the private sector ● Bring about changes in cooperatives to make them true representatives of farmers instead of functioning as parastatals. ● Support to dairying as an enterprise to encourage commercial dairy farming and encourage production and productivity by extension and breed development ● Enhance packaged milk distribution in more areas ● Strengthen dairy farmer cooperatives to enable farmers to get a higher price for milk ● Create rational export policy to enable farmers to take advantage of higher prices ● Strictly implement quality regulations and improve infrastructure and training for quality ● Strengthen the breed development programmes ● Strengthen extension facilities ● Increase access to credit through dairy farmer organizations and other agencies

Contd...

Table 1.3–*Contd...*

Opportunities	*How to Pursue Them*
● Increased farmer income by exploiting the high demand ● Increased consumer sophistication and awareness of quality reception of quality packaged products (though slowly) ● Entry of large corporations in retailing, which can lead to more investment ● Immense scope to enhance governance of dairy farmer organizations and thus enable dairy farmers to demand higher prices ● Potential for exports due to low cost of production ● Overall positive growth environment, which is triggering the Government to enhance infrastructure	● Create policies and activities geared towards enhancing dairy farming activity by increasing, production, productivity and ensuring fair farmer price of milk ● Establish enabling policy environment to enhance investment ● Create policy support to enhance governance of producer companies ● Focus on quality issues that are a barrier to exports ● Encourage private sector to increase investment in dairying
Threats	*How to Avert Them*
● Large portion of the population does not care about quality issues in milk ● Because of high price sensitivity for dairy products, people are not willing to pay for quality ● Significant increase in feed prices ● Large informal markets that extend credit are constraining farmers ● Low productivity and scattered production leading to high cost of transportation ● Emphasis on milk fat and not on SNF content maintaining relatively lower prices of milk	● Initiate consumer education about the negative health impacts of unpackaged products ● Develop packaging in small quantities to meet the needs of the poor ● Increase milk prices in accordance with feed prices ● Support expansion of dairy farmer organizations ● Enhance productivity by breed improvement and extension ● Enforce price setting of milk based on both *i.e.* fat and SNF content to encourage production of cow milk

opportunities and threats derive from the external environment. As evident in Table 1.3, there are a large number of weaknesses in the sector, implying considerable scope for interventions. This SWOT analysis entailed matching each of these elements with an appropriate action (Punjabi, 2009).

Conclusion

The potential impact of increasing the efficiency of the dairy value chain in India has to be seen in view of the 70 million rural households relying on dairy animals. However, the difficulties in replicating the considerable success of co-operative dairy systems in individual states on a national scale highlight the challenges in reaching these households.

Based on planning commission's estimate and subsequent corrections on account of consistently higher growth in GDP, it is expected that demand for milk is likely to be about 155 million tonnes by 2016-17 and around 200 million tonnes by 2021-22. With such a mammoth volume of milk to be dealt with, the organized dairy sector needs to look within, and convert its challenges into capabilities.

The objective of doubling per animal milk productivity could be achieved by interventions at all stages of the value chain. The improved quality and cost-efficiency of inputs and services will lead to an increase in their use. The integration of producers into producer communities will enable a better utilization of available resources and an increased investment into dairy production. The gains in efficiency and quality of milk collection, processing and marketing, as well as the improvements in the policy framework will encourage various stakeholders in the value chain to profitably increase their involvement in the dairy sector. In particular, attractive and reliable marketing opportunities will be the basis for small-scale dairy producers to intensify their dairy production and to increase their market integration.

Doubling dairy animal productivity will lead to considerable income increases for the poorest households, especially if the efficiency of input delivery and of milk marketing is improved simultaneously. Improved awareness of nutritional requirements of children together with increased household milk supply will contribute to a reduction in child undernutrition. Improved milk supply to markets and more efficient marketing systems will allow poor consumers to increase the share of milk and milk products in their diets without exposing themselves to increased health risks. The growth of the dairy value chain will create additional employment opportunities, both in the formal and the informal sectors. Updated policies will allow the informal sector to grow through supporting improvements in food safety and environmental impacts while retaining its efficiency advantage.

References

Addisu B, Mesfin B, Kindu M, Duncan A 2012. Production aspects of intensification and milk market quality in Amhara region, Ethiopia. Livestock Research for Rural Development 24 (9): 154.

Bachmann F 2004. Livelihood and livestock: Lessons from Swiss livestock and dairy development programmes in India and Tanzania. IC series no. 4.Intercooperation (IC). Bern, Switzerland. 40 pp.

Chaudhary U, Singh KHR, Datta KK 2010. Successful Dairy Value Chain: The Organized Sector Approach. Agricultural Economics Research Review 23: 561.

Hemme T, Garcia O, Saha A 2003. A Review of Milk Production in India with Particular Emphasis on Small Scale Producers.Pro-Poor Livestock Policy Initiative (PPLPI) working paper no. 02, International Livestock Research Institute, Rome, pp 1-58.

ILRI (International Livestock Research Institute), 2011. Dairy value chain in India: Background proposals for the CGIAR Research Program on Livestock and Fish. http: //cgspace.cgiar.org/bitstream/handle/10568/16886/LivestockFish_DairyVCIndia.pdf?sequence=1. website visited on November 15, 2013.

ILRI (International Livestock Research Institute), 2013. Knowledge to action: Enhancing traditional dairy value chains in Assam, India. http: //www.ilri.org/knowledgetoaction. website visited on November 15, 2013.

Karmakar KG, Banerjee GD 2006. Opportunities and Challenges in the Indian Dairy Industry. Technical Digest 9: 24-27.

Naik G, AbrahamM 2009. Interventions in the Food Value Chain to ImproveQuality and Competitiveness: A Case Study of Dairy Cooperative in India. Bangalore: Centre for Public Policy, Indian Institute of Management Bangalore. (http: //www.eoq.hu/iama/conf/1007_case.pdf)

Punjabi M, 2009. India: Increasing demand challenges the dairy sector. In: Smallholder dairy development: Lessons learned in Asia. Morgan N (eds). FAO Regional Office for Asia and the Pacific (RAP), Maliwan Mansion, Bangkok, Thialand, pp 44-62.

Staal SJ, Pratt AN, Jabbar M 2008.Dairy Development for the Resource Poor. Part 3: Pakistan and India Dairy Development Case study. Pro-Poor Livestock Policy Initiative (PPLPI) working paper no. 44-3, International Livestock Research Institute, Rome.

2016, Dairy and Food Product Technology *Pages 19–38*
Editors: **Birendra Kumar Mishra and Subrota Hati**
Published by: **BIOTECH BOOKS, NEW DELHI**

Chapter 2

Smallholder Dairying in India: A Living from Livestock

A.M. Shendurse[1] *and S.K. Roy*[2]

[1]*Assistant Professor,*
Shri GN Patel Dairy Science and Food Technology College,
[2]*Dean,*
Faculty of Dairy Science and Food Technology,
Sardarkrushinagar Dantiwada Agricultural University,
Sardarknagar – 385 506, Dist. Banaskantha, Gujarat

Introduction

India is approaching a population inflection point as the numbers are projected to rise sharply. Urgent attention is required to provide food for this growing demand. Much of the demand for dairy products will be concentrated in the urban and peri-urban area (Azage and Wold, 1998). Given suitable government policy support and access to market and services, there is a great potential to develop small-scale dairy householder dairy schemes in peri-urban and urban areas. Small-scale dairy farmers are at the centre of concerns about globalization, and rightly so because they are the largest employment and small business group among the world's poor (von Braun, 2004). The majority of the dairy cattle are in the hands of smallholder dairy producers. Dairying is also considered a strong tool to develop a village micro economy in order to improve rural livelihoods and to alleviate rural poverty. Potentially, therefore, small-scale dairying is a viable tool to spur economic growth and alleviate poverty and malnutrition. Among other reasons, low agricultural productivity, and high population growths not matching with the available resources to support them are associated with high incidences of poverty in many countries (Mwankemwa, 2004; Mason and Lee, 2005). Smallholder farming has been characterized by low productivity. This situation is partly attributed to lack of capital and uses of poor farming technologies by smallholder farmers, drought, and lack of market for the produce (Mwankemwa, 2004).

Dairying is a part of mixed farming system in India and contributes to the livelihoods of many small-scale farmers in our country through income, employment and food. Smallholder dairy production has thrived since independence owing to supportive subsidized services, and guaranteed milk markets and prices for farmers. In order to take advantage of emerging market demands for reducing their poverty, smallholders have to face challenges to improve production costs and productivity (Uddin *et al.*, 2012). The recent historical rise in world food prices has further aggravated the situation of dairy input prices which has also increased farm costs and ultimately affects farm profitability. In addition, there is a lack of institutional support, research and training, which would be beneficial to the farming environment (Sriri *et al.*, 2011). As in many other parts in India, there is a growing need for information about detail householders and small-scale dairy production parameters to enhance household life styles. Furthermore, dairying represents one of the fastest returns for livestock keepers. It provides regular returns to farmers, especially to women, enhances household nutrition and food security and creates off-farm employment, as many as one job for each 20 kg milk processed and marketed (Hooten, 2008).

Livestock production makes an important contribution to economic development, rural livelihoods, poverty alleviation and meeting the fast growing demands for animal protein in developing countries. Livestock provide over half the value of global agricultural output and one-third in developing countries. Rapid growth in demand for livestock products in developing countries is viewed as a 'food revolution'. Because livestock products are more costly than staple foods, their consumption levels are still low in developing countries, although they are increasing as incomes rise. Increased dairy production and greater self-sufficiency save on foreign exchange. Livestock also contribute to rural livelihoods through employment and poverty relief by integrating with and complementing crop production, embodying savings and providing a reserve against risks (Moran, 2009).

India is rich in its livestock wealth. India has one of the largest livestock population in the world. Fifty percent of the buffaloes and twenty percent of the cattle in the world are found in India, most of which are milk cows and buffaloes. The livestock sector occupies an important place in the economies of our country. It contributes to the economy in four different ways, as it provides: (a) energy in the form of draft and traction power for various activities and fuel for cooking and other heating purposes; (b) food in the form of milk, milk products and meat; (c) raw materials in the form of wool, hair, skins, hides, bones, hoof and horns and a number of other products of pharmaceutical and industrial use; and (d) manure for crops. The livestock sector in the country is characterized by the preponderance of smallholders typically possessing only one or two milch animals, low productivity, lack of proper feeding and animal health care, an inadequate supporting infrastructure for supply of feed and veterinary medicines, procurement, processing, storage, transport and marketing of milk (Singh and Pundir, 2002).

Milk Production

India has the largest dairy industry in the world. To help get this in perspective, the quantity of milk handled annually in the Indian informal market alone is greater than the annual world trade of exports of all dairy products, measured in liquid milk equivalents. Dairy development in India has been acknowledged the world over as one of modern India's most successful developmental programme. Milk and milk products is rated as one of the most promising sectors which deserves appreciation in a big way. In India, smallholder dairying has become a good income-earning occupation for farmers. With further improvement in productivity and reduction in production costs, Chantalakhana and Skunmun (2002) believe that smallholder dairying can become a very sound and sustainable enterprise.

Over the last few years, India has emerged as one of the largest producer of milk in the world. The rise in milk production in India comes coupled with two important facts.

First, India has the world's largest concentration of bovine population in the world. It has more than 40 per cent of world's total buffaloes, and more than one sixth of world's total cows1. This implies that though our aggregate milk production may be high, the productivity of milch cattle in the country remains abysmally low.

Second, despite being the largest producer of milk in the world, India is also the largest consumer of milk. The per capita availability of milk in India, though increasing over the last 20 years, still falls below the world's average milk availability per person per day. The per capita availability of the milk has reached a level of 281 grams per day during year 2010-11, but it is still lower than the world average of 284 grams per day.

Dairy Sector in India

The dairy sector in India is characterized as follows: by small-scale, widely dispersed and unorganised milch animal holders; low productivity; lack of assured year-round remunerative producer price for milk; inadequate basic infrastructure for provision of production inputs and services, and for procurement, transportation, processing and marketing of milk; and lack of professional management. Other important characteristics of the dairy sector are the predominance of mixed crop–livestock farms and the fact that most of the milch animals are fed on crop by-products and residues, which have a very low opportunity cost. In addition, dairy development policies and programmes followed in India, including those relating to foreign trade, are not congenial to promoting sustainable and equitable dairy development.

Low productivity of milch animals is a serious constraint to dairy development in India. This is due mostly to low genetic potential of the milch animals, and inadequate and inappropriate feeding and animal health care. The productivity of dairy animals could be increased substantially through crossbreeding of the low yielding nondescript cows with high yielding selected indigenous purebreds or suitable exotic breeds in a phased manner and by better feeding, disease control and management. The cattle breeding policy should also provide for the production of good quality bullocks to meet the draft power requirement of agriculture. Upgrading of nondescript buffalo through selective breeding with high yielding purebreds should be given a high priority in all areas where buffalo are well adapted to the agroclimatic conditions. While fixing procurement price, producers' interest should receive the utmost attention. The producer price should at least cover the long-run average cost of milk production and provide a reasonable mark-up (Singh and Pundir, 2002).

Dairy plants, cattle feed factories, technical inputs and services should be managed professionally and run as commercial enterprises and not as social welfare schemes. The role of governments should be to direct, co-ordinate and regulate the activities of various organisations engaged in dairy development, to establish and maintain a level playing field for all stakeholders and to create and maintain a congenial socio-economic, institutional and political environment for smallholder dairy development.

The new world trade regime ushered in by the World Trade Organization (WTO) poses several challenges and opens up many opportunities for smallholder milk producers. There is need to enhance the competitive economic advantage in dairy products in terms of both quality and cost. There exists a vast untapped potential for increasing the multifarious contributions of the dairy sector to the economy of the country. What is needed to realize this potential is, among other things, a comprehensive and integrated dairy development policy and determination and total commitment from politicians and bureaucracy at all levels in order to effectively implement the policy (Singh and Pundir, 2002).

Salient Characteristics of the Indian Dairy Sector

The dairy sector in India has emerged as an important source of livelihood for a vast majority of the rural population, especially the poor. Besides being a source of supplementary income and nutrition, the sector also provides draft power, fuel and organic manure. More importantly, the sector contributes significantly to the national economy of our country; for example, milk is the single largest contributor (in the agricultural sector) to the national GDPs of India.

Milk is the cheapest source of all nutrients when compared with other food items. Thus, it has an important role in national nutritional programmes, particularly for those below poverty line, children and expectant mothers. It is a boon for India, where per capita incomes are low and for about 40 per cent of the region's population incomes are below the level necessary to ensure adequate nutrition (Singh and Pundir, 2002).

In India, some 70 per cent of the cows and 60 per cent of the buffalo are nondescript and have very low productivity. To convert this huge population of low producing milch animals into high yielding milch animals, India needs a sound breeding policy. The breeding policy needs to consist of: (i) selective breeding of Indian dairy cattle for milk production; (ii) upgrading of the nondescript Indian cattle through breeding with selected Indian donors; (iii) selective breeding of the major buffalo breeds for milk production; and (iv) upgrading of nondescript and minor breeds of buffalo through breeding with the Murrah buffalo breed. Crossbreeding, as a tool to improve the quality of milch animals, is a time-tested technique in the country. However, organised breeding operations, mainly artificial insemination services under the government departments, reach only about 20 per cent of the breeding animals among cattle and less than 5 per cent of the buffalo (Singh and Pundir, 2002).

Smallholder milk producers are vulnerable to fluctuations in the prices of both milk and the inputs that go into milk production. Due to these characteristics, they have very low or practically no bargaining power vis-a-vis those to whom they sell their produce and from whom they buy their supplies. Consequently, they are exploited on both fronts, *i.e.* selling their milk and buying their production inputs. This heightens the need for government intervention in the sector through policies aimed at equalizing opportunities, at strengthening the bargaining power of milk producers in rural areas and at restraining the powerful from exploiting the weak. In fact, the governments have intervened in the dairy sector by launching the rural development programmes.

Milk production is less vulnerable than crop production to weather-induced risks and hence serves as an informal means of insurance for milk producers.

SWOT Analyses of the Dairy Sectors

The dairy sector in India has many strengths and weaknesses. The new era characterized by privatization and globalization has opened up many opportunities but also poses many threats to the smallholders. This section attempts a brief Strength, Weaknesses, Opportunities, Threats (SWOT) analysis of the dairy sector (Singh and Pundir, 2002).

Strengths

Sizeable population of high yielding cows and buffalo; huge domestic market for milk and milk products; good infrastructural and institutional support for dairying; high producer's share (89 per cent) in the consumer's price of milk; availability of all kinds of machinery and equipment for dairy plants at the most competitive rates in the world; a well-developed and professionally managed system of dairy co-operatives set up under Operation Flood; and the largest network of artificial insemination (AI) centres in the world.

Weaknesses

Small and scattered animal holdings; low milk yields; a large population of unproductive cattle; socio-cultural constraints on culling less productive/ unproductive animals; shortages of feed and fodder in many milksheds; competition between man and animals for scarce land and water resources; undue interference by the government in the affairs of dairy co-operatives; lack of strict regulation by the government of the unethical practices of unscrupulous private operators; lack of access for smallholders to institutional credit; lack of professional management; and lack of a well-defined national policy for dairy development.

Opportunities

Potential for increasing the productivity of milch animals and export of high quality dairy products since the new world trade regime came into effect; scope for dairy sector reforms by restructuring the Departments of Animal Husbandry in Indian states and reorienting their mandates; good scope for problem-solving and action-oriented research funded by private agencies; and good scope for privatization of animal health care services in selected areas.

Threats

Unregulated competition from national and multinational private companies; dumping of cheap dairy products on Indian markets by developed countries; unethical practices by unscrupulous private dairy operators; and inadequate public and private investment in modernization of the sector.

Major Concerns of Dairy Farmers

Dairy farmers around the world have the same goal of running a profitable and sustainable business. These farmers, however, face different challenges in different countries, to achieve their primary goal. IFCN (2005) reported on a survey undertaken with dairy farmers from 30 different countries throughout the world, to prioritize major concerns affecting their business future. The concerns were grouped in various categories with percentages as follows:

- ☆ 85 per cent reported that the joint market forces of low output returns and high input prices mattered most.
- ☆ 82 per cent considered policy factors (such as global market, local/national market, environmental/animal welfare issues) to be important.
- ☆ 80 per cent considered production factors (such as milk quota, labour, capital, land and animals) to be important.
- ☆ 50 per cent expressed concern that despite recent advancements in their dairy sectors, many of the farm strategic factors (such as skilled management, optimal farm size, reducing production costs, and diversification) lag below optimal levels for the majority of their farmers.
- ☆ 50 per cent expressed similar concerns about the direct farm factors (such as feeding, breeding, animal health).

When the findings for China, India and Pakistan were grouped together to represent Asia, the key issues were prioritized as follows:

1. Low milk yield
2. High feed prices and shortages of feed
3. Insufficient veterinary and breeding services
4. Access to credit
5. Access to markets (except in China)
6. Strong informal sector (especially in India and Pakistan)
7. Low adoption of technology

Improving the productivity, profitability and sustainability of smallholder farming is then a major pathway out of poverty. World Bank (2007) considers that this will require:

- ☆ Improving price incentives and increasing the quality and quantity of public investment
- ☆ Making product markets work better
- ☆ Improving access to financial services and reducing exposure to uninsured risks
- ☆ Enhancing the performance of producer organizations
- ☆ Promoting innovation through science and technology
- ☆ Making agriculture more sustainable and a provider of environmental services.

The shift from labour intensive towards capital intensive practices, both on-farm and in the market, is primarily due to the increased opportunity costs for labour. This shift to higher productivity of labour can be used as a measure of dairy development, reflecting change in all parts of dairy value chain.

In fact, Psilos (2008) has suggested that the rural wage rate is a key determinant of the optimum size of smallholder farms. Rising rural wages make other activities more attractive and tend to divert smallholders away from dairy farming. Because such dairy systems are labour intensive, their competitiveness relies on the low opportunity cost for labour. From surveys across the world's tropical smallholder dairy industries, dairy herd sizes grow as rural wage rates rise, when mixed farmers respond with capital investments (such as land for grazing as well as milking equipment and other labour saving devices) thus introducing economies of scale into their long-term plans for profitability and sustainability. Rising rural wages also provide the farming family with greater opportunities for more remunerative off-farm wages, hence they are less likely to contribute their labour to dairy farming activities. Such farms may transition to specialized, small commercial farms where the emphasis is more on cash remuneration rather than various forms of income and asset building that makes dairying attractive to smallholder multi enterprise farmers.

The role of cow manure as an income generator varies considerably around Asia. In country such as India, sun dried manure is an important domestic fuel for the kitchen where it can be an important contributor to farm profits. Some countries use manure in constructing houses, although its major sale benefit throughout Asia is as a fertilizer. In dense, subsistence regions in Kenya, Psilos (2008) noted that its sale value on smallholder farms was 130 per cent of the value of raw milk. Financial benefits from manure are reduced when rural wages rise because manure handling is a labour intensive process.

The attractiveness of smallholder dairying then depends on low labour costs and lack of access to other farm investments. Where opportunities for other uses of labour are low and where soil nutrients and land are scarce, smallholder mixed dairy producers can successfully outcompete larger more specialized producers locally because they require lower formal financial returns from sale of milk.

Infrastructure, in the form of roads and milk collection and handling facilities, can also greatly influence the milk marketing sector. In fact, it partially sets the farm gate price for milk as Staal *et al.* (2008) noted that poor feeder roads can reduce milk prices paid to farmers by 3 per cent for each additional km separating farm from market. This is not just due to the simple costs of transport, but also to the seasonal risks that such roads can impose.

Countries that do not have a strong tradition of milk production and consumption are particularly vulnerable to import competition and tend to be less self-sufficient in dairy products. This is more a function of demand rather than a lack of domestic supply. Where there are strong dairy traditions, most demand is for raw milk and traditional products, for which imports cannot easily be substituted, if at all. Supporting the development of traditional markets thus takes on the added feature of helping buffer domestic production from imports.

Development of Smallholder Dairy Farming

Smallholder dairy development provides opportunities to address the persistent problem of rural poverty by transferring income from affluent urban households to their poorer rural counterparts, and improve food and nutritional security for poor rural and urban households. Commercialization and intensification are frequently used as synonyms for development (Moran, 2009).

Objectives of Dairy Development

The three major objectives of any dairy development programs are to:

- ✰ Raise the living standards of traditional small-scale farmers and dairy market agents.
- ✰ Improve the nutrition of poor consumers; and, at the same time
- ✰ Sustain the natural resource base to ensure their long-term impact.

Such development can be achieved through:

- ✰ Creating employment in rural and peri-urban areas, both on-farm and along market distribution and value chains.
- ✰ Generating reliable income and asset accumulation for resource-poor farmers.
- ✰ Providing low cost and safe dairy products to poorly resourced (or using a recently coined term, resource-poor) consumers.
- ✰ Improving natural resource management and sustained farming systems through recycling dairy farm nutrients.
- ✰ Improving infant nutrition and social development in resource-poor households.

Dairy Development and Farm Technology

Dairy development is generally associated with technical changes to improve milk yield per milch animal. However it should be noted that:

- ☆ The use of exotic cattle is a rapid and potentially sustainable path to higher productivity, even for small-scale resource-poor farmers and in warm, semi arid or humid climates. However, there have been many repeated failures of such schemes for obvious but often ignored reasons.
- ☆ National and local breeding strategies need to address the realities of climate and disease risk to increase the likelihood of successful crossbreeding programs.
- ☆ Fodder technology should be an integral part of any dairy development program, particularly if it incorporates importation of stock of high genetic merit (or quality).
- ☆ The success or otherwise of intensive fodder production schemes is more likely to depend on availability of cheap labour, scarcity of land and good access to milk markets than it is on agro climatic setting. Where labour is scarce, intensive fodder cultivation practices and the feeding of crop residues to cattle are unlikely to be taken up unless mechanized. Promotion of such schemes should pay very close attention to labour opportunity costs (Moran, 2009).

Requisites for Long-term Sustainable Dairy Development

For dairy development to be sustainable, there must be:

- ☆ Adequate infrastructure and marketing opportunities.
- ☆ Access to reliable markets for increased milk production.
- ☆ Promotion through government policy.
- ☆ Availability of credit for purchasing of livestock and planting pastures.
- ☆ Available productive and adapted forage species.
- ☆ Ready access to information.
- ☆ A farm management system which ensures adequate feed throughout the year.
- ☆ Management of animal wastes.
- ☆ Disease control measures.
- ☆ Adequate hygiene for milk collection.

Improving Small-scale Farm Productivity

After several decades of dairy development in many Asian countries, average milk yields per cow per day still range between 8–10 kg as compared to average yields of 20–30 kg in developed countries. In addition, the average calving interval of dairy cows in smallholder farms is commonly as long as 16–20 months, when it could be reduced to 14–15 months. This clearly shows their low levels of farm productivity. Some technical solutions are available but they must be carefully selected so they will be suitable for small farmers and their socio-economic conditions. This means that scientists and extension workers must be able to understand factors influencing the acceptance of technology by farmers (Moran, 2009). Scientific knowledge alone cannot solve small-scale farm problems.

Some of the major challenges to such service providers are:

- Effective delivery of appropriate technology to benefit small-scale farmers at farm level
- Fair price policy and efficient rural livestock marketing systems promoted by national governments
- Promoting active and workable farmer groups or cooperatives
- Involving farmer participation in research and extension
- Linking public institutions and the private sector in technology delivery.

Chantalakhana and Skunmun (2002) recommended new strategies to facilitate dairy development which include:

- Establishing a national dairy board, consisting of representatives of all the industry stakeholders, to formulate and oversee national dairy policies to promote the smallholder dairy industry.
- Putting major inputs into strengthening dairy training for farmers, for example mobile extension units to provide on-farm advice.
- Establishing a national herd improvement program, firstly to select and multiply superior quality dairy sires and cows, and secondly to cull cows with below average milk yields. These cull cows could either be used for beef production in other areas or if slaughtered, the farmers should be provided with some compensation.
- Continued support for dairy research with highly selected topics aimed at solving 'real farmer's problems'.

Dairying is an important source of subsidiary income to small/marginal farmers and agricultural labourers. The manure from animals provides a good source of organic matter for improving soil fertility and crop yields. The gober gas from the dung is used as fuel for domestic purposes as also for running engines for drawing water from well. The surplus fodder and agricultural by-products are gainfully utilised for feeding the animals. Almost all draught power for farm operations and transportation is supplied by bullocks. Since agriculture is mostly seasonal, there is a possibility of finding employment throughout the year for many persons through dairy farming. Thus, dairy also provides employment throughout the year. The main beneficiaries of dairy programmes are small/marginal farmers and landless labourers (www.keralaagriculture.gov.in, 2013).

Dairy development helps the rural poor in having additional regular income. At present, unorganized milk traders put a stiff competition to the organized milk sector. Therefore, organized milk marketing has a dual task to attract more and more producer-sellers to its fold by offering good price, and to help producers to produce more milk. Co-operatives provide their members bargaining power, fair deal, and assistance in improving the productivity of the dairy business. But, the co-operatives have to face some problems also like low literacy of member-farmers, lack of their ownership over productive assets, lack of credit/finance, lack of training facilities, etc. Today in India, there are about 75,000 dairy cooperative societies, spread all over the country with a membership of 10 million. There are nearly 70 million households engaged in milk production, of which more than 10 million are in the co-operative sector. However, except for the brand name 'Amul', most of the state federation brands are regional. These include 'Verka' in Punjab, 'Nandini' in Karnataka, 'Vijaya' in Andhra Pradesh, 'Saras' in Rajasthan, 'Anchal' in Uttarakhand, 'Mother Dairy' in Delhi and Kolkata (Sharma *et al.*, 2007).

Dairy cooperatives account for a major share of processed liquid milk marketed in the country. As far as dairy industry is concerned, forecasting of demand, procurement of raw milk, and transportation of processed milk are the vital components of supply chain management. On this front, Indian co-operatives are little susceptible and lag behind the multinational corporations and other private firms. Non-observance of timeliness, inefficient distribution of milk, frequent break-downs of milk vans, delays in loading and unloading at depots lead to disturbed supply of milk and its products (Sharma *et al.*, 2007).

Package of Common Management Practices Recommended for Dairy Farmers

Modern and well established scientific principles, practices and skills should be used to obtain maximum economic benefits from dairy farming. Some of the major norms and recommended practices are as follows (www.keralaagriculture.gov.in, 2013):

Housing

1. Construct shed on dry, properly raised ground.
2. Avoid water-logging, marshy and heavy rainfall areas.
3. The walls of the sheds should be 1.5 to 2 meters high.
4. The walls should be plastered to make them damp proof.
5. The roof should be 3-4 metres high.
6. The cattle shed should be well ventilated.
7. The floor should be pucca/hard, even non-slippery impervious, well sloped (3 cm per metre) and properly drained to remain dry and clean.
8. Provide 0.25 metre broad, pucca drain at the rear of the standing space.
9. A standing space of 2 x 1.05 metre for each animal is needed.
10. The manger space should be 1.05 metre with front height of 0.5 metre and depth of 0.25 metre.
11. The corners in mangers, troughs, drains and walls should be rounded for easy cleaning.
12. Provide 5-10 sq. metre loaf space for each animal.
13. Provide proper shade and cool drinking water in summer.

14. In winter keep animals indoor during night and rain.
15. Provide individual bedding daily.
16. Maintain sanitary condition around shed.
17. Control external parasites (ticks, flies etc.) by spraying the pens, sheds with Malathion or Copper sulphate solution.
18. Drain urine into collection pits and then to the field through irrigation channels.
19. Dispose of dung and urine properly. A gobar gas plant will be an ideal way. Where gobar gas plant is not constructed, convert the dung alongwith bedding material and other farm wastes into compost.
20. Give adequate space for the animals. (The housing space requirement of crossbred cattle in various categories/age-groups is given in Table 2.1).

Table 2.1: Housing Space Requirements for Crossbred Cattle

Age-group	*Manger Space (mtr.)*	*Standing or Covered Area (sq.mtr.)*	*Open Space (sq.mtr.)*
4-6 months	0.2-0.3	0.8-1.0	3.0-4.0
6-12 months	0.3-0.4	1.2-1.6	5.0-6.0
1-2 years	0.4-0.5	1.6-1.8	6.0-8.0
Cows	0.8-1.0	1.8-2.0	11.0-12.0
Pregnant cows	1.0-1.2	8.5-10.0	15.0-20.0
Bulls*	1.0-1.2	9.0-11.0	20.0-22.0

*To be housed individually

Selection of Animal

1. Immediately after release of the loan purchase the stock from a reliable breeder or from nearest livestock market.
2. Select healthy, high yielding animals with the help of bank's technical officer, veterinary/animal husbandry officer of State government/Zilla Parishad, etc.
3. Purchase freshly calved animals in their second/third lactation.
4. Before purchasing, ascertain actual milk yield by milking the animal three times consecutively.
5. Identify the newly purchased animal by giving suitable identification mark (ear tagging or tattooing).
6. Vaccinate the newly purchased animal against disease.
7. Keep the newly purchased animal under observation for a period of about two weeks and then mix with the general herd.
8. Purchase a minimum economical unit of two milch animals.

9. Purchase the second animal/second batch after 5-6 months from the purchase of first animal.
10. As buffaloes are seasonal calvers purchase them during July to February.
11. As far as possible purchase the second animal when the first animal is in its late stage of lactation and is about to become dry, thereby maintaining continuity in milk production vis-a-vis income. This will ensure availability of adequate funds for maintaining the dry animals.
12. Follow judicious culling and replacement of animals in a herd.
13. Cull the old animals after 6-7 lactations.

Feeding of Milch Animals

1. Feed the animals with best feeds and fodders. (Feeding schedule is given in Table 2.2).
2. Give adequate green fodder in the ration.
3. As far as possible, grow green fodder on your land wherever available.
4. Cut the fodder at the right stage of their growth.
5. Chaff roughage before feeding.
6. Crush the grains and concentrates.
7. The oil cakes should be flaky and crumbly.
8. Moisten the concentrate mixture before feeding.
9. Provide adequate vitamins and minerals. Provide salt licks besides addition of mineral mixture to the concentrate ration.

Table 2.2: Feeding Schedules for Dairy Animals (Quantity in Kgs.)

Sl.No.	Type of Animal	Feeding During	Green Fodder	Dry Fodder	Concentrate
1	2	3	4	5	6
(A)	CROSSBRED COW				
	a) 6 to 7 litres milk per day	Lactation days	20 to 25	5 to 6	3.0 to 3.5
		Dry days	15 to 20	6 to 7	0.5 to 1.0
	b) 8 to 10 litres milk per day	Lactation days	25 to 30	4 to 5	4.0 to 4.5
		Dry days	20 to 25	6 to 7	0.5 to 1.0
(B)	BUFFALOES				
	a) Murrah (7 to 8 litres milk per day)	Lactation days	25 to 30	4 to 5	3.5 to 4.0
		Dry days	20 to 25	5 to 6	0.5 to 1.0
	b) Mehasana (6 to 7 litres milk per day)	Lactation days	15 to 20	4 to 5	3.0 to 3.5
		Dry days	10 to 15	5 to 6	0.5 to 1.0
	c) Surti (5 to 6 litrs milk per day)	Lactation days	10 to 15	4 to 5	2.5 to 3.0
		Dry days	5 to 10	5 to 6	0.5 to 1.0

10. Provide adequate and clean water.
11. Give adequate exercise to the animals. Buffaloes should be taken for wallowing daily. In case this is not possible sprinkle sufficient water more particularly during summer months.
12. To estimate the daily feed requirements remember that the animals consume about 2.5 to 3.0 percent of their body weight on dry matter basis.

Milking of Animals

1. Milk the animals two to three times a day.
2. Milk at fixed times.
3. Milk in one sitting within eight minutes.
4. As far as possible, milking should be done by the same person regularly.
5. Milk the animal in a clean place.
6. Wash the udder and teat with antiseptic lotions/luke-warm water and dry before milking.
7. Milker should be free from any contagious diseases and should wash his hands with antiseptic lotion before each milking.
8. Milking should be done with full hands, quickly and completely followed by stripping.
9. Sick cows/buffaloes should be milked at the end to prevent spread of infection.

Protection against Diseases

1. Be on the alert for signs of illness such as reduced feed intake, fever, abnormal discharge or unusual behaviour.
2. Consult the nearest veterinary aid centre for help if illness is suspected.
3. Protect the animals against common diseases.
4. In case of outbreak of contagious disease, immediately segregate the sick, in-contact and the healthy animals and take necessary disease control measures. (Vaccination schedule is given in Table 2.3).
5. Conduct periodic tests for Brucellosis, Tuberculosis, Johne's disease, Mastitis etc.
6. Deworm the animals regularly.
7. Examine the faeces of adult animals to detect eggs of internal parasites and treat the animals with suitable drugs.
8. Wash the animals from time to time to promote sanitation.

Breeding Care

1. Observe the animal closely and keep specific record of its coming in heat, duration of heat, insemination, conception and calving.
2. Breed the animals in time.

Table 2.3: Programme for Vaccination of Farm Animals against Contagious Diseases

Sl.No.	*Name of Disease*	*Type of Vaccine*	*Type of Vaccination*	*Duration of Immunity*	*Remarks*
1.	Anthrax (Gorhi)	Spore vaccine	Once in an year premonsoon vaccination	One season	–
2.	Black Quarter (Sujab)	Killed vaccine	– do –	– do –	–
3.	Haemorrhagic Septicaemia (Galghotu)	Ocladjuvant vaccine	– do –	– do –	–
4.	Brucellosis (Contagious abortion)	Cotton strain 19 (live bacteria)	At about 6 months of age	3 or 4 calvings	To be done only in infected herds
5.	Foot and Mouth disease (Muhkhar)	Polyvalent tissue culture vaccine 4 months later	At about 6 months of age with booster dose One season	After vaccination repeat vaccination every year in Oct./Nov.	
6.	Rinderpest (Mata)	Lapinised avianised vaccine for exotic and crossbred catte, caprinised vaccine for zebu cattle.	At about 6 months of age	Life long	It is better to repeat after 3 to 4 years

3. The onset of oestrus will be within 60 to 80 days after calving.
4. Timely breeding will help achieving conception within 2 to 3 months of calving.
5. Breed the animals when it is in peak heat period (*i.e.* 12 to 24 hours of heat).
6. Use high quality semen preferably frozen semen of proven sires/bulls.

Care during Pregnancy

Give special attention to pregnant cows two months before calving by providing adequate space, feed, water etc.

Marketing of Milk

1. Marketing milk immediately after it is drawn keeping the time between production and marketing of the milk to the minimum.
2. Use clean utensils and handle milk in hygienic way.
3. Wash milk pails/cans/utensils thoroughly with detergent and finally rinse with chloride solution.
4. Avoid too much agitation of milk during transit.
5. Transport the milk during cool hours of the day.

Care of Calves

1. Take care of new born calf.
2. Treat/disinfect the navel cord with tincutre of iodine as soon as it is cut with a sharp knife.
3. Feed colostrum to calf.
4. Assist the calf to suckle if it is too weak to suckle on its own within 30 minutes of calving.
5. In case it is desired to wean the calf immediately after birth, then feed the colostrum in bucket.
6. Keep the calf separately from birth till two months of age in a dry clean and well ventilated place.
7. Protect the calves against extreme weather conditions, particularly during the first two months.
8. Group the calves according to their size.
9. Vaccinate calves.
10. Dehorn the calves around 4 to 5 days of age for easy management when they grow.
11. Dispose of extra calves not to be reared/maintained for any specific purpose as early as possible, particularly the male calves.
12. The female calves should be properly reared.

Challenges and Opportunities for Small Livestock Holders under the New World Trade Regime

India had signed an agreement that led to the establishment of the WTO and therefore, is obliged to follow the dictates of the new world trade regime spearheaded by the WTO. The new regime concerning dairy products became effective on 1 July 1995. Liberalization of world trade in dairy products under the new trade regime poses new challenges and has opened up new export opportunities for the dairy industry. There is need to enhance the competitive economic advantage in dairy products, in terms of both quality and cost, and to enhance the credibility in international markets. The role of governments should be to direct, co-ordinate and regulate the activities of various organisations engaged in dairy development, to establish and maintain a level playing field for all stakeholders and to create and maintain a congenial socio-economic, institutional and political environment for smallholder dairy development through appropriate policies and programmes (Singh and Pundir, 2002).

The new trade regime is not expected to affect the overall world trade in milk and milk products. However, there will be some redistribution in terms of regions of origin and destination. It is expected that the decreased volume of subsidised exports of dairy products from several developed countries will be offset, to some extent, by increased export from countries like India, which do not subsidise their exports of dairy products (Singh and Pundir, 2002).

In order to benefit from the new trade opportunities, India will need to set and enforce high quality standards for various dairy products through an independent non-governmental authority and to improve the basic infrastructure (particularly the ports) and the air transport system. India will also need to improve the competitive advantage in milk production by improving milk yields to reduce the per litre cost of production and by improving the quality of the products by adopting the latest processing and packaging technologies and professional management. Compliance with phytosanitary specifications will also be necessary in order to increase the export of dairy products. A general switch to higher-value dairy products consequent upon increased access to high-priced markets in developed countries is also likely to occur (Sharma 2002).

In today's context, trade between neighbours is the harbinger of goodwill and economic uplift. In this context, the South Asian Preferential Trade Arrangement (SAPTA), which became operational in December 1995, is a welcome and significant development. It aims to facilitate trade among South Asian countries through preferential tariffs. The South Asian countries have identified a substantial number of commodities for preferential trading among themselves. Previously, they used to import/export some of these commodities indirectly from their neighbours through distant third parties. Besides the preferential tariff, the other gain to South Asian trade from SAPTA should be a drastic reduction in transportation costs. It is hoped that, following SAPTA, trade in dairy products among South Asian countries will usher in an era of prosperity in the region (Malhotra 1997).

Conclusion

Milk production in India has shown remarkable growth, but the potential role of dairy farming as a tool to increase household incomes, create rural employment and increase the regional competitiveness at producing milk are still to be realized. For dairy to play such a development role, there is an urgent need to provide the vast majority of small-scale dairy farmers with quality livestock services packaged in manners that are affordable and have maximum impacts on the key production and economic factors of their farms.

Smallholder dairy farmers are generally competitive and are likely to endure for many years to come, particularly where the opportunity costs of family labour and wages remain low. Furthermore, dairying is a viable enterprise even among the landless and socially marginalized groups. Policy makers should resist the all too common assumption that development efforts should move from smallholders towards supporting larger-scale, 'more efficient' milk producers to meet growing consumer demand. Instead, growing demand should be used as a stimulus to help continue and sustain smallholder dairy enterprises, particularly when they face increasing barriers to participate in value chain markets. One good model to encourage is 'colony farming', which is established in centralized governed societies such as China. With colony farming, smallholders house their herds together in a large dairy shed but are still responsible for feeding and maintaining their animals. These innovations require a large investment in buildings but they do allow smallholders to own and manage their own stock in a well-constructed durable shed and with the benefits of magnitude of size. This allows for communal forage production, large-scale silage making and bulk purchases of concentrates together with specialized labour undertaking machine milking and rearing of young stock.

Finally, it can be concluded that smallholder dairy production have the potential to poverty alleviation, food security, improved family nutrition and income and employment generation. However, disease, unpredictable milk market, high prices of drugs, feed concentrates and failure of AI are main constraints limiting small-scale dairy production. Therefore, in order to improve small-scale householders life style by the way of improving dairy production, there is a need for technical and institutional intervention to alleviate the constraints through dissemination of appropriate technologies for better disease prevention strategy, establishing the reliable milk market, availability of drugs with convenient price, feeding, artificial insemination service, improved dairy animals supply and awareness, which will significantly increase milk production and animal performance.

References

Azage T, Wold AG 1998. Prospects for peri-urban dairy development in Ethiopia. In: ESAP (Ethiopian Society of Animal Production), fifth national conference of Ethiopian Society of Animal Production, Addis Ababa, Ethiopia, pp 28-39.

Chantalakhana C, Skunmun P 2002. Sustainable smallholder animal systems in the tropics. Kasetsart University Press, Bangkok.

Hooten N 2008. Dairy development for the resource poor. Lessons for policy and planning strategies. In: Developing an Asian regional strategy for sustainable small holder dairy development. Proceedings of an FAO/APHCA/CFC funded workshop, Chiang Mai, February 2008. pp 48–51.

IFCN 2005. Dairy report 2005. For a better understanding of milk production worldwide. International Farm Comparison Network, Braunschweig, Germany.

Malhotra R 1997. SAARC dairy grid. In: Gupta P.R. (ed), Dairy India 1997. 5th edition. A-25 Priyadarshini Vihar, New Delhi, India.

Mason A, Lee S 2005. The demographic dividend and poverty reduction. In Proceedings of the United Nations Seminar on the Relevance of Population Aspects for the Achievement of the Millennium Development Goals. 17-19 November 2004, New York, USA http: //www.un.org/esa/population/publications/PopAspectsMDG/19_MASONA.pdf

Moran J 2009. Smallholder dairy farming in Asia. In: Business Management for Tropical Dairy Farmers. John Moran (ed). Landlinks Press, CSIRO Publishing, PO Box 1139, Collingwood, Victoria 3066, Australia, pp 25-42.

Mwankemwa ASA 2004. Performance of saving and credit co-operative societies and their impact on rural livelihoods: A case study of Morogoro rural and Mvomero districts, Tanzania. M. Sc. Dissertation, Sokoine University of Agriculture, Morogoro, Tanzania. pp 132.

Psilos P 2008. Competitiveness framework for Asian small holder dairy development. In: Developing an Asian regional strategy for sustainable small holder dairy development. Proceedings of an FAO/APHCA/CFC funded workshop, February 2008, Chiang Mai. pp 27–37.

Sharma ML, Saxena R, Mahato T, Das D 2007. Potential and Prospects of Dairy Business in Uttarakhand: A Case Study of Uttaranchal Cooperative Dairy Federation Limited. Agri Eco Res Review 20: 489-502.

Sharma VP 2002. Implications of international trade regulations (World Trade Organization agreement on agriculture and Codex Standards) for smallholder dairy development. In: Rangnekar D. and Thorpe W. (eds), Smallholder dairy production and marketing—Opportunities and constraints. Proceedings of a South–South workshop held at NDDB, Anand, India, 13–16 March 2001. NDDB (National Dairy Development Board), Anand, India, and ILRI (International Livestock Research Institute), Nairobi, Kenya.

Singh K, Pundir RS 2002. Problems and prospects of smallholder dairy production and marketing in South Asia: An overview. In: Rangnekar D. and Thorpe W. (eds), Smallholder dairy production and marketing—Opportunities and constraints. Proceedings of a South–South workshop held at NDDB, Anand, India, 13–16 March 2001. NDDB (National Dairy Development Board), Anand, India, and ILRI (International Livestock Research Institute), Nairobi, Kenya.

Sriri MT, Jaouhari MEl, Saydi A, Kuper M, Le Gal PY 2011. Supporting small-scale dairy farmers in increasing milk production: evidence from Morocco. Trop Anim Health Prod 43: 41–49.

Staal SS, Pratt AN, Jabbar M 2008. Dairy development for the resource-poor. A comparison of dairy policies and development in South Asia and East Africa. Pro-Poor Livestock Policy Initiative Working Paper No. 44–1, FAO Rome.

Uddin MN, Uddin MB, Al Mamun M, Hassan MM, Khan MMH 2012. Small Scale Dairy Farming for Livelihoods of Rural Farmers: Constraint and Prospect in Bangladesh. J Anim Sci Adv 2(6): 543-550.

von Braun J 2004. Small-scale farmers in liberalized trade environment. Proceedings of the seminar on October 2004 in Haikko, Finland. pp 21-52.

World Bank 2007. World development report 2008. Agriculture for development – an overview. The World Bank, Washington.

www.keralaagriculture.gov.in 2013. Dairy Farming for Small Farmer. http: // www.keralaagriculture.gov.in/htmle/bankableagriprojects/ah per cent 5Cdairyfarming.htm. website visited on December 05, 2013.

2016, Dairy and Food Product Technology Pages 39–48
Editors: Birendra Kumar Mishra and Subrota Hati
Published by: BIOTECH BOOKS, NEW DELHI

Chapter 3
Health Claims of Probiotic Foods in Human Life

D.C. Rai[1], *P.R. Patel*[1] *and M. Verma*[2]

[1]*Department of Animal Husbandry and Dairying,*
[2]*Department of Home Science,*
Banaras Hindu University, Varanasi – 221 005, Uttar Pradesh

Introduction

Probiotics mean "for life" and are defined as "living microorganisms which upon ingestion in certain numbers exert health benefits in humans and animals beyond inherent basic nutrition". The large number of probiotics currently used and available in dairy fermented foods, especially in yoghurt, dahi and shrikand. Lactic acid bacteria constitute a various group of organisms providing considerable benefits to humankind, some as natural inhabitants of the intestinal tract and others as fermentative lactic acid bacteria used in food industry, imparting flavor, texture and possessing preservative properties. Beyond these, some species are administered to humans as live microbial supplements, which positively influence our health mainly by improving the composition of intestinal microbiota. The term "probiotics foods" comprises some bacterial strains and products of animal origin containing physiologically active compounds benecial for human health and reducing the risk of chronic diseases, lactose maldigestion, diarrhea, irritable bowel syndrome, constipation certain cancers (colorectal, bladder, cervical, breast), coronary heart disease, urinary tract disease, upper respiratory tract and related infection, reduce serum cholesterol and blood pressure. But some harmful effect of the human being like infectious diarrhea or dysentery, inflammatory bowel disease (*e.g.*, ulcerative colitis and Crohn's disease),chronic stomach inflammation, tooth decay and periodontal disease, vaginal infections, stomach and respiratory infections that children acquire in daycare and skin infections.

The concept of probiotics (which means, "for life") was introduced in early 20th century by Nobel Prize laureate Metchnikoff (1907). According to the currently adopted definition by World Health Organization (2001) and the Food and Agriculture Organization of the United Nations, probiotics are: "Live microorganisms which when administered in sufficient amounts confer a health benefit on the host". There are a large number of probiotics currently used and available in dairy fermented foods, especially in yoghurt and dahi. Lactic acid bacteria like *Lactobacillus, Bifidobacterium* and *Saccaromyces* etc. constitute various group of organisms providing considerable benefits to mankind, some as natural inhabitants of the intestinal tract and others (*in vitro*) as fermentative lactic acid bacteria used in food industry, imparting flavor, texture and possessing preservative properties.

Studying longevity and general health of a Bulgarian population dwelling in the Rhodopes Mountains and fed basically on dairy products, the scientist introduced the idea that lactic acid bacteria in yogurt may neutralize deleterious effects of gut pathogens thus extending life span. He further contributed to the adoption of the name of the species, *Lactobacillus bulgaricus*, one of the two essential yoghurt starter microorganisms. This also meant the birth of modern dairy industry.

Strictly speaking, however, the term "probiotic" should be reserved for live microbes that have been shown in controlled human studies to impart a health benefit. Fermentation of food provides characteristic taste profiles and lowers the pH, which prevents contamination by potential pathogens. Fermentation is globally applied in the preservation of a range of raw agricultural materials (cereals, roots, tubers, fruit and vegetables, milk, meat, fish etc.).

Lactic Acid Bacteria (LAB)

A functional classification of nonpathogenic, nontoxigenic, Gram-positive, fermentative bacteria that are associated with the production of lactic acid from carbohydrates, making them useful for food fermentation. Species of Lactobacillus, Lactococcus, and Streptococcus thermophilus are included in this group. Since the genus Bifidobacterium is not associated with food fermentation and is taxonomically distinct from the other LABs, it is not usually grouped as a member of the LABs. Many probiotics are also LABs, but some probiotics (such as certain strains of E. coli, spore-formers, and yeasts used as probiotics) are not.

Human-Associated Microbial Ecosystems

The microbes that colonize humans are dynamic components of the body's ecosystem, both gaining and providing nourishment. In the process, these microbes aid in the development of intestinal cells and participate in the maturation and function of the innate immune system. Although much is still unknown about the microbes that colonize humans, some important points can be made.

- ✫ Each individual has his or her own unique population of microbes, even if there are commonalities of species among people.

- The microbes colonizing different regions of the human body (skin, mouth, gastrointestinal tract, vaginal tract of women) are both diverse and numerous, and they differ according to their habitat.
- Intestinal microbes are fairly stable through time, although transitions occur at weaning and again in the elderly. Colonizing microbiota can be impacted by antibiotics, diet, immunosuppression, intestinal cleansing, and other factors; however, the populations generally return to normal after being disturbed, with no intervention.
- Most colonizing microbes are not harmful in their natural body habitat, but some may generate undesirable metabolic end products.
- The composition of the "normal, healthy intestinal microbiota" is not defined currently. Likewise, the characteristics of the intestinal microbiota that may lead to many different disease states also are not well understood. The nature of the end products of growth of these microbes may be as important as which specific microbes are present.
- Activities such as regulating immune function, enhancing the intestinal barrier to prevent unwanted microbes from entering the blood stream, colonization resistance, and digestion are important functions of colonizing microbes.

Health Benefits

The most convincing evidence for use of probiotics is in reducing the duration of infectious infant viral diarrhea and dysentery a frequent cause of infant morbidity in developing countries. Usually managed by rehydration therapy, there is growing evidence to support the use of probiotics as a complementary therapy. Specific probiotic strains or mixtures have beenshown to reduce the disease time frame by as much as two days, a significant effect.

Areas of future interest for the application of probiotics include colon and bladder cancers, diabetes, heat-inactivated potential, neonatal intestinal infection and rheumatoid arthritis (Goldin and Gorbach 2008). Although scientific literature reports some conflicting data, the importance of strain, viable cell populations, matrix, host health status and method of production of the probiotic have been identified as critical to functionality and efficacy.

Probiotics are intended to assist the body's naturally occurring gut microbiota. Some probiotic preparations have been used to prevent diarrhea caused by antibiotics, or as part of the treatment for antibiotic-related dysbiosis. Studies have documented probiotic effects on a variety of gastrointestinal and extraintestinal disorders, including inflammatory bowel disease (IBD), irritable bowel syndrome (IBS), vaginal infections, and immune enhancement. Some probiotics have also been investigated in relation to atopic eczema, rheumatoid arthritis, and liver cirrhosis. Although there is some clinical evidence for the role of probiotics in lowering cholesterol, the results are conflicting. In general, the strongest clinical evidence for probiotics is related to their use in improving gut health and stimulating immune function as fallow:

Keeping Healthy People Healthy

This product concept developed further, the value of probiotics to prevent, rather than treat, disease was appreciated more fully. Toward this end, studies have been conducted in healthy populations, with end points such as decreasing the incidence of colds (de Vrese *et al.*, 2005), winter infections (Turchet *et al.*, 2003), or even absences from work (Tubelius *et al.*, 2005) or day care (Weizman *et al.*, 2005). These controlled human studies provide support that certain probiotic strains consumed as part of a daily diet will in-crease the number of illness-free days. Infants were helped by *Lactobacillus reuteri,* which decreased crying time due to colic (Savino *et al.*, 2007).

Lactose Maldigestion

Lactose is a sugar found in milk, composed of a glucose molecule linked to a galactose molecule. The people consume *dairy products with lactose*, they can develop gastrointestinal symptoms such as abdominal bloating, pain, flatulence, and diarrhea (de Vrese *et al.*, 2001). Yogurt (*Streptococcus thermophilus* and *Lactobacillus delbrueckii* subsp. *bulgaricus*) also produce lactase, and when consumed with dairy products can improve lactose digestion and symptoms in these individuals (Kolars *et al.*, 1984).

Bowel Transit

Daily consumption of one to three servings of fermented milk containing a probiotic strain, *Bifido bacterium* animalis, decreased the amount of time it took food to travel from the mouth to the anus for people who had longer-than-desired transit time (Marteau *et al.*, 2002).

Irritable Bowel Syndrome

Symptoms of abdominal pain, bloating, and flatulence commonly occur in patients with IBS. These symptoms may result in part from fermentations taking place in the colon that generate gas. The two other placebo-controlled trials have shown relief of abdominal bloating in patients with IBS treated with *Lactobacillus plantarum* (Kim *et al.*, 2003; Nobaek *et al.*, 2000). In children, *L. rhamnosus*decreased perceived ab-dominal distension but not abdominal pain (Bausserman and Michail 2005). The effect was more pronounced in elderly subjects and in women. A mixture of eight different strains of *Lactobacilli*, *Bifidobacteria*, and *S. thermophilus* had no effect on gastro-intestinal transit time in *irritable bowel syndrome* (IBS) subjects (Kim *et al.*, 2003).

Gastrointestinal Infections

The clinical trials have tested the efficacy of probiotics in the prevention of acute diarrhea, including antibiotic-associated diarrhea. Probiotics given along with antibiotic therapy have been shown to decrease the incidence of antibiotic-associated diarrhea in children and in adults. Different strains have been tested including *L. rhamnosus*, the yeast *Saccharomyces cerevisiae*, and undefined strains of *Lactobacillus acidophilus* and *L. delbrueckii* subsp. *Bulgaricus* (McFarland 2006; Sazawal *et al.*, 2006).

Probiotics have been tested as a strategy for eradication of *H. pylori* infection of the stomach. Some strains of lactic acid bacteria are known to inhibit the growth of *H.*

pylori in laboratory experiments. But results in human studies with different probiotics are mixed. Eradication of *H. pylori* was attempted by feeding yogurt containing probiotic strains selected for their ability to inhibit *H. pylori* in laboratory studies. (Wendakoon *et al.*, 2002; Lionetti *et al.*, 2006; Myllyluoma *et al.*, 2005; Sheu *et al.*, 2006 and Sykora *et al.*, 2005).

Prevention of Systemic Infections

The Bacterial translocation is the passage of bacteria through the lining of the intestine, which can lead to infection of organs or the blood. This passage can occur when patients have undergone surgical procedures or are seriously ill with critical conditions, such as severe acute pancreatitis, advanced liver cirrhosis, or multisystem organ failure. Probiotic organisms rarely translocate, even though a disturbed epithelium (Daniel *et al.*, 2006). In a study of patients with severe acute pancreatitis, treatment with *L. plantarum* significantly decreased the incidence of infection (Olah *et al.*, 2002).

Necrotizing enterocolitis resulting from immaturity and poor function of the gut mucosal barrier is a severe clinical condition that may occur in low birth weight premature infants. Two controlled studies have demonstrated that the use of probiotic mixtures in these infants significantly decreases the incidence and severity of necrotizing enterocolitis and prevents death (Bin-Nun *et al.*, 2005; Lin *et al.*, 2005).

Inflammatory Bowel Diseases

Ulcerative colitis, pouchitis, and *Crohn's disease* are chronic conditions of evidence suggests that abnormal activation of the *mucosal immune system* against the gut microbiota is the key event that triggers this abnormal inflammatory response that in turn causes ulcers in the gut that fail to heal, leading to chronic intestinal disease. Three studies investigated the effectiveness of an oral preparation of *Escherichia coli* compared with mesalazine, the standard treatment for maintenance of remission in patients with ulcerative colitis (Kruis *et al.*, 1997, 2004; Rembacken *et al.*, 1999).

The probiotic mixture proved highly effective for maintaining remission of chronic relapsing pouchitis (Gionchetti *et al.*, 2000; Mimura *et al.*, 2004). In Crohn's disease, however, clinical studies with *L. rhamnosus*GG failed to show efficacy in preventing postoperative recurrence of the disease (Prantera *et al.*, 2002) or as maintenance therapy (Bousvaros *et al.*, 2005).

Allergy

The prevalence of allergic diseases in western societies is increasing at an alarming rate. The effectiveness of *L. rhamnosus* in the prevention of atopic dermatitis has been reported in randomized, controlled trials (Kalliomaki *et al.*, 2001, 2003). In a subsequent study, this same strain was combined with three other probiotic strains, *L. rhamnosus, B. breve,* and *P. freuden-reichii* subsp. *shermanii*JS, and a *prebiotic* to determine the impact on the cumulative incidence of allergic dis-eases (Kukkonen *et al.*, 2007). Effectiveness in the management of cow's milk allergy in children is associated with the use of probiotics (Kirjavainen *et al.*, 2003).

Colon Cancer

Human intervention trials to confirm these animal studies are intrinsically difficult because of the natural history of the disease. A 4-year studied found that *L. casei* Shirota decreased the recurrence of atypical colonic polyps (Ishikawa *et al.*, 2005). The European Union (EU)-sponsored "Synbiotics and Cancer Prevention in Humans" project tested a synbiotic (oligofructose plus *L. rhamnosus* and *B. animalis* subsp. *lactis*) in patients at risk for colonic polyps. Among several intermediate end points that were used as biomarkers of colon cancer risk, the study found that the synbiotic decreased uncontrolled growth of intestinal cells (Van Loo *et al.*, 2005).

Vaginal Infection

Vaginal infections are caused mostly by fecal microbes ascending into the vaginal tract and displacing the normal *Lactobacilli microbiota*. The potential of using probiotic *Lactobacilli* to decrease the risk of bacterial or yeast vaginal infections or to improve the clinical outcome during treatment for these infections has captured the interest of researchers for decades (Reid and Bocking 2003 and Reid *et al.*, 2001). These studies have provided the best evidence to date for successful pro-biotic intervention to improve vaginal health. Some other recent studies have not shown positive results (Eriksson *et al.*, 2005).

Harmful/Unfriendly to Pathogen

"Unfriendly" microorganisms such as disease-causing bacteria, yeasts, fungi, and parasites can also upset the balance. Researchers are exploring whether probiotics could halt these unfriendly agents in the suppress their growth and activity in conditions like

- Infectious diarrhea or dysentery
- Irritable bowel syndrome
- Inflammatory bowel disease
- Infection with *Helicobacter pylori* (*H. pylori*), a bacterium that causes most ulcers and many types of chronic stomach inflammation
- Tooth decay and periodontal disease
- Vaginal infections
- Stomach and respiratory infections that children acquire in daycare
- Skin infections

Recent Findings

Recent studies have demonstrated that prophylactic administration of probiotics to preterm neonates decreases both the incidence and severity of subsequent necrotizing enterocolitis, mannanoligosaccharide, haematological and immunological.

Probiotics represent a therapeutic effort to bolster natural host defenses via the 'normalization' of abnormal gut microflora of the premature infant at risk, thereby reducing the subsequent threat of necrotizing enterocolitis, rotavirus-associated diarrhearadiation intestinal disease, vaginosis, ulcerative colitis,gastrointestinal tract's and the irritable bowel syndrome. The appeal of probiotics in neonatology is threefold.

- ✰ Safety record renders them an attractive alternative to many of the more aggressive therapeutic options
- ✰ They represent a simple, noninvasive attempt to recreate a natural or normal flora rather than a disruption of nature.
- ✰ The probiotics are used mainly for disease prevention and are naturally occurring. As such, they are not considered to be drugs, but rather food supplements.
- ✰ To date, very few other strategies have been proven definitively to be efficacious in decreasing the incidence of necrotizing enterocolitis.

References

Bausserman, M. and S. Michail (2005). The use of *Lactobacillus* GG in irritable bowel syndrome in children: A double-blind randomized con-trol trial. *J Pediatr*147: 197–201.

Bin-Nun, A., R. Bromike, M. Wilschanski, M. Kaplan, B. Rudensky, M. Caplan, and C. Hammerman (2005). Oral probiotics prevent necrotizing enterocolitis in very low birth weight neonates. *J Pediatr*147: 192–196.

Bousvaros, A., S. Guandalini, R. N. Baldassano, C. Botelho, J. Evans, G. D. Ferry, B. Goldin, L. Hartigan, S. Kugathasan, J. Levy, K. F. Mur-ray, M. Oliva-Hemker, J. R. Rosh, V. Tolia, A. Zholudev, J. A. Vanderhoof, and P. L. Hibberd (2005). A randomized, double-blind trial of *Lactobacillus* GG versus placebo in addition to standard maintenance therapy for children with Crohn's disease.*Inflamm Bowel Dis* 11: 833–839.

Daniel, C, S. Poiret, D. Goudercourt, V. Dennin, G. Leyer, and B. Pot (2006).Selecting lactic acid bacteria for their safety and functionality by use of a mouse colitis model.*Appl Environ Microbiol*72(9): 5799–5805.

deVrese, M., A. Stegelmann, B. Richter, S. Fenselau, C. Laue, and J. Schrezenmeir (2001). Probiotics—compensation for lactase insufficiency. Am J ClinNutr 73(Suppl): 421– 429.

de Vrese, M., P. Winkler, P. Rautenberg, T. Harder, C. Noah, C. Laue, S. Ott, J. Hampe, S. Schreiber, K. Heller, and J. Schrezenmeir (2005). Effect of Lactobacillus gasseri PA 16/8, Bifidobacteriumlongum SP 07/3, B. bifidum MF 20/5 on common cold episodes: A double blind, randomized, controlled trial. ClinNutr 24(4): 481–491.

Eriksson, K., B. Carlsson, U. Forsum, and P. G. Larsson (2005).A double-blind treatment study of bacterial vaginosis with normal vaginal lactobacilli after an open treatment with vaginal clindamycin ovules.*ActaDermVenereol*85(1): 42–46.

FAO/WHO (2001).Evaluation of Health and Nutritional Properties of Probiotics in Food, Cordoba, Argentina.

Gionchetti, P., F. Rizzello, A. Venturi, P. Brigidi, D. Matteuzzi, G. Bazzocchi, G. Poggioli, M. Miglioli, and M. Campieri (2000). Oral bacteriotherapy as maintenance treatment in patients with chronic pouchitis: A double-blind, placebo-controlled trial. *Gastroenterol- ogy*119: 305–309.

Gionchetti, P., F. Rizzello, U. Helwig, A. Venturi, K. M. Lammers, P. Brigidi, B. Vitali, G. Poggioli, M. Miglioli, and M. Campieri (2003). Prophylaxis of pouchitis onset with probiotic therapy: A double-blind, placebo-controlled trial. *Gastroenterology* 124: 1202– 1209.

Ishikawa, H., I. Akedo, T. Otani, T. Suzuki, T. Na-kamura, I. Takeyama, S. Ishiguro, E. Miyaoka, T. Sobue, and T. Kakizoe (2005).Randomized trial of dietary fiber and *Lactobacillus casei*administration for prevention of colorectal tumors.*Int J Cancer* 116: 762–767.

Kalliomaki, M., S. Salminen, H. Arvilommi, P. Kero, P. Koskinen, and E. Isolauri (2001). Probiotics in primary prevention of atopic dis-ease: A randomised placebo-controlled trial. *Lancet* 357: 1076–1079.

Kalliomaki, M., S. Salminen, T. Poussa, H. Ar-vilommi, and E. Isolauri (2003). Probiotic and prevention of atopic disease: 4-year follow-up of a randomised placebo-controlled trial. *Lancet* 361: 1869–1871.

Kim, H. J., M. Camilleri, S. McKinzie, M. B. Lempke, D. D. Burton, G. M. Thomford, and A. R. Zinsmeister (2003).A randomized controlled trial of a probiotic, VSL#3, on gut transit and symptoms in diarrhoea-pre-dominant irritable bowel syndrome.*Aliment Pharmacol Ther*17: 895–904.

Kim, H. J., M. I. Vazquez Roque, M. Camilleri, D. Stephens, D. D. Burton, K. Baxter, G. Thomforde, and A. R. Zinsmeister (2005). A randomized controlled trial of a probiotic combination VSL# 3 and placebo in irritable bowel syndrome with bloating. *Neurogastro-enterolMotil*17: 687–696.

Kirjavainen, P. V., S. J. Salminen, and E. Isolauri (2003). Probiotic bacteria in the management of atopic disease: Underscoring the impor-tance of viability. *J PediatrGastroenterolNutr* 36: 223–227.

Kolars, J. C., M. D. Levitt, M. Aouji, and D. A. Savaiano (1984). Yogurt—Anautodigesting source of lactose. *New Engl J Med* 310: 1–3.

Kruis, W., E. Schutz, P. Fric, B. Fixa, G. Judmaier, and M. Stolte (1997). Double-blind compari- son of an oral *Escherichia coli* preparation and mesalazine in maintaining remission of ulcerative colitis.*Aliment PharmacolTher*11: 853–858.

Kruis, W., P. Fric, J. Pokrotnieks, M. Lukas, B. Fixa, M. Kascak, M. A. Kamm, J. Weismueller, C. Beglinger, M. Stolte, C. Wolff, and J. Schulze (2004). Maintaining remission of ulcerative colitis with the probiotic *Esch-erichia coli* Nissle 1917 is as effective as with standard mesalazine. *Gut* 53: 1617–1623.

Kukkonen, K., E. Savilahti, T. Haahtela, K. Jun-tunen-Backman, R. Korpela, T. Poussa, T. Tuure, and M. M. Kuitunen (2007). Probiotics and prebiotic galacto-oligosaccharides in the prevention of allergic diseases: A randomized, double-blind, placebo-controlled trial. *J Al-lergy ClinImmunol*119(1): 192–198.

Lin, H. C., B. H. Su, A. C. Chen, T. W. Lin, C. H. Tsai, T. F. Yeh, and W. Oh (2005). Oral probiotics reduce the incidence and severity of necrotizing enterocolitis in very low birth weight infants. *Pediatrics* 115: 1–4.

Lionetti, E., V. L. Miniello, S. P. Castellaneta, A. M. Magista, A. de Canio, G. Maurogiovanni, E. Ierardi, L. Cavallo, and R. Francavilla (2006). *Lactobacillus reuteri*therapy to reduce side-effects during anti-*Helicobacter pylori* treatment in children: A randomized placeb P controlled trial. *Aliment PharmacolTher*24(10): 1461–1468.

Marteau, P., E. Cuillerier, S. Meance, M. F. Ger-hardt, A. Myara, M. Bouvier, C. Bouley, F. Tondu, G. Bommelaer, and J. C. Grimaud (2002).*Bifidobacteriumanimalis*strain DN- 173 010 shortens the colonic transit time in healthy women: A double-blind, randomized, controlled study. *Aliment PharmacolTher*16: 587–593.

Marteau, P., P. Seksik, and R. Jian (2002). Probiot-ics and intestinal health effects: A clinical perspective. *Br J Nutr*88(Suppl 1): S51–S57.

McFarland, L. V. (2006). Meta-analysis of probiotics for the prevention of antibiotic associated diar-rhea and the treatment of *Clostridium difficile*disease. *Am J Gastroenterol* 101(4): 812–822.

Mimura, T., R. Rizzello, U. Helwig, G. Poggioli, S. Schreiber, I. C. Talbot, R. J. Nicholls, P. Gionchetti, M. Campieri, and M. A. Kamm (2004).Once daily high dose probiotic therapy (VSL#3) for maintaining remission in recur-rent or refractory pouchitis.*Gut* 53: 108–114.

Myllyluoma, E., L. Veijola, T. Ahlroos, S. Tynk-kynen, E. Kankuri, H. Vapaatalo, H. Rautelin, and R. Korpela (2005). Probiotic supplemen-tation improves tolerance to *Helicobacter pylori* eradication therapy—a placebo-con-trolled, double-blind randomized pilot study. *Aliment PharmacolTher*21: 1263–1272.

Nobaek, S., M. L. Johansson, G. Molin, S. Ahrne, and B. Jeppsson (2000). Alteration of intesti- nalmicroflora is associated with reduction in abdominal bloating and pain in patients with irritable bowel syndrome. *Am J Gastroenterol*95: 1231–1238.

Olah, A., T. Belagyi, A. Issekutz, M. E. Gamal, and S. Bengmark (2002).Randomized clinical trial of specific *Lactobacillus* and fibre supple-ment to early enteral nutrition in patients with acute pancreatitis.*Br J Surg*89: 1103–1107.

Prantera, C., M. L. Scribano, G. Falasco, A. Andreoli, and C. Luzi (2002). Ineffective-ness of probiotics in preventing recurrence after curative resection for Crohn's disease: A randomised controlled trial with *Lactobacillus* GG. *Gut* 51: 405–409.

Reid, G. and A. Bocking (2003).The potential for probiotics to prevent bacterial vaginosis and preterm labor.*Am J ObstetGynecol*189(4): 1202–1208.

Reid, G., D. Beuerman, C. Heinemann, and A. W. Bruce (2001). Probiotic *Lactobacillus* dose required to restore and maintain a normal vaginal flora. *FEMS Immunol Med Microbiol* 32(1): 37–41.

Rembacken, B. J., A. M. Snelling, P. M. Hawkey, D. M. Chalmers, and A. T. Axon (1999). Non- pathogenic *Escherichia coli* versus mesala-zine for the treatment of ulcerative colitis: A randomised trial. *Lancet* 354: 635–639.

Sanders, M. E., Gibson, G., Gill, H. S. and F. Guarner (2007). Probiotics: Their Potential t P Impact Human Health. CAST 36(5): 1-20.

Savino, F., E. Pelle, E. Palumeri, R. Oggero, and R. Miniero (2007).*Lactobacillus reuteri* (American Type Culture Collection Strain 55730) versus simethicone in the treatment of infantile colic: A prospective randomized study. *Pediatrics* 119(1): e124–130.

Sazawal, S., G. Hiremath, U. Dhingra, P. Malik, S. Deb, and R. E. Black (2006). Efficacy of probiotics in prevention of acute diarrhoea: A meta-analysis of masked, randomised, placebo-controlled trials. *Lancet Infect Dis* 6(6): 374–382.

Stamatova I. and J. H. Meurman (2009). Probiotics: Health benefits in the mouth. Amer J Dent 22(6): 329-337.

Tubelius, P., V. Stan, and A. Zachrisson (2005).Increasing work-place healthiness with the probiotic *Lactobacillus reuteri*: A randomised, double-blind placebo-controlled study. *Envi-ron Health* 4: 25.

Turchet, P., M. Laurenzano, S. Auboiron, and J. M. Antoine (2003). Effect of fermented milk containing the probiotic *Lactobacillus casei*DN-114001 on winter infections in free-liv- ing elderly subjects: A randomised, controlled pilot study. *J Nutr Health Aging* 7(2): 75– 77.

Van Loo, J., Y. Clune, M. Bennett, and J. K. Collins (2005). The SYNCAN project: Goals, set-up, first results and settings of the human interven-tion study. *Br J Nutr*93(Suppl 1): 91–98.

Weizman, Z., G. Asli, and A. Alsheikh (2005). Ef-fect of a probiotic infant formula on infections in child care centers: Comparison of two probiotic agents. *Pediatrics* 115(1): 5–9.

Wendakoon, C. N., A. B. Thomson, and L. Ozimek (2002).Lack of therapeutic effect of a specially designed yogurt for the eradication of *Helico-bacter pylori* infection.*Digestion* 65: 16–20.

2016, Dairy and Food Product Technology Pages 49–59
Editors: **Birendra Kumar Mishra and Subrota Hati**
Published by: **BIOTECH BOOKS, NEW DELHI**

Chapter 4

Biofunctional Significance of Soybean Based Food Products

***Brij Pal Singh*[1], *Shilpa Vij*[1], *Subrota Hati*[2], *Deependra Singh*[1] *and Bharat Bhushan*[1]**

[1]***Dairy Microbiology Division,***
National Dairy Research Institute, Karnal – 132 001, Haryana
[2]***Dairy Microbiology Department,***
S.M.C. College of Dairy Science, Anand Agricultural University, Anand – 388 001, Gujarat

Introduction

Soybean-based foods have generated much interest because of the evidence that consumption of large amounts of soybean can lower the risk of chronic diseases such as cardiovascular disease and cancer. In addition, consumption of soy foods may reduce the risk of osteoporosis and help alleviate menopausal symptoms which are major health concerns for women (Genovese *et al.*, 2002). Soy food products are perceived as healthy foods and are considered an important part of the diet. More than 50 per cent consumers in the USA agreed that soy foods are healthy foods. Fermented soymilk is a good source of bioactive peptides such as anti-ACE, antioxidative, anti-cancer and immunomodulatory (Vij *et al.*, 2011). Many fermented soy milk based products such as soy cheese, soymilk-kefir, soy yoghurt etc. are produced. The major soy proteins are β-conglycinin and glycinin which constitute up to 90 per cent of the total soy protein (Gianazza *et al.*, 2003). Evaluation of these dietary proteins is very interesting because their hydrolysis by proteases produces peptides with biological activities. Many bioactive peptides have been isolated from soybean. Numerous studies of the enzymatic or chemical hydrolysis of soybean proteins have demonstrated their functional properties and physiological effects such as antimicrobial, antifungal, anticancer, antiobesity, antihypertensive, anti-inflammatory, hypocholesterolemic, immunostimulatory and antioxidant activities. Additionally, soybean is a rich source of isoflavone, which are reported to have beneficial estrogenic effects with potential antioxidant properties. The isoflavones present in soy milk deal with many health issues; most important being the prevention of various cancers, heart disease, osteoporosis, antioxidant etc. As it doesn't contain galactose, soy milk can safely replace breast milk in children with galactosemia. It is also good source of lecithin and vitamin E. It also helps in fighting the symptoms of menopause and in promoting eye health or anticatract. Soy milk is safe for people with lactose intolerance, or milk allergy, it can conveniently be used as a weaning formula instead of cow's milk for infants, and it also shows antidiabetic and antiobesity property. It has the highest protein content amongst plant products. As animal protein contains all the essential amino acids, lacking in pulse protein, soy is often used to replace the animal proteins in an individual's diet. Soybean is the only vegetable food that contains all eight essential amino acids.

Soy Based Products

Soybeans and its fermented and unfermented products are traditionally part of diet in the peoples of many countries from log times such as in China, Japan, and Korea, the bean and products made from it are a popular part of the diet (Table 4.1).

Soy Milk

Soy milk (also called soya milk, soymilk, soybean milk, or soy bean juice) and sometimes referred to as soy drink/beverage is a beverage made from soybeans. A stable emulsion of oil, water and protein, it is produced by soaking dry soybeans and grinding them with water. Soymilk produced by the traditional process presents the following composition, in average: 3.4 per cent of protein, 1.8 per cent of lipids, 1.5 per cent of carbohydrate and 0.4 per cent of ash. Soymilk provides a plentiful and inexpensive supply of protein and calories. It is considered as a suitable economical substitute for cow's milk and an ideal nutritional supplement for lactose-intolerant population.

Soy Flour

It is made entirely from defatted soy meal and is currently used worldwide by commercial processors. Soy flour is also a common ingredient in blended food aid products and can also be fortified with various micronutrients.

Soy Protein Concentrate

It is made wholly from defatted soy meal. Soy protein is flour like product consisting of about 70 per cent protein and is being used in a variety of meat systems, baked foods and dairy applications.

Soy Protein Isolates

It is made wholly from defatted soy meal and is used as an ingredient in high protein foods including dairy foods, nutritional supplements, meat systems, infant formulas, nutritional beverages, cream soups, sauces and snacks. It is also a good source of protein in milk replacers.

Table 4.1: Fermented Soy Products Originated From Different Countries

Product	Description	Origin
Cheonggukjang	A fermented soybean paste, contain whole as well as ground soybeans.	Korea
Doenjang	A traditional Korean fermented soybean paste.	Korea
Doubanjiang	A spicy, salty paste made from fermented soybeans.	China
Douchi	Used in the cuisine of China, where it is most widely used for making black bean sauce.	China
Mis	A traditional Japanese seasoning produced by fermenting rice, barley, and/or soybeans with salt and the fungus *kôjikin*, the most typical miso being made with soy. The result is a thick paste used for sauces and spreads, pickling vegetables or meats, and mixing with dashi soup stock to serve as miso soup called *misoshiru*), a Japanese culinary staple.	Japan
Natt	A traditional Japanese food made from soybeans fermented with *Bacillus subtilis*. It is especially popular as a breakfast food. As a rich source of protein, nattô and the soybean paste miso formed a vital source of nutrition in feudal Japan.	Japan
Soy sauce	A condiment produced from a fermented paste of boiled soybeans, roasted grain, brine, and *Aspergillus oryzae* or *Aspergillus sojae* molds. After fermentation, the paste is pressed, producing a liquid.	China
Tempeh	A traditional soy product originally from Indonesia. It is made by a natural culturing and controlled fermentation process that binds soybeans into a cake form, similar to a very firm vegetarian burger patty.	Indonesia
Tauc	A paste made from preserved fermented yellow soybeans in Chinese Indonesian cuisine. The name comes from the pronunciation in the Hokkien dialect and it is originated from China.	Indonesia

Douchi

Douchi is a type of fermented and salted soybean product. It is most famous in China, where they are most widely used for making black bean sauce. It is made by fermentation and salting of soybeans. It contain sharp flavor and spicy in smell, with a taste that is salty and somewhat bitter and sweet.

Soy Sauce

Soy sauce is a condiment made from paste of boiled soybeans, roasted grain with fermentation by *Aspergillus oryzae* or *Aspergillus sojae* molds. After fermentation the paste is pressed, producing a liquid, which is the soy sauce, and a solid by product. Soy sauce is a traditional ingredient in East and Southeast Asian cuisines, where it is used in cooking and as a condiment.

Tempeh

Tempeh is a traditional soy product that is originally from Indonesia. It is made by a natural culturing and controlled fermentation process that binds soybeans into a cake form, similar to a very firm vegetarian burger patty.

Tofu

Tofu also called bean curd is a food made by coagulating soy milk and then pressing the resulting curds into soft white blocks. There are many different varieties of tofu, including fresh tofu and tofu that has been processed in some way. Tofu has a low calorie count, relatively large amounts of protein, and little fat. It is high in iron and depending on the coagulant used in manufacturing, may also be high in calcium or magnesium.

Bioactive Components of Soy products

Various bioactive components present in soy foods like isoflavones, saponins, phytosterols, bioactive peptides etc. which contribute to various biofunctional properties.

Isoflavones

Isoflavones are a group of naturally occurring heterocyclic phenols found mainly in soybean. They are present in soybean at the level of 0.1 to 5 mg/g. Isoflavones are phytoestrogens because they are found in plant foods and appear to have estrogen-like activity. They are structurally similar to estrogen and bind to estrogen receptors. Many potential health benefits of isoflavones in soy products have been investigated, including effects on cancer, vascular disease, osteoporosis, menopausal symptoms, and cognitive function (Anderson *et al.*, 1997; Sirtori 2001). The three major groups of isoflavones found in soybeans are genistein, daidzein and glycitein. Isoflavones generally exist in soybeans and soy foods as aglycones (daidzein, genistein, and glycitein), β-glycosides (daidzin, genistin, and glycitin), acetylglycosides (63-*O*-acetyldaidzin, 63-*O*-acetylgenistin, and 63-*O*-acetylglycitin), and malonylglycosides (63-*O*-malonyldaidzin, 63-*O*-malonylgenistin, and 63-*O*-malonylglycitin) (Wang *et al.*, 1994).

Saponins

Saponins are compounds containing a steroid or triterpenoid aglycone linked to one or more oligosaccharide moieties (Liener, 1994). Saponins have antiviral activity against HIV, cholesterol level lowering activity (Sugano *et al.*, 1990) and antioxidative activity. Anti-tumor-promotion and growth inhibition of tumors or tumor cell lines by soy saponins have also been reported (Konoshima *et al.*, 1992). At high concentrations, saponins cause cell damage by disrupting the cell membrane or inducing apoptosis. The membranolytic activity of soybean saponins as indicated by their interaction with human colon carcinoma cells explains their role as anticarcinogens. The blood cholesterol-lowering properties of soybean saponins are also of particular interest in human nutrition. Saponins cause a depletion of body cholesterol by preventing its reabsorption, thus increasing its excretion. Studies on rats have shown soybean saponins to have an anabolic effect on bone components, suggesting its role as a nutritional factor in the prevention of osteoporosis. Soybeans contain 0.5–2 per cent saponins and are a significant source of saponins in human diets.

Phytosterols

Soybean phytosterols consist mainly of β-sitosterol, campesterol, and stigmasterol; β-sitosterol content being highest among them. It was reported that phytosterols such as β-sitosterol offer anticancer effects (Awad *et al.*, 2000), prostatic hyperplasia-lowering effects (Klippel *et al.*, 1997) and stimulation of a plasminogen activating factor (Kojima *et al.*, 1986). It has been demonstrated that soybean phytosterols are able to reduce serum cholesterol by inhibiting the absorption of cholesterol. Therefore, phytosterols are often supplemented into functional foods to improve human health. The FDA has also authorized a food claim for phytosterols: "Food containing at least 0.65 g per serving of plant sterol esters, eaten twice a day with meals for a daily total intake of at least 1.3 g, as part of a diet low in saturated fat and cholesterol, may reduce the risk of heart disease".

Bioactive Peptides

Bioactive peptides have been defined as specific protein fragments that have a positive impact on body functions or conditions and may ultimately influence health. Bioactive peptides are inactive within the sequence of the parent protein and can be released in three ways: (a) enzymatic hydrolysis by digestive enzymes, (b) food processing and (c) proteolysis by enzymes derived from microorganisms or plants. Once bioactive peptides are liberated, they may act as physiological modulators with hormone-like activity. Thus, these peptides represent potential health-enhancing nutraceuticals for food and pharmaceutical applications. The beneficial health effects of bioactive peptides may be several like antihypertensive, antioxidative, antiobesity, immunomodulatory, antidiabetic, hypocholesterolemic and anticancer. Fermentation is considered to be an efficient way to produce bioactive peptides. By soymilk fermentation, proteins are degraded into simpler form like oligopeptides and di, tri-peptides, thus relieving protein allergy problem and serve as good source of bioactive peptides. Calcium availability occurs at low pH, so by fermenting soymilk and utilizing complex sugars present in it, calcium availability also increases. Also because of low pH, harmful or pathogenic bacteria in the intestine get eliminated

Health Benefits of Soy Products

Antihypertensive

Hypertension is reported to affect 25 per cent of the population (National Centre for Health Statistics). In the U.S. alone, hypertension and its associated complications led to 35 million medical consultations in 2002 (Cherry *et al.*, 2002). ACE, widely distributed in mammalian tissues, nonspecific dipeptidyl carboxypeptidase associated with the regulation of blood pressure by modulating the rennin-angiotensin system. ACE is important for blood pressure regulation, because it catalyses the hydrolysis of inactive pro-hormone angiotensin I (decapeptide) to angiotensin II (octapeptide), which increase the blood pressure through vasoconstriction. Soy based products have shown excellent antihypertensive activity, it is may be due to bioactive peptide release during fermentation or hydrolysis. Several antihypertensive peptides studied in soy foods, they show their activity by inhibiting angiotensin-converting enzyme. Several ACE inhibitory bioactive peptides have been found in enzyme hydrolysates of soy proteins. These peptide fractions show blood pressure lowering effect when addminsterd orally to rats (SHR) at a level of 2.0 g/kg body (Chen *et al.*, 2003). Several ACE inhibitory peptides, such as Val-Ala-His-Ile-Asn-Val-Gly-Lys or Tyr-Val-Trp-Lys, were isolated from fermented soy products with *Bacillus natto* or *B. subtilis* (Kimura *et al.*, 2000). ACE inhibitory peptides have also been found in many traditional Asian fermented soy foods, such as soybean paste (His-His-Leu) (Shin *et al.*, 2001), soy sauce natto, and tempeh (Gibbs *et al.*, 2004).

Antoxidative

An antioxidant is a molecule that inhibits the oxidation of other molecules. Oxidation is a chemical reaction that transfers electrons or hydrogen from a substance to an oxidizing agent. Oxidation reactions can produce free radicals. In turn, these radicals can start chain reactions. When the chain reaction occurs in a cell, it can cause damage or death to the cell. Antioxidants terminate these chain reactions by removing free radical intermediates, and inhibit other oxidation reactions. Cell culture studies suggest that isoflavones may enhance the cellular antioxidant network by increasing metallothione in mRNA levels, inhibiting peroxynitrite-mediated LDL oxidation by delaying tyrosine nitration and activating glutathione peroxidase or inhibiting superoxide production and thus, cell-mediated LDL modification (Hwang *et al.*, 2003). During hydrolysis, the soy protein structure will be altered and more active amino acid R group will be exposed. Therefore, soybean peptides can have higher antioxidant activity than intact protein (Chen *et al.*, 1998). After enzyme digestion of β-conglycinin and glycinin the radical-scavenging activities were increased 3 to 5 times. The antioxidant capacity of soy peptides is dependent on its structure and therefore affected by the hydrolysis procedures. Many antioxidative peptides have been isolated from fermented soy product when it ferments with *Lactobacillus plantarum Lp 6* (Amadou *et al.*, 2010).

Hypcholestromic

The beneficial effects of soybean on cardiovascular diseases were first considered through its impact on blood cholesterol. When soy proteins administered to animals or human subjects by orally these subjected to protease digestion in the GI tract releasing the bioactive peptides, which then may lower cholesterol levels (Potter., 1995). Dietary intakes of soy protein (20 g/d) for 5 week would be effective in reducing CHD risk among high-risk, middle-aged men (Sagara *et al.*, 2004). Soybean phytosterols are able to reduce serum cholesterol by inhibiting the absorption of cholesterol.

Immunomodulatory

Innate and adaptive responses of immune system play important role in host defense against microorganisms. The innate immune response, forming the first line of defense against attack by pathogens, is mediated by macrophages, dendritic cells

(DC) and natural killer (NK) cells. Soy-based products has been shown to modulate some aspects of the immune (Maskarinec *et al.*, 2009), possibly due to its isoflavone content. Also consumption of fermented soy products has been suggested to be of possible benefit for allergic conditions. Recently, some soy proteins and their digests have been shown to possess immunomodulatory activities. Soymetide-13 (Met- Ile-Thr-Leu-Ala-Ile-Pro-Val-Asn-Lys-Pro-Gly-Arg) from trypsin digests of soybean protein as a peptide stimulating phagocytosis in vitro by human polymorphonuclear leukocytes (Tsuruki, 2003). Liu *et al.* (2002) report that soymilk kefir elevates intestinal IgA levels as well as inhibits sarcoma tumor cell growth in an in *vivo* murine mode. Antimicrobial activity of soy yoghurts was also carried out in our lab, it showed significantly higher (P<0.0001) inhibition against all the pathogens, with maximum against *S.typhi, L. monocytogenes, Shigella spp.* and *S. enteritidis* (Yadav., 2012).

Anticancer

Genistein and daidzein are the two principal components in soy products and have been the focus of studies investigating the role of soy isoflavones in cancer prevention. Numerous biological activities have been documented for genistein and daidzein *in vitro*. Genistein is a known tyrosine kinase inhibitor inhibits topoisomerase I and II and inhibits angiogenesis (Fotsis *et al.*, 1993). The evidence of an effect of soy peptides on oncogenesis is preliminary, however, the newly discovered peptide lunasin has shown promise as an anticancer agent. Other soy peptides with anticancer properties include Kunitz trypsin inhibitor, a peptide that was reported to suppress ovarian cancer cell invasion by blocking urokinase upregulation. Lunasin, a 43-amino acid peptide naturally present in soy protein, has been found to suppress transformation of mammalian cells induced by carcinogens and viral oncogenes E1A and RAS (Jeong *et al.*, 2007). This novel peptide can be found in amounts ranging from 0.10 to 1.33 g/100 g flour in different soybean varieties and in commonly available soy proteins. Lunasin internalizes into mammalian cells within minutes of exogenous application, and localizes in the nucleus after 18 hours. It inhibits acetylation of core histones in mammalian cells (Lumen., 2005). Epidemiological evidence suggests that diets rich in soybean products are associated with lower cancer mortality rates, particularly for cancers of the colon, breast, and prostate.

Antidiabetic

The incidence of type 2 diabetes is increasing in Western populations in an alarming rate compared to Asian populations, because Asians consume fermented soybean products which are being unique to the traditional Asian diet. However, soy peptides and dietary phytoestrogens in fermented soybean foods consumed in traditional Asian diets may help prevent and slow the progression of type 2 diabetes. Type 2 diabetes is a heterogeneous metabolic disorder initiated by both insulin deficiency and peripheral insulin resistance. The pancreatic β cell adapts to increased nutrient availability and insulin resistance by increasing its function and mass. These processes are driven by signals derived from nutrient metabolism, hormones and cytokines. The failure of β-cell adaptation results in type 2 diabetes. Thus, β-cell dysfunction and deficient mass pancreatic mass play crucial roles in the development of insulin resistance and subsequent β-cell compensation failure and diabetes (Kwon *et al.*, 2010). Soybean products are considered a good substitute for animal protein and their nutritional value is almost equivalent to that of animal protein because soy proteins contain most of the essential amino acids for human nutrition. Soybean products have less saturated fat and various phytochemicals, which possess numerous health benefits. In particular, the association of high-quality protein and phytochemicals, especially isoflavones, is unique among plant-based proteins because isoflavones are not widely distributed in plants other than legumes. The large protein, lipid and carbohydrate molecules in raw soybean are broken down by enzymatic hydrolysis during fermentation to small molecules such as peptides, amino acids, fatty acids, and sugars, which are responsible for the unique sensory and functional properties of the final products. Short-term fermented soy foods such as chungkookjang and tempeh, which are fermented with *Bacillus subtilis* and *Rhizopus oligosporus* (Wang *et al.*, 2008). respectively, for less than 72 hours have a much greater concentration of large molecules than do long-term fermented foods including doenjang and miso, which are fermented for more than 6 months with *Aspergilus* and *Bacillus* species from rice straw and koji, respectively. Of this hydrolyzed protein, 65 per cent remains in the fermented products as amino acids and peptides, 25 per cent is assimilated into the mold biomass, and 10 per cent is oxidized. After the initial stage of fermentation, however, soy proteins are rapidly degraded and only 9 per cent to 17 per cent of the crude protein remains at the end stage of long-term fermentation. In addition a G-protein coupled receptor called cholecystokinin receptor-type 1 (CCK1R), is activated by the natural peptide ligand cholecystokinin (CCK).There are indications that bioactive peptides from food protein have similar effect as CCK, (Staljanssens *et al.*, 2012) therefore it can be used as an ingredient for functional foods has relevance for the treatment and prevention of obesity.

Menstrual Cycle and Osteoporosis

The ingestion of high concentrations of isoflavones has adversely affected reproduction in several animal species and in premenopausal women daily ingestion of soy protein has been reported to lengthen the menstrual cycle and to suppress the usual midcycle surge in pituitary gonadotropins, effects that epidemiological evidence suggests are beneficial in decreasing risk of breast cancer. As isoflavones are weak estrogens, ingestion of soybean foods was earlier proposed as an alternative to hormone replacement therapy for post-menopausal women. It is estimated that by the year 2010, 35 million women in the United States either will have osteoporosis or be at risk of developing the disease if appropriate preventive measures are not taken (Arjmandi *et al.*, 2005). In women, osteoporosis most often occurs after menopause when the ovaries stop producing estrogen. Animal studies, as well as a double blind placebo controlled trials in humans suggest that genistein can help restore bone protection (Morabito *et al.*, 2002).

Conclusion

Soy products especially fermented has many nutritional benefits including reducing cardiovascular disease, reducing menopausal symptoms, weight loss, arthritis and brain function. They contain phytochemicals such as isoflavones, saponins, phytosterols as well as bioactive peptides that promote health. When we consume food these bioactive peptides can be release in GI system and exert their action on specific target organs after absorption. Commercial interest in production and use of bioactive peptides has been increased recently but on industrial-scale production of such peptides is still not well established. So more work is required to develop a large-scale of protein hydrolysates to obtain products enriched with biologically active peptides of the specific function that could be used as additives in functional foods.

References

Amadou, I., Gbadamosi, O., Shi, Y.H., Kamara, M., Jin, S., and Le, G.W. 2010. Identification of Antioxidative peptides from *Lactobacillus plantarum* Lp 6 fermented soybean protein meal. *Research J. of Microiology*. **5(5)**: 372-380.

Anderson, J., Garner S. 1997. Phytoestrogens and human function. *Nutrition Today* **32(6)**: 232.

Arjmandi, B., Lucas, E., Khalil, D., Devareddy, L., Smith, B., McDonald, J. 2005. One year soy protein supplementation has positive effects on bone formation markers but not bone density in postmenopausal women. *Nutrition Journal* **4(1)**: 8.

Awad, A, . Fink, C. 2000. Phytosterols as anticancer dietary components: evidence and mechanism of action. *The Journal of Nutrition* **130(9)**: 2127.

Chen, H.M., Muramoto, K., Yamauchi, F., Fujimoto, K., Nokihara, K., 1998. Antioxidative properties of histidine-containing peptides designed from peptide fragments found in the digests of a soybean protein. *J Agric Food Chem* **46**: 49–53.

Chen, J., Okada, T., Muramoto, K., Suetsuna, K., and Yang, S. 2003. Identification of angiotensin I-converting enzyme inhibitory peptides derived from the peptic digest of soybean protein. *J Food Biochem* **26(6):** 543–54.

Cherry, D.K., and Woodwell, D.A. 2002. National ambulatory medical care survey: 2000 summary. Advanced data from vital and health statistics. *National Center for Health Statistics Hyattsville*, MD. No. **328.**

Fotsis, T., Pepper, M., Adlercreutz, H., Fleischmann, G., Hase, T., Montesano, R. 1993. Genistein, a dietary-derived inhibitor of *in vitro* angiogenesis. *Proceedings of the National Academy of Sciences* **90(7)**: 2690.

Genovese, M.I., and Lajolo, F.M. 2002. Isoflavones in soy-based foods consumed in Brazil: levels, distribution and estimated intake. *Journal of Agricultural and Food Chemistry*, Vol. 50, No. **21**, 5987-5993. ISSN: 0021-8561.

Gianazza, E., Eberini, I., Arnoldi, A., Wait, R., and Sirtori, C.R. 2003. A proteomic investigation of isolated soy proteins with variable effects in experimental and clinical studies. *The Journal of Nutration*, Vol. 133, No. **1**, 9–14. ISSN: 0022-3166.

Gibbs, B.F., Zougman, A., Masse, R., and Mulligan, C. 2004. Production and characterization of bioactive peptides from soy hydrolysate and soy-fermented food. *Food Research International*. **37**: 123–131.

Hwang, J., Sevanian, A., Hodis, H., Ursini, F. 2000. Synergistic inhibition of LDL oxidation by phytoestrogens and ascorbic acid. *Free Radical Biology and Medicine* **29(1)**: 79-89.

Jeong, H.J., Jeong, J.B., Kim, D.S., Park, J.H., Lee, J.B., Kweon, D.H., Chung, G.Y., Seo, E.W., And Lumen, B.O. 2007. The cancer preventive peptide lunasin from wheat inhibits core histone acetylation. *Cancer Letters*. Vol. **255, No. 1**, 42–48. ISSN: 0304- 3835.

Kimura, A., Takada, A., Okada, T., and Yamada, H, 2000. Microbial manufacture of angiotensin I-converting enzyme inhibiting peptides. *Japan: Toyo Hatsuko K.K.: Jpn. Kokai Tokkyo Koho JP 2000229996* A2 22.

Klippel, K., Hiltl, D., Schipp, B. 1997. A multicentric, placebo-controlled, double-blind clinical trial of and beta-sitosterol (phytosterol) for the treatment of benign prostatic hyperplasia. *British Journal of Urology* **80(3)**: 427-432.

Kojima, S., Soga, W., Hagiwara, H., Shimonaka, M., Saito, Y., Inada, Y. 1986. Visible fibrinolysis by endothelial cells: effect of vitamins and sterols. *Bioscience Reports* **6(12)**: 1029-1033.

Konoshima, T., Kokumai, M., Kozuka, M., Tokuda, H., Nishino, H., Iwahima, A. 1992. Anti-tumor-promoting activities of afromosin and soyasaponin I isolated from Wistaria brachybotrys. *Journal of Natural Products* **55(12)**: 1776-1778.

Kwon, D.Y., Daily, J.W., Kim, H.J., Park, S. Antidiabetic Effects of Fermented Soybean Products on Type 2 Diabetes. *Nutrition Research*. **2010**, 30, 1-13.

Liener, I. 1994. Implications of antinutritional components in soybean foods. *Critical Reviews in Food Science and Nutrition* **34(1)**: 31-67.

Lumen, B.O. 2005. Lunasin: A Cancer-preventive soy peptide. *Nutr Rev* **63**: 16-21.

Maskarinec, G., Steude, J., Franke, A., Cooney, R. 2009. Inflammatory markers in a 2-year soy intervention among premenopausal women. *Journal of Inflammation* **6(1)**: 9.

Morabito, N., Crisafulli, A., Vergara, C., Gaudio, A., Lasco, A., Frisina, N. 2002. Effects of Genistein and Hormone-Replacement Therapy on Bone Loss in Early Postmenopausal Women: A Randomized Double-Blind Placebo-Controlled Study. *Journal of Bone and Mineral Research* **17(10)**: 1904-1912.

Potter, S.M. 1995. Overview of proposed mechanisms for the hypocholesterolemic effect of soy. *J Nutr* **125(3):** 606S.

Sagara, M., Kanda, T., Njelekera, M., Teramoto, T., Armitage, L., Birt, N., Birt, C., and Yamori, Y. 2004. Effects of dietary intake of soy protein and isoflavones on cardiovascular disease risk factors in high risk, middle-aged men in Scotland. *J Am Coll Nutr* **23(1):** 85–91.

Shin, Z.I., Yu, R., Park, S.A., Chung, D.K., Ahn, C.W., Nam, H.S., Kim, K.S., and Lee, H.J. 2001. His-His-Leu, an angiotensin I converting enzyme inhibitory peptide derived from Korean soybean paste, exerts antihypertensive activity in vivo. *J Agric Food Chem* **49:** 3004–9.

Sirtori, C. 2001. Risks and benefits of soy phytoestrogens in cardiovascular diseases, cancer, climacteric symptoms and osteoporosis. *Drug Safety* **24(9)**: 665-682.

Staljanssens, D., Camp, J. V., Billiet, A., Meyerc, T.D., Shukora, N.A., Vos, W.H., Smagghe, G. Screening of soy and milk protein hydrolysates for their ability to activate the CCK1 receptor. *Peptides*. **2012**, 34(1), 226–231.

Sugano, M., Goto, S., Yamada, Y., Yoshida, K., Hashimoto, Y., Matsuo, T., and Kimoto, M. 1990. Cholesterol-lowering activity of various undigested fractions of soybean protein in rats. *J Nutr* **120:** 977–85.

Tsuruki, T., Takahata, K., Yoshikawa, M. 2005. Anti-alopecia mechanisms of soymetide-4, an immunostimulating peptide derived from soy [beta]-conglycinin. *Peptides* **26(5)**: 707-711.

Vij, S., Hati, S., and Yadav, D. 2011. Biofunctionality of Probiotic Soy Yoghurt. *Food and Nutrition Sciences*, 2011, **2**, 502-509.

Wang, H., Murphy, P. 1994. Isoflavone content in commercial soybean foods. *Journal of Agricultural And Food Chemistry* **42(8)**: 1666-1673.

Wang, S.H., Zhang, C., Yang, Y.L., Diao, M., Bai, M.F. Screening of a High Fibrinolytic Enzyme Producing Strain and Characterization of the Fibrinolytic Enzyme Produced from *Bacillus subtilis* LD-8547. *World Journal of Microbiology and Biotechnology*. 2008, 24, 475-482.

Yadav, D. 2012. Preparation and evaluation of Biofunctional attributes of soy based probiotic yoghurt. *Ph.D thesis. N.D.R.I. Karnal 2012.*

2016, Dairy and Food Product Technology Pages 61–75
Editors: Birendra Kumar Mishra and Subrota Hati
Published by: BIOTECH BOOKS, NEW DELHI

Chapter 5

Probiotic, Prebiotic and Synbiotic Rich Fermented Food and Human Health: An Overview

D.C. Rai and Dilip Kumar

Department of A. H. and Dairying, Institute of Agricultural Sciences, Banaras Hindu University, Varanasi, Uttar Pradesh

Introduction

Fermentation is an ancient method of food preservation and nutrition enhancement that humans discovered thousands of years ago through careful observation. Modern scientists have determined that the benefits of fermented foods and beverages include digestive support and improved mental health.

The term "fermentation" is sometimes used to specifically refer to the chemical conversion of sugars into ethanol, a process which is used to produce alcoholic beverages such as wine, beer, and cider. Fermentation is also employed in the leavening of bread (CO_2 produced by yeast activity), in preservation techniques to produce lactic acid in sour foods such as sauerkraut, dry sausages, kimchi, and yogurt; and in pickling of foods with vinegar (acetic acid).

History of Fermented Food

Fermented products with live organisms have existed for centuries and were developed because of the need to extend the shelf-life of common and precious commodities like dairy and became one of the oldest forms of preservation. The ability of beneficial bacteria to transform milk into a longer lasting food was recorded as far back as 6,000 years ago. Long before the existence of micro-organisms was even known or recognized, fermented products were used therapeutically, to treat colds, fevers, and ailments of the gastro- intestinal tract such as constipation and diarrhoea.

Natural fermentation precedes human history. Since ancient times, however, humans have been controlling the fermentation process. The earliest evidence of an alcoholic beverage, made from fruit, rice, and honey, dates from 7000–6600 BCE, in the Neolithic Chinese village of Jiahu (McGovern *et al.*, 2004) and winemaking dates from 6000 BCE, in Georgia, in the Caucasus area (8,000-year-old wine unearthed in Georgia, 2003). Seven-thousand-year-old jars containing the remains of wine, now on display at the University of Pennsylvania, were excavated in the Zagros Mountains in Iran (7,000 Year-old Wine Jar, wikipedia). There is strong evidence that people were fermenting beverages in Babylon circa 3000 BC (Fermentation in food processing, wikipedia), ancient Egypt circa 3150 BC (Cavalieri, 2003), pre-Hispanic Mexico circa 2000 BC (Fermentation in food processing, wikipedia), and Sudan circa 1500 BC (Dirar, 1993).

French chemist Louis Pasteur was the first known *zymologist*, when in 1856 he connected yeast to fermentation. Pasteur originally defined fermentation as "respiration without air". Pasteur performed careful research and concluded:

I am of the opinion that alcoholic fermentation never occurs without simultaneous organization, development, and multiplication of cells,. If asked, in what consists the chemical act whereby the sugar is decomposed. I am completely ignorant of it.

In the last 100 years, most traditional fermented foods have practically been eliminated from our diet. In this country, about the only fermented food we continue to eat with any regularity is pickles made from fermented cucumbers. But commercial pickles are made using vinegar instead of just salt and water—and then they are pasteurized, which kills all the lactic acid bacteria. This process, in effect, renders the product nearly useless when it comes to improving health. This fact should be fairly evident when you consider the 35 million people in this country alone who suffer from irritable bowel syndrome, and the millions more who suffer from ulcers, indigestion, recurring vaginal infections, chronic constipation or diarrhoea, and dozens of other related health problems.

Other History shows that:

- ☆ During the Roman era, people consumed sauerkraut because of its taste and health benefits.
- ☆ In ancient India, it was common to enjoy lassi, a pre-dinner yogurt drink. This traditional practice is anchored on the principle of using sour milk as a probiotic delivery system to the body.
- ☆ Bulgarians are known for their high consumption of fermented milk and kefir, and for their high level of health.

Ukrainians consumed probiotics from a fermented food list that included raw yogurt, sauerkraut, and buttermilk. Various Asian cultures ate pickled fermentations of cabbage, turnips, eggplant, cucumbers, onions, squash, and carrots, and consume these fermented treats until today.

Probiotics: The Good Bacteria

The World Health Organization defines *probiotics* as "live microorganisms, which, when administered in adequate amounts, confer a health benefit on the host."

Probiotics are usually defined as microbial food supplements with beneficial effects on the consumers. Most probiotics fall into the group of organisms known as lactic acid-producing bacteria and are normally consumed in the form of yogurt, fermented milks or other fermented foods. Some of the beneficial effect of lactic acid bacteria consumption include:

- Improving intestinal tract health;
- Enhancing the immune system, synthesizing and enhancing the bioavailability of nutrients; reducing symptoms of lactose intolerance, decreasing the prevalence of allergy in susceptible individuals; and
- Reducing risk of certain cancers.

The mechanisms by which probiotics exert their effects are largely unknown, but may involve modifying gut pH, antagonizing pathogens through production of antimicrobial compounds, competing for pathogen binding and receptor sites as well as for available nutrients and growth factors, stimulating immunomodulatory cells, and producing lactase. Selection criteria, efficacy, food and supplement sources and safety issues around probiotics are reviewed. Recent scientific investigation has supported the important role of probiotics as a part of a healthy diet for human as well as for animals and may be an avenue to provide a safe, cost effective, and 'natural' approach that adds a barrier against microbial infection. This paper presents a review of probiotics in health maintenance and disease prevention.

Probiotics are the fastest growing component of the functional food industry. They have been aiding human health since biblical times, through their role in increasing the number of beneficial bacteria in the gut. These living microorganisms are primarily derived from lactic acid bacteria, comprising multiple strains from the genera Lactobacillus and Bifidobacterium. Some foods combine several different probiotics based on the unique attributes associated with each strain. Probiotics are most often found in yogurt, fermented milk, cheese, ice cream and other dairy desserts, infant formulas, and fermented soymilks. More recently, scientists have devised ways to incorporate probiotics into fermented cereal products, juices, fermented sausages, hams, and confectionaries like chocolate (Lee and Salminen, 2009).

History of Probiotics

The word "probiotic" (origins: Latin pro meaning "for" and Greek bios meaning "life") was first used in 1954 to indicate substances that were required for a healthy life. Out of a number of definitions, the most widely used and accepted definition is that proposed by a joint FAO/WHO panel (FAO/WHO, 2010), "Live micro-organisms which, when administered in adequate amounts, confer a health benefit on the host".

Metchnikoff (1907) suggested that "The dependence of the intestinal microbes on the food makes it possible to adopt measures to modify the flora in our bodies and to replace the harmful microbes by useful microbes". Tissier, (1906) suggested that these bacteria could be administered to patients with diarrhoea to help restore a healthy gut flora.

The works of Metchnikoff and Tissier were the first to make scientific suggestions concerning the probiotic use of bacteria, even if the word "probiotic" was not coined until 1960, to name substances produced by microorganisms which promoted the growth of other microorganisms. Fuller (1989), in order to point out the microbial nature of probiotics, redefined the word as "A live microbial feed supplement which beneficially affects the host animal by improving its intestinal balance".

It is clear that these definitions have:

1. Restricted the use of the word probiotic to products which contain live microorganisms;
2. Pointed out the need for providing an adequate dose of probiotic bacteria in order to the desirable effects.

The probiotic concept was therefore regarded as scientifically unproven and it received minor interest for decades, with some research involving animal feeding, in order to find healthy substitutes for growth promoting agents. In the last 20 years however, research in the probiotic area has progressed considerably and significant advances have been made in the selection and characterization of specific probiotic cultures and substantiation of health claims relating to their consumption.

Probiotics and Human GI

Members of the genera Lactobacillus and Bifidobacterium are mainly used, but not exclusively, as probiotic microorganisms and a growing number of probiotic foods are available to the consumer. Some ecological considerations on the gut flora are necessary to understand the relevance, for human health, of the probiotic food concept.

Bacteria are normal inhabitants of humans (as well as the bodies of upper animals and insects) including the gastrointestinal tract, where more than 400 bacterial species are found (reviewed by Tannock, 1999): half of the wet weight of colonic material is due to bacterial cells whose numbers exceed by 10-fold the number of tissue cells forming the human body. Normally the stomach contains few bacteria (10^3colony forming units per ml of gastric juice) whereas the bacterial concentration increases throughout the gut resulting in a final concentration in the colon of 10^{12} bacteria/g. Bacterial colonization of the gut begins at birth, as new-borns are maintained in a sterile status until the delivery begins, and continues throughout life, with notable age-specific changes. Bacteria, forming the so-called resident intestinal microflora, do not normally have any acute adverse effects and some of them have been shown to be necessary for maintaining the wellbeing of their host.

As an example of the beneficial role of intestinal microflora, it is possible to cite what has been referred to as "colonization resistance" or "barrier effect" meaning the mechanism used by bacteria already present in the gut to maintain their presence in this environment and to avoid colonization of the same intestinal sites by freshly ingested microorganisms, including pathogens. Therefore, it could be assumed that dietary manipulation of gut microflora, in order to increase the relative numbers of "beneficial bacteria" could contribute to the wellbeing of the host. This was also the original assumption of Metchnikoff who however, cautioned that:

"Systematic investigations should be made on the relation of gut microbes to precocious old age, and on the influence of diets which prevent intestinal putrefaction in prolonging life and maintaining the forces of the body."

Applications

Application of probiotics in food Probiotic organisms are used in a variety of foods, the main category being dairy products, but they are also present as food supplements in capsule or tablet form.

Since viability is an essential property of a probiotic, the final product must contain an adequate amount of living probiotic(s) until the end of its shelf life. A health claim for the addition of probiotics to foods or food supplements should only be made if there are documented benefits based on good quality human trials conducted with the relevant food product containing the specific strain that is the subject of the claim and using relevant endpoints.

These studies should also be able to demonstrate the safe, effective dose of the probiotic organism in food. Like legislation on food safety, regulation of health claims for foods varies by country or region and any claims on commercial products containing probiotics must adhere to requirements, which in some cases include pre-market approval of the claim by the regulatory authorities.

Prebiotics: Fiber Food for Bacteria

Prebiotics are basically food for probiotics. Taking prebiotics helps probiotics work better and more efficiently. Prebiotics not digestible by humans, but they stimulate the growth of beneficial bacteria. Common prebiotics are inulin and carbohydrate fibers called oligosaccharides. Prebiotics are found in fruits, vegetables, and whole grains. Prebiotics are an emerging field of study, and many researchers believe their potential to help you stay healthy may be as important as probiotics.

Prebiotics are a more recent discovery, utilized to promote the survival of probiotics. Prebiotics are non-digestible carbohydrates that are not absorbed in the intestine. They travel to the colon where they promote the growth of specific advantageous microbiota (probiotics) by supplying food/energy, while simultaneously influencing the microbiota's gene expression (Lee and Salminen, 2009). Additionally, fermentation of the prebiotics by the probiotics results in the production of beneficial by products. Prebiotics can be found in natural sources such as whole grains, onions, bananas, garlic, honey, leeks, and artichokes, to name a few. In addition, many foods are fortified with prebiotics (such as inulin or fructo-oligosaccharide) that are either directly extracted from plants or synthesized through chemical methods (Lee and Salminen, 2009). This has led to the incorporation of prebiotics in foods like soft drinks and croissants.

Table 5.1: Top 10 Foods Containing Prebiotics (Moshfegh *et al.*, 1999)

Food	*Prebiotic Fiber Content by Weight*
Raw Chicory Root	64.6 per cent
Raw Dandelion Greens	24.3 per cent
Raw Garlic	17.5 per cent
Raw Leek	11.7 per cent
Raw Onion	8.6 per cent
Cooked Onion	5 per cent
Raw Asparagus	5 per cent
Raw Wheat bran	5 per cent
Whole Wheat flour, Cooked	4.8 per cent
Raw Banana	1 per cent

History of Prebiotic

As mentioned, the Japanese were the first to recognise the value of fermentable oligosaccharides, initially in feeding piglets and later, during the 1980s, with the identification of human milk oligosaccharides. However, it was not until 1995 that the prebiotic concept for modulation of gut micro biota was introduced. Although a number of definitions have been proposed, there is as yet no full agreement on a single definition of a prebiotic. The International Scientific Association of Probiotics and Prebiotics (ISAPP) was agreed in a meeting at the 2010 (Gibson *et al.*, 2011):

"A dietary prebiotic is a selectively fermented ingredient that results in specific changes, in the composition and/or activity of the gastrointestinal microbiota, thus conferring benefit(s) upon host health."

Synbiotics: Combining Probiotics and Prebiotic

A synbiotic is a supplement that contains both probiotics and prebiotics. It makes sense to make sure any supplement you take contains both pro- and prebiotics, because the two work in tandem to make sure your system has enough of the healthy, beneficial bacteria it needs.

Synbiotics refer to nutritional supplements combining probiotics and prebiotics in a form of synergism, hence synbiotics. The concept of synbiotic is "the mixtures of probiotics and prebiotics that beneficially affect the host by improving the survival and implantation of live microbial dietary supplements in the gastrointestinal tract, by selectively stimulating the growth and/or by activating the metabolism of one or a limited number of health-promoting bacteria, thus improving host welfare" (McGovern *et al.*, 2004).

Probiotics are live bacteria which are intended to colonize the large intestine and confer physiological health benefits to the host. A prebiotic is a food or dietary supplement product that confers a health benefit on the host associated with modulating the microbiota.

Table 5.1: Examples of Common Probiotics, Prebiotics, and Synbiotics

	Probiotic	Prebiotic	Synbiotic
Lactobacillus sps.	*L. acidophilus*	Fructo-oligosaccharides	Lactobacilli + lactitol
	L. amylovorus	(FOS)	Lactobacilli+ inulin
	L. bulgaricus	Inulin	Lactobacilli + FOS or inulin
	L. brevis	Lactulose	Lactobacillus rhamnosus
	L. casei	Lactitol	GG + inulin
	L. cellobiosus	Galactooligosaccharides	Bifidobacteria + FOS
	L. crispatus	(GOS)	Bifidobacteria + GOS
	L. curvatus	Isomaltooligosaccharides	Bifidobacteria and
	L. fermentum	Xylooligosaccharides	Lactobacilli + FOS
	L. gallinarum	Lactosucrose,	or inulin
	L. johnsonii	Cereals fibres	
	L. lactis	Soy oligosaccharides	
	L. paracasei	Raffinose	
	L. plantarum	Fructo-oligosaccharides	
	L. salivarius	(FOS)	
	L. sporogenes		
	L. reuteri		
	L. rhamnosus		
Bifidobacterium spp.	*B. adolescentis*		
	B. animalis		
	B. bifidum		
	B. breve		
	B. infantis		
	B. longum		
	B. thermophilum		
Streptococcus spp.	*S. thermophilus*		
	S. salivarius subsp. *thermophilus*		
	S. diaacetylactis		
Saccharomyces sps.	*S. cerevisiae, S. boulardii*		
Others	*Bacillus cereus*		
	Bacillus coagulans		
	Escherichia coli		
	Enterococcus faecium		
	Propionibacterium freudenreichii		
	Homeostatic Soil Organisms (HSO's)		

Prebiotics are not drugs, not functioning because of absorption of the component, not due to the component acting directly on the host, and are due to changes to the resident bacteria – either changing the proportions of the resident bacteria or the activities thereof. Measurable changes to the microbiota in the absence of a desirable physiological consequence in the host does not qualify as a prebiotic. A prebiotic may be a fibre, but a fibre is not necessarily a prebiotic.

Using prebiotics and probiotics in combination is often described as synbiotic, but the United Nations Food and Agriculture Organization (FAO) recommends that the term "synbiotic" be used only if the net health benefit is synergistic. Probiotic bacteria may colonise the upper part of the intestine to avoid the adhering of pathogens to the intestinal tract and may help in digestion. A prebiotic is a fibre such as fructose oligosaccharide, galactose oligosaccharide, etc., and is consumed that is intended to stimulate the microflora in the large intestine. The combination thus works separately in the small and large intestine, but synergistically as they increase the overall gut health. A common mistake is to require that the prebiotic be shown to increase the population and/or function of the probiotic it is paired with, as the probiotic is an external species, whereas prebiotics stimulate the flora which is already present.

Role of Pro, Pre and Synbiotics for Healthy Human Gut

Many factors affect the composition of the large intestinal microbiota in humans. These include the age, susceptibility to infections, nutritional requirements, and immunologic status of the host and the pH, transit time, interactions between flora components, and presence and availability of fermentable material in the gut.

The gastrointestinal tract of new born is inoculated primarily by organisms originating from the mother's vagina and from the environment. For new born delivered by caesarean birth this latter factor is of particular importance. Bacterial populations develop during the first day of life. Not surprisingly, facultative anaerobic strains such as *Escherichia coli* and Streptococci initially exist in highest numbers (Rotimi and Duerden, 1981). These bacteria may subsequently create a highly reduced environment that allows the growth of strictly anaerobic species. Soon after delivery, the infant may be weaned from breast milk to formula. Differences in the faecal flora of breast-fed and bottle-fed infants exist and have been associated with a lower risk of gastrointestinal infection in breast-fed infants. Although the data are conflicting, there is some evidence that the microflora of breast-fed infants is dominated by populations of bifidobacteria (Benno *et al.*, 1984), an observation first promoted of Tissier (Tissier, 1905). In contrast, formula-fed infants are thought to have a more complex flora with no one bacterial genus showing a numerical predominance. The high incidence of bifidobacteria has been cited as one possible explanation for the purported health advantages. Although the ecosystem development is undoubtedly influenced by both host and environmental factors, some postulations directly attribute this development to the feeding regime as follows:

1. Oligosaccharides, including Nacetyl-glucosamine (Gyorgy, 1953), glucose, galactose, and fructose oligomers or certain glycoproteins which form a significant proportion of human breast milk, may be specific growth factors for bifidobacteria.

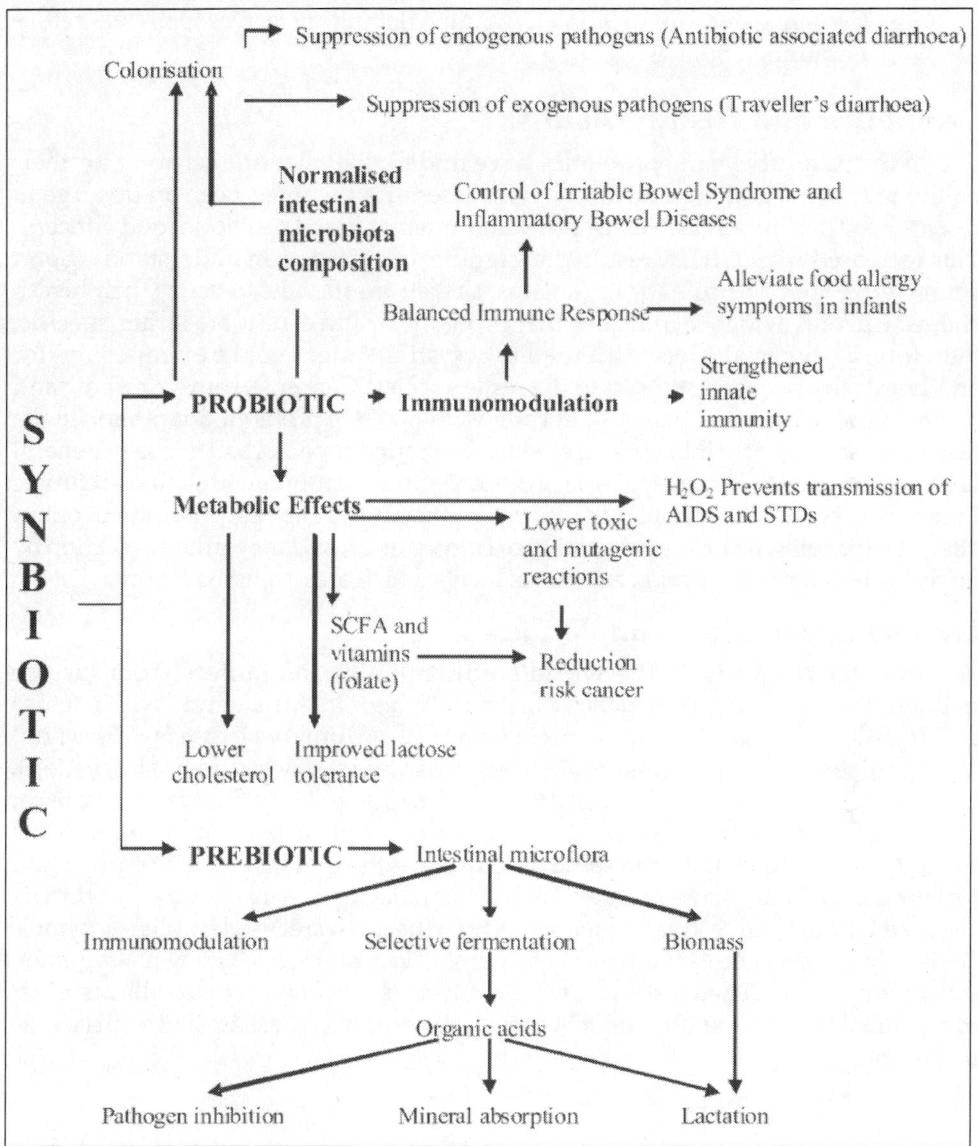

Figure 5.1: An Overview of Prebiotics, Probiotics and Synbiotics (Adopted from Sekhon and Jairath, 2010).

2. Low protein content and reduced buffering capacity of human milk may allow elevated growth of bifidobacteria (Bullen *et al.*, 1977; Willis *et al.*, 1973). The nature, type, absorption, and quantity of milk proteins present in the feed may also exert an effect.
3. Certain compounds, including lactoferrin and some lipids, inhibit microorganisms (Raibaud, 1988).

4. Certain bacteria may stimulate immunoglobulin molecules such as secretory immunoglobulin A.

Probiotics and Health Claims

In the past fifty years, probiotics have undergone scientific scrutiny for their ability to prevent and cure a variety of diseases and there has been an upsurge in research for probiotics, as well as profound interest in the probiotic food concept. This increased research has resulted in significant advances in understanding and characterize specific probiotic organisms, as well as attempts to verify their health claims. Current evidence indicates that probiotic health claims are strain-specific, therefore, a beneficial effect attributed to one strain might not be extrapolated for another strain, even if it belongs to the same species. Genomic, transcriptomic and proteomic studies reveal genes and proteins involved in probiotic adaptation in the host while exerting their beneficial effects. Probiotics have a broad scope in general health interventions like managing lactose intolerance, immunomodulation in chronic inflammatory and infectious conditions, managing oral absorption syndromes, allergies, diabetes, arthritis, prevention of colon cancer, and for cardio protection by lowering blood pressure and cholesterol levels which are discussed below:

Lactose-maldigestion/Intolerance

About two-thirds of the world's adult population suffers from lactose maldigestion, with the prevalence particularly high in Africa and Asia. Lactose maldigestion occurs frequently, especially in adults (primary lactose maldigestion) and in persons with bowel resection or enteritis (secondary lactose maldigestion). Symptoms include loose stools, abdominal bloating, pain, flatulence and nausea. Individuals with lactose maldigestion can tolerate lactose present in yoghurt to a much greater degree than the same amount of lactose in raw milk. Yoghurt and probiotic lactic acid bacteria contain high levels of lactase (β-galactosidase), which is released within the intestinal lumen when these bacteria are lysed by bile secretions. Lactase/β-galactosidase then acts on the ingested lactose, thus relieving maldigestion symptoms. The reduced intestinal transit time of yoghurt might also allow slower digestion of lactose, thereby reducing the symptomatology associated with lactose intolerance.

Intestinal Disorders

Gastrointestinal disorders occur when the digestive tract (gastrointestinal) does not function properly due to microbial imbalance and disturbance in host microbe interaction. GI infections particularly diarrheal diseases are the major cause of morbidity and mortality in children and adults across the world particularly in the third world countries. Several probiotic strains have proven effective against gastrointestinal abnormalities however benefits are dose dependent. Various meta-analyses showed an overall reduction in Gastric inflammation during administration of different strains of *L. casei, L. reuteri, L. plantarum, L. acidophilus, B. animalis, B. infantis* and *S. boulardii.*

Inflammatory Bowel Diseases

Inflammatory bowel diseases, such as ulcerative colitis and Crohn's disease are serious intestinal disorders that can eventually necessitate the surgical removal of the colon. The causes are diverse, but are often related to depletion of beneficial gut bacteria. IBD refers to a group of disorders of unknown aetiology that are characterized by chronic or recurrent mucosal inflammation. Probiotic administration, either through regulation of the inflammatory response or modulation of gut microbiota composition and/or activity might bring about relief in IBD symptoms or maintain remission from symptoms. Formula VSL#3 is a mixture of four lactobacilli (*L. acidophilus, L. bulgaricus, L. casei* and *L. plantarum*), three bifidobacteria (*B. breve, B. infantis* and *B. longum*) and *S. thermophilus*. The mixture has proven quite effective in reducing the recurrence of chronic relapsing paucities.

Allergies

Probiotics direct modulation of the immune system through the induction of anti-inflammatory cytokines or through switching the production of secretory IgA. Probiotics can potentially modulate the toll-like receptors and the proteoglycan recognition proteins of enterocytes, leading to activation of dendritic cells and a Th1 response. The resulting stimulation of Th1 cytokines can suppress Th2 responses. Paediatric studies suggest that probiotic use in children with atopic conditions such as atopic dermatitis results in enhancement of IFN-production and decrease d IgE and antigen-induced TNF-α, IL-5, and IL-10 secretion. Studies involving *L. casei, L. reuteri, L. paracasei, L. acidophilus, B. animalis* and *B. infantis* reported that probiotics shows variable effects in different clinical trials. Probiotic effects in allergies are limited to initiation of improvement in mild allergic diseases like atopic dermatitis and rhinitis.

Metabolic Syndrome

Metabolic syndrome is a group of metabolic abnormalities, increases an individual's risk of developing obesity, type 2 diabetes and cardiovascular disease. This syndrome is influenced by more than one factors *i.e.* genetic, lifestyle and environmental factors (gut microbiota). Altered dietary intake not only affects our energy balance but also has a major impact upon gut microbial composition, which results in altered gut microbiota, increased energy harvest, increased inflammatory tone and altered intestinal endocrine function.

Obesity

Different studies show the relation between body weight and gut microbiota. In obese people the ratio of Bacteroidetes, Firmicutes and Archea is altered in obese individuals. However human studies gave controversial data on this fact. In overweight human adults lower ratio of Firmicutes to Bacteroidetes is also found as well as weight loss diets, found no proof of the link between the proportion of Bacteroidetes and Firmicutes and human obesity.

Type 2 Diabetes

Compared to non-diabetics, reduction in the relative abundance of Firmicutes is observed with increase in the proportion of Bacteroidetes and Proteobacteria in diabetics. The level of probiotic *Bifidobacterium* significantly and positively correlated with improved glucose-tolerance and low-grade inflammation in mice model.

Cardiovascular Diseases

It is an inflammatory metabolic syndrome because high-fat diet has mainly been attributed to the inflammatory properties of dietary fatty acids (*i.e.*, palmitic acid). It trigger an inflammatory response by acting via LPS receptor [TLR-4]) signalling in adipocytes and macrophages, inflammation of adipose tissue in obesity. Metabolic endo-toxaemia increases adipose TNF-α and IL-6 concentrations, insulin resistance and stimulates vascular inflammation which results in hypertension and CVD. Probiotics strongly promote SCFA production from indigestive carbohydrates (prebiotics) and are also associated with changes in ingestive behaviour by mechanisms linked to the modulation of gut peptide production and secretion which changes the food intake behaviour.

Respiratory Infections

A few studies have shown that certain strains of probiotic bacteria may prevent viral respiratory tract infections (common cold and influenza), alleviate complaints and/or shorten the duration of the disease. *L. rhamnosus L. casei* strain, a vitamin-mineral-mixture plus *L. gasseri, B. longum* and *B. bifidum* have been found to be effective in winter-infections in children and elderly subjects. *L. acidophilus* NCFM and *Bifidobacterium animalis ssp. lactis* Bi 07 have also been shown to reduce the incidence of fever, cough and sneezing, their duration and the incidence of application of antibiotics. Meta-analysis on probiotics and respiratory infections showed a significant decrease in the number of days with fever, clinical visits and antibiotic prescriptions. However, the efficacy of probiotics was found to be strain and dose dependent and also exhibited variable results in different studies.

Urogenital Infections

A number of human clinical studies have demonstrated that probiot-ics strains including *Lactobacillus acidophilus, Lactobacillus rhamnosus, Lactobacillus reuteri, L. gasseri, L. jhonsenii* and *Lactobacillus cris-patus* cured bacterial vaginosis. Intravaginally administration of probiotics were found to be more effective in these studies. Curing of vulvuvaginal candidiasis, urogenital tract infection and bacterial vaginosis is related to acid and H_2O_2 production and concentration of applied probiotic.

Cancer

In vitro systems and animal models provide considerable evidence that probiotics, prebiotics and synbiotics exert anti-neoplastic effects *e.g.* - the polysaccharide fraction of *B. bifidum* BGN4 inhibited the growth of HT-29 and HCT-116 cells but did not inhibit the growth of Caco-2 cells. The probiotic strains *L. rhamnosus, B. lactis* and *L. reuteri* ATCC 6475 activate caspase proteins in cancerous cell line. The consumption of probiotics may be beneficial in preventing the onset of cancer, but also in the treatment of existing tumours. However, evidence from human studies is still limited.

Generic Health Claims

Probiotics are beneficial microorganisms and habitat in our body with other commensal flora. Beside all specific health effects, they have role in maintenance of normal health status of host. Probiotics are involved in different processes such the improvement of nutrient bioavailability and degradation of non-digestible dietary compounds, the supply of new nutrients, and the removal of harmful, toxic and non-nutritional compounds and suppress the growth of pathogens besides enhancing immune status.

Why Eat Fermented Foods?

Besides the fact that they taste great and really grow on you, there are several great reasons to start making and eating fermented foods:

- ☆ **Probiotics**- Eating fermented foods and drinking fermented drinks like Kefir and Kombucha will introduce beneficial bacteria into your digestive system and help the balance of bacteria in your digestive system. Probiotics have also been shown to help slow or reverse some diseases, improve bowel health, aid digestion, and improve immunity!
- ☆ **Absorb Food Better**- Having the proper balance of gut bacteria and enough digestive enzymes helps you absorb more of the nutrients in the foods you eat. Pair this with your healthy real food diet, and you will absorb many more nutrients from the foods you eat. You won't need as many supplements and vitamins, and you'll be absorbing more of the live nutrients in your foods.
- ☆ **Budget Friendly**- Incorporating healthy foods into your diet can get expensive, but not so with fermented foods. You can make your own whey at home for a couple of dollars, and using that and sea salt, ferment many foods very inexpensively. Drinks like Water Kefir and Kombucha can be made at home also and cost only pennies per serving. Adding these things to your diet can also cut down on the number of supplements you need, helping the budget further.
- ☆ **Preserves Food Easily**- Homemade salsa only lasts a few days in the fridge-Fermented homemade salsa lasts months. The same goes for sauerkraut, pickles, beets and other garden foods. Lacto-fermentation allows you to store these foods for longer periods of time without losing the nutrients like you would with traditional canning.
- ☆ **Condiments and Sauces**- Many traditional condiments are fermented foods, including sauerkraut, Korean kimchi, miso, tamari, chutneys, vinegars and some types of pickles. Fermented pickles are sour pickles made in a salt-water brine instead of being made with vinegar, so check the label if you aren't sure whether a particular type of pickle is fermented.
- ☆ **Dairy Products**- Dairy products other than milk are usually produced through fermentation. These include yogurt, kefir and cheeses. The

fermentation process helps break down the lactose in the milk used to make these products, making them easier to digest. This is why some lactose-intolerant people can consume these products at least in small amounts without digestive complaints.

- **Beverages**- Beverages can also be fermented, including alcoholic beverages like beer, which is made by fermenting grains, and wine, which is made by fermenting grape juice. Another fermented beverage is Kombucha, made by fermenting tea and considered by some to be a healthier alternative to soda for those who want something fizzy to drink, according to a March 2010 article published in "The New York Times."
- **Other Fermented Foods**- Any food made with the use of a starter is a fermented food. This includes sourdough bread and Amish friendship bread. Tempeh, made by fermenting soybeans, is often crumbled up and used as an alternative to ground meat by vegetarians and people wanting to lower their fat and cholesterol consumption.

References

"7, 000 Year-old Wine Jar [Object of the Day #23]". Retrieved 2014-10-14.

"8, 000-year-old wine unearthed in Georgia". The Independent. 2003-12-28. Retrieved 2014-10-14. http: //www.stonepages.com/news/archives/000498.html

Benno Y, Sawada K, Mitsuoka T. The intestinal microflora of infants: composition of fecal flora in breast-fed and bottle-fed infants. *Microbiology Immunology*, 1984. 28(9): 975-86.

Bhupinder Singh Sekhon and Saloni Jairath. Prebiotics, probiotics and synbiotics: an overview. *J Pharm Educ Res*. (2010). 1(2),

Bullen CL, Tearle PV, Stewart MG. The effect of "humanized" milks and supplemented breast feeding on the faecal flora of infants. *J. Med. Microbiology*, 1977, 10(4): 403-13.

Cavalieri, D., McGovern P.E., Mortimer R., Polsinelli M. Evidence for *S. cerevisiae* Fermentation in Ancient Wine. *Journal of Molecular Evolution*, (2003). 57(1): S226-S232.

Dirar, H. The Indigenous Fermented Foods of the Sudan: A Study in African Food and Nutrition, CAB International, UK, (1993).

FAO/WHO Joint FAO/WHO Expert Consultation on evaluation of health and nutritional properties of probiotics in food including powder milk with live lactic acid bacteria. Cordoba, Argentina, October 2001.

http: //www.who.int/foodsafety/publications/fs_management/en/probiotics.pdf

Fermentation in food processing. From Wikipedia. Retrieved 2014-10-15. http: //en.wikipedia.org/wiki/Fermentation_in_food_processing#cite_note-FAO-4

Fuller R, ed. Probiotics: The Scientific Basis. London, UK: Chapman and Hall, 1992.

Gibson, G.R. Dietary prebiotics: current status and new definition. IFIS Functional Foods Bulletin, (2011). 7: 1–19.

Gyorgy P. A hitherto unrecognized biochemical difference between human milk and cow's milk. Pediatrics, 1953, 11(2): 98-108.

http: //www.penn.museum/blog/collection/125th-anniversary-object-of-the-day/7000year-old-wine-jar-object-of-the-day-24/

Lee YK, Salminen S. Handbook of probiotics and prebiotics. 2nd ed. Hoboken, N.J.: John Wiley and Sons, Inc. (2009).

Lee YK, Salminen S. Use of probiotics as replacements for antibiotics in pigs and poultry. Handbook of probiotics and prebiotics. 2nd Ed Wiley-Interscience, 2009.

McGovern, P.E., Zhang, J., Tang, J., Zhang, Z., Hall, G.R., Moreau, R.A., Nunez, A., Butrym, E. D., Richards, M.P., Wang, C.S., Cheng, G., Zhao, Z., Wang, C. "Fermented beverages of pre- and proto-historic China". *Proceedings of the National Academy of Sciences*, (2004). **101**(51): 17593–17598.

http: //en.wikipedia.org/wiki/Digital_object_identifier

Moshfegh AJ, Friday JE, Goldman JP, Ahuja JK. Presence of inulin and oligofructose in the diets of Americans. *J. Nutr.*, 1999. 129(7): 1407S-11S

Raibaud P. Factors Controlling the Bacterial Colonization of the Neonatal Intestine. In: Hanson LA. (ed.) Biology of Human Milk, Nestle Nutrition Workshop Series, Vol.15, Vevey/Raven Press: New York, 1988, 205-19.

Rotimi VO, Duerden BI. The development of the bacterial flora in normal neonates. *J. Med. Microbiology*, 1981. 14(1): 51-62.

Tannock GW, ed. Probiotics: A Critical Review. Wymondam, UK: Horizon Scientific Press, 1999, 85–111.

Tissier H. Repartition des microbes dans l'intestin du nourisson. Ann. Inst. Pasteur (Paris), 1905, 19: 109-23.

Willis AT, Bullen CL, Williams K, Fagg CG, Bourne A, Vignon M. Breast milk substitute: a bacteriological study. *Br. Med. J.*, 1973, 4(5884): 67-72.

2016, Dairy and Food Product Technology Pages 77–86
Editors: Birendra Kumar Mishra and Subrota Hati
Published by: BIOTECH BOOKS, NEW DELHI

Chapter 6

Structural Changes of Indian Dairy Farming: Smallholder Prospects

Shiv Raj Singh[1], K.K. Datta[2] and Rohit Charan[3]

[1]Department of Dairy and Food Business Management, SDAU, Banaskantha – 385 506, Gujarat
[2]Dairy Economics Statistics and Management, National Dairy Research Institute, Karnal, Haryana
[3]M.V.Sc Research Scholar, College of Veterinary Science and Animal Husbandry, SDAU, S.K. Nagar, Gujarat

Introduction

Dairying is an important activity in Indian economy contributing about 27 per cent of the agricultural gross domestic product (GDP) and which around 4.35 per cent of the national GDP (CSO 2007-08). The total milk production has increased from 48.40 million tons in 1988-89 to 116.9 million tons in 2009-10. Dairying in India is more inclusive compared to crop production in the sense that it involves a majority of the vulnerable segments of the society for livelihoods. Livestock population is more equitably distributed than the land (Kumar and Singh, 2008). The income from livestock sector helps in alleviating poverty and smoothening of income distribution (Birthal *et al.*, 2002).

In case of livestock sector, targeted growth rate during the Eleventh Five Year Plan Period was 6 per cent per annum but its achievement was 4.17 per cent (Mid

Term Appraisal of XIth Plan). Over the plan periods, the annual growth rate of livestock sector maintained a moderate pace and their contributions towards the total growth process in agricultural sector helped to achieve around 3 per cent in the first three years of the Eleventh Five Plan Period. The outcomes of the growth are very important in the democratic country like India because in the one billion plus populated country there is a need to address the inclusiveness and poverty alleviation through the growth process. The association between rural poverty and agriculture suggested that there is a direct relationship between the growth of agriculture and reduction in the rural poverty. Recent study indicated that the benefits of growth in agriculture have trickled down to the rural poor and the distributive growth has been inclusive (Kumar *et al.*, 2011).

Over the years operational land holdings have increased nearly two folds from 56 Million to 101 million. The reason of increase in number of operational holdings was land fragmentation. As a result of land fragmentation average land holdings was reduced from 1.67 ha. to 1.06 ha within the period 1981-82 to 2002-03 (NSSO, 2006). Land fragmentation also impacts distribution of dairy animals because of integration of agricultural land with dairying. The results of different studies indicated positive and negative relationship between landholdings size and productivity in crop production systems (Ram *et al.*, 1999, Wattanutchariya and Jitsanguanet, 1992). The result of recent study shows that dairy farms can increase their income from 9 to 14 per cent by reducing the degree of land fragmentation (Corral *et al.*, 2011).

Increase in agricultural holdings and their continuous subdivision among the family siblings seemed to be affecting the consolidation of milch animal holdings. As a result of land fragmentation, number of operational holdings across the landless, marginal and small categories have increased over the years resulting in reduction in average size (NSSO, 2006). Therefore, in these categories households dependence on the crop sector reduced over the years and off-farm and dairying income share would have increased. The income from off-farm and dairy farming reduces the income equality across the different categories (Mandal *et al.*, 2010). It is with this background that this paper has studied the structural transformation and current structure of Indian dairy sector, along with its contribution to the household income.

Data Sources and Analytical Tools

The study is based on the secondary data, available from National Sample Survey Organization(NSSO) unit level data on *"Situation Assessment Survey of Farmers"*,2003 (visit-2), Annual Survey of Industries report of *"Livestock Ownership Across Operational Land Holding Classes in India, 2002-03"* survey round and milk production statistics from Department of Animal Husbandry, Dairying and Fisheries. It may be mentioned that the *Situation Assessment survey of Farmers* was conducted in rural area of India with the sample size of 51,105 households.

Data on distribution pattern of land holders and dairy animals across the different category of households has been analysed through tabular analysis and percentage method.

Results and Discussion

Milk Production Scenario

Production of milk in Indian dairy sector has shown tremendous growth from 17 million tons (1950-51) to 121.8 million tons (2009-10). This transition from deficiency to sufficiency has been achieved by a series of policy interventions by the government. Dairy development has been transformed from the earlier phase (1970-80) of slow growth rate of milk production and its utilization in the form of value-added in milk products to a high growth rate of milk production and value-added milk products. It was due to the higher allocations under 'Operational Flood' programme in the Fourth Five-Year Plan, which was considered as the turning phase from deficiency to sufficiency in the Indian dairy history. This programme aimed at replicating the Amul Model of milk cooperative, which essentially ensured a favourable price regime to milk producers and mode of disbursal of prices through a fair and transparent system. The impact of this programme was reflected in the annual growth rates of milk production which increased from 2.80 per cent per annum in 1970s to 6.72 per cent per annum during 1980s but declined subsequently to 4.04 per cent per annum (Singh and Datta, 2010).

If we look at the decadal growth process of different promising milk producing states of India, it clearly indicates that there are some states such as Andhra Pradesh, Gujarat, Orissa and Uttar Pradesh where consistent and systematic growth process was achieved during 1990s and 2000s, the growth process continued during the Eleventh Five Year Plan period also (Table 6.1). In other states the growth process was found to be inconsistent and volatile.

Stylization of Districts in Terms of Milk Productivity

At all India level, productivity of milch animals (Table 6.2) of 391 districts across different states (mainly 13 states contributing more than 90 per cent of total Indian milk production) indicate that average realized productivity is about 3.03 kg/day whereas the figures very widely with as high as 15.39 kg/day in Mumbai Suburban and 15.03 kg/day in Amritsar district of Punjab to a low of 0.33 kg/day in Katihar district of Bihar and 0.26 kg/day in Purualia district of West Bengal. Considering variation as 154 per cent, these 391 districts were clubbed into following three categories as 1) High milk producing districts with average productivity of more than 6.27 kg/day (62 districts); 2) Moderate but promising districts of 3.81 to 6.27 kg/day (113 districts) and 3) Low milk producing districts with lesser than 3.81 kg/day (216 districts). Summarily, the distribution pattern clearly indicates that 62, 113 and 216 districts, respectively could be classified into high, moderate and low milk producing districts. The district level productivity profile clearly reflects that there is a scope to enhance milk production in the districts where the production is at moderate level. If at least another 113 districts can join the high productivity level districts, it may prove to be a breakthrough in the Indian way of White Revolution.

Table 1: Annual growth rate of milk production in major states (per cent)

States	*Decadal Growth Rate*		
	1990-00	*2000-10*	*XIth Plan (2007-10)*
Andhra Pradesh	6.22	7.01	9.29
Kerala	4.65	–0.79	6.40
Gujarat	4.93	5.55	5.54
Karnataka	7.33	–0.04	5.51
Orissa	5.96	8.64	4.68
J and K	10.22	1.68	4.17
Chhattisgarh	–	2.03	4.12
Madhya Pradesh	1.94	4.45	4.01
Haryana	3.71	2.04	3.99
Uttarakhand	–	2.67	3.95
Bihar	1.33	12.13	3.83
Uttar Pradesh	4.30	4.26	3.73
Maharashtra	5.45	3.00	3.26
West Bengal	2.20	2.47	2.55
Jharkhand	–	6.65	1.47
Tamil Nadu	3.09	2.40	1.32
Punjab	4.53	2.34	0.83
Rajasthan	5.92	3.17	0.50
Himachal Pradesh	3.21	1.75	–1.04
All India	4.33	3.83	3.69

Source: Authors' estimates from Annual Reports (Various Issues) of Department of Animal Husbandry, Dairying and Fisheries, GOI.

Dairy Income as a Source of Rural Livelihood

In the rural area there were four major sources of income (receipts) like crop, off-farm, dairy farming and livestock farming (excluding dairy). Correlation analysis suggested that there was a low (positive) correlation between the income from off-farm and dairy farming (0.028), other livestock (0.022) and crop sector (0.021) at household level. Somehow, this relationship suggested that by promoting the basket of income generating activities (off-farm and agricultural) at household level we can increase the household income without conflict. From Table 6.3, it is clear that landless households earned about 43 per cent of their income from livestock out of which about 41 per cent income was exclusively from dairying. If we compare the non-livestock keeping farm households vs. livestock keeping farm households, then we observe that majority of the non-livestock keeping households depended on off-farm source of income which were uncertain and seasonal.

Table 6.2: Number of Districts of different Level of Milk Productivity during 2007-08 (kg/day)

States	*Low*	*Moderate*	*High*
Andhra Pradesh	17	5	0
Karnataka	22	5	0
Kerala	1	4	9
Tamil Nadu	7	18	4
Gujarat	9	12	4
Maharashtra	27	5	0
UP	46	17	7
Bihar	31	6	1
Haryana	1	10	8
Punjab	1	1	15
Rajasthan	14	12	6
MP	24	16	8
West Bengal	16	2	0
All India	216	113	62

Source: Authors' estimates from Integrated Sample Survey Report of different states (2007-08).

Another observation from Table 6.3 is that at the household level, landless and marginal farmers spend maximum amount in dairying, while small, medium and large farmers do the same for crop production. This is related to relative asset distribution across different types of productive assets and their uses and it is the livestock including dairy animals that ensure higher stability in family incomes and hence relatively larger expenditure. Table 6.3 also indicates that those households who keep animals in marginal category earn four times higher monthly incomes. The income from dairy farming varies widely across the different farm categories *e.g.*, in landless category contribution around Rs. 412 per month (40.58 per cent) of total household income while farms in large category is Rs. 2545 per month (7.54 per cent). In comparison of landless farmers, large farmers earn more than six times higher income from the dairying, but income from dairying sustains livelihood of landless families. Therefore, dairying has the potential to reduce rural poverty, as demonstrated through the analysis of unit level data of the NSSO. Three important points emerge out from the unit level data of the 59th NSSO (i) proportion of expenditure on livestock and dairying to total production expenditure of the households is inversely related to land ownership (ii) proportion of household income from livestock and dairying to total family incomes of the households is also inversely related to land ownership, and (iii) dairying helps in poverty alleviation. Essentially, the NSSO data corroborates the significance of livestock and dairy incomes in the vulnerable segments of the farming community compared to large land owning classes of farmers.

Table 6.3: Different Sources of income and Expenses per Month by the Dairy Farmers for their Livelihood Activities during 2002-03

Particulars	*Landless.1*		*Marginal*		*Small*		*Medium*		*Large*	
	LKH	*NLKH*	*LKH*	*NLKH*	*LKH*	*NLKH*	*LKH*	*NLKH*	*LKH*	*NLKH*
Expenses (per cent)										
Off farm	1.17	44.00	0.43	1.57	0.16	0.39	0.06	0.18	0.03	0.11
Dairy farming	81.64		48.71		33.08		27.67		16.34	
*Livestock farming	12.26		8.33		6.93		4.71		3.36	
Crop	4.93	56.00	42.53	98.43	59.82	99.61	67.56	99.82	80.27	99.89
Total(Rs.)	359	12	1011	339	2089	999	4395	1841	10999	3092
Receipts (per cent)										
Off farm	53.84	98.21	35.30	62.72	16.42	30.6	22.30	13.32	26.22	46.57
Dairy farming	40.58		26.79		20.56		12.21		7.54	
*Livestock farming	2.25		2.06		1.43		1.04		0.22	
Crop	3.33	1.79	35.85	37.28	61.59	69.94	64.45	86.66	66.01	53.43
Total (Rs.)	1014	1168	8667	2094	4684	3552	9832	5155	33755	14005

Note: *Livestock farming excluded dairy farming

LKH: Livestock Keeping Households; NLKH: Non Livestock Keeping Households.

Source: Authors' estimates based on unit level data of NSSO 59th round on Situation Assessment Survey of Farmers

Attempt to Link Milk Producers with Milk Processing Units

Most of the milk producers in the country belong to the categories of small and marginal farmers and landless households. So, any strategy for increasing milk production as well as forward linkage with dairy factories must aim at small and marginal holders who are in advantageous position in terms of cost of production and their geographical location. Therefore, it can be assumed beyond any doubt that the future course of growth in the form of value-added products will be completely guided by the small and marginal holders. To link the smallholder milk producers with the organised dairy industry, it is important to visualise the structural changes in the Indian dairy industry. This information is helpful to infer whether industrial dynamics has been in favour of smallholder dairy farming or not. In this regard, Table 6.4 provides the results for structural changes in Indian dairy sector over the post liberalized period of 1994 to 2008. The study period can be classified into the two phases *i.e.* 1994-95/2000-01 is Pre MMPO-1999 and 2000-01/2008-09 Post MMPO-1999. Due to the growing pressure of competition from global players in the dairy sector, the tightening of the WTO Agreements as well as the anomalies in the license structure, the government made an amendment (in the year 1999) in the MMPO in 1992. The amendment allowed the dairy players to setup dairy processing units wherever and whenever they want to. MMPO-1992 was actually introduced in India to protect the interest of the cooperative as well as domestic small and medium size dairy plants. So, this amendment is one of the major policy amendments in the Indian dairy sector from government front in the post liberalized period.

Table 6.4: Dynamics of Organized Dairy Industry with Respect of Types of Ownership (in per cent)

Type of Ownership	*1994-95*		*2000-01*		*2007-08*	
	Quantity of Milk Processed	*No. of Factories*	*Quantity of Milk Processed*	*No. of Factories*	*Quantity of Milk Processed*	*No. of Factories*
Public	17.48	10.23	21.52	12.46	24.44	7.65
Private	4.62	11.13	4.15	18.28	16.25	18.20
Co-operative	45.98	33.39	53.53	17.62	44.85	21.00
Others	31.92	45.24	20.80	51.64	14.45	53.14

Source: Authors' estimates based on unit level data of ASI (1994-95, 2000-01 and 2007-08).

Others: Individual Proprietorship, Joint family, Partnership, Govt. Departmental Enterprise, Public Corporation by Special act of Parliament/legislator/PSU, Khadi and village industries commission, Handlooms and Others (incl Trusts, wakf board, etc)

From Table 6.4 it can be inferred that at that time (1994-95), Indian dairy sector was mainly dominated by the cooperatives and others (mainly small and medium size dairy plants) in terms of ownership of dairy plant. These two subsectors constituted around 78 and 79 per cent, respectively of the total quantity of milk processed and number of dairy plants in the organized dairy sector. The private sector dairy plants were very less both in terms of numbers as well as milk handled by them. In the Pre MMPO-1999 period their share did not change much. At the same time, cooperative sector increased their share in terms of quantity of milk handled despite the fact that their number reduced drastically. It means that in the Pre MMPO-1999 period, the cooperative sector kept their reliance on consolidation of milk procurement and handling capacities and capabilities as even though they reduced their numbers they continued to increase their share in Indian dairy sector.

In the starting phase of Post MMPO-1999 period, the organized sector was mainly dominated by the cooperatives sector (53.53 per cent in the year 2000-01). Similar trend continued till the end of 2008-09 (Table 6.4). At the same time, the private sector processed 4.15 per cent of total milk handled in the organized sector with 18.28 per cent of the dairy factories. But at the end of 2007-08, this sector increased their share from 4.15 to 16.25 per cent with almost same proportion of dairy factories. Similar but marginal incremental trend was observed in the public sector operated dairy industry. One important observation from Table 6.4 is dairy plants operating in the "Others" category, they increased their numbers but at the same time they drastically lost their share in the milk processing in the organized sector (as per the unit level data average milk handing in Other category factories reduced from around 8000 tons/annum to 3000 tons/annum between the period 1994-95 to 2007-08). This infers that plants operating in the private sector category increased their milk handling capacity primarily at the expense of those in the "Others" who were marginalized in terms of quantity of milk processed by them in the organized sector over the study period. This means that private sector owned dairy factories must be providing some kinds of sops to the farmers and at the same time have consolidated their milk handling capacity as there is no proportional increase in number of their dairy factories.

Conclusions

The study observed that fragmentation of land has led to increase in operational holdings across the different categories in the last four decades. The operational land holdings are more concentrated (90 per cent) in smallholder (less than 2 hectare operational landholding). The smallholders account for around 52 per cent of land resources with about 83 per cent of milch animals. They are contributing around 77 per cent in the national milk production. This production system indicates dominance of smallholder dairy farming system in India and moving towards more intensification. Therefore, it puts more pressure on feed and fodder resources. Study point out that in the early 1990s organized dairy industry was mainly dominated by cooperative and other (small and medium size dairy plants) sectors but as a result of amendment in MMPO-1992 dairy industry is consolidated in the form of cooperative and private sector.

Operation Flood program emphasis on developing smallholder-based dairy sector in the pre liberalized era is justified on the ground that it realized the needs of the production base by the masses. Need of the day is to provide quality of efficient input and output support services as provided by the co-operatives (Amul model at Gujarat, Nandani Milk Federation at Karnataka Model), private sector (*Nestlé*) and contract dairy farming. In the recent years some new dairy development models have been implemented and scaled up by cooperative sector like New Generation

Cooperatives (Dairy Producer Companies) in Junagadh (Gujarat) initiative taken by NDDB and Punjab Progressive Dairy Farmers Association started by a group of progressive farmers. In the liberalized economy, the replication and scaling up of these models largely depends on the governance, institutional support and market forces.

The study also revealed that the proportional expenses on dairying to total production expenditure at the household level is inversely related to land ownership whereas as income from dairying to total family income of the households is also inversely related to land ownership. Therefore, dairying has capacity to reduce poverty at the household level and it should be an integral part of poverty alleviation programs. The study revealed that 1 per cent incremental increase in crop income will trigger total income inequality by 1.38 per cent with a caveat that other things are unchanged. On the other hand, the income from off farm, dairy farm and livestock source has a equalizing effect on the distribution of total income for all categories of farm households, which otherwise corroborates the hypothesis of relative income equalizing effect through dairying and other livestock farming compared to distribution of incomes through crop. The income from dairy farming reduces the income inequality. It also confirms that growth through inclusive dairying does not worsen income distribution, but helps in reducing absolute poverty and inequality. Promotion of economic development and reduction of poverty will depend on the capacity of dairy farming to contribution to smallholder income and employment. In the liberalized era the researchable questions is therefore to search whether smallholder-based dairy sector is justifiable and if it is "yes", then in what forms it will work? Shall it follow the combination of co-operative or contractual or corporate format or each should work independently according to the needs and priority of the region/states?

References

Birthal PS, Joshi, PK and Kumar A 2002. Assessment of research priorities for livestock sector in India. Policy Paper No. 15, National Centre for Agricultural Economics and Policy Research (ICAR), New Delhi.

Corral JD, Perez JA, and Roibas D, 2011. The impact of land fragmentation on milk production. Journal of Dairy Science, Vol. 94, pp.517-525.

CSO, 2007-08. "National Accounts Statistics" Ministry of Statistics and Programme Implementation, Government of India, New Delhi.

Kumar A, Kumar S, Singh DK and Shivjee 2011. Rural employment diversification in India: trends, determinants and implications on poverty. Agricultural Economics Research Review Vol. 24, pp. 361-372.

Kumar A, and Singh, Dhiraj K 2008. Livestock production system in India: An appraisal across agro-ecological regions. Indian Journal of Agriculture Economics, Vol.63, No.4, pp. 577-597.

Ram KA, Tsunekawa A, Sahad DK and Miyazaki T 1999. Subdivision and fragmentation of land holdings and their implication in desertification in the Thar Desert, India. Journal of Arid Environments 41, 463–477.

Mandal S, Datta KK and Lama TD 2010. Economic evolution of farming system research in NEH region: some issues. Indian Journal of Agricultural Economics 65(1), 118-134.

NSSO, 2006. Livestock ownership across operational land holding classes in India, 2002-03. National Sample Survey 59thRound, Ministry of Statistics and Programme Implementation, Govt. of India.

Singh SR and Datta KK (2010) Understanding Value Addition in Indian Dairy Sector: Some

Perspectives, Agricultural Economics Research Review, Vol. 23 (Conference Number): 487-493.

Wattanutchariya S and Jitsanguan T 1992. Increasing the scale of smallfarm operations in Thailand. Food and Fertilizer Technology Centre. http: / /www.fftc.agnet.org/ library/article/eb344a.html.

2016, Dairy and Food Product Technology Pages 87–104
Editors: Birendra Kumar Mishra and Subrota Hati
Published by: BIOTECH BOOKS, NEW DELHI

Chapter 7

Meat Based Snacks

Parminder Singh, Tarun Pal Singh and Pavan Kumar

Department of Livestock Products Technology,
College of Veterinary Science,
Guru Angad Dev Veterinary and Animal Sciences University,
Ludhiana – 141 004, Punjab

Historical Background

Snack foods have always been a significant part of modern life; they represent a distinct and constantly widening and changing group of food items. It has been known for centuries that salting, drying and smoking meats and fish is a very effective way to preserve them. Long before Europeans arrived in the Americas, pemmican and charqui (from whence the word "Jerky" is derived) were dietary staples. These dry meat products were a convenient source of sustenance for hunters, as they were constantly on the move. According to a rough working estimate, an annual worldwide sale of snacks including the US was $30-35 billion. American Heritage dictionary defines snack as "hurried or light meal" or "food eaten between meals." Traditional snack foods appeal to consumers on a number of levels (Riaz, 2004).

Introduction

Meat and meat products have provided a valuable contribution to our diets since the time of hunter-gatherer societies. The importance of meat is reflected not only in terms of a concentrated source of proteins with high biological value but also due to containment of almost all essential amino acids, vitamins, minerals, fatty acids etc. Besides it is well known for its palatability due to the inherent sensory attributes. At present total meat production of India is about 7.2 million tonnes of meat which accounts for 33.8 per cent share of total export from the livestock industry

(DGCIS, 2009). So, Indian meat industry is growing at a very fast pace from the last one decade. The processing of meat into different ready-to-eat value added products is also increasing due to urbanization and the growth of fast food sector. However, our traditional meat products due to their limited shelf life at room temperature and bulkiness find very limited role in our diets. Recently the outburst of dietary fat related health problems such as obesity, hypertension, high blood cholesterol level, coronary heart diseases and cancer set the trigger for various recommendations to minimize the dietary fat content to the 30 per cent of total energy intake, out of which not more than 10 per cent from the saturated fatty acids (Department of Health, 1994). The concept of functional foods includes exploiting the properties of certain foods to treat, mitigate or prevent diseases in addition to their nutritional value. For these reasons, various organizations have recommended a total fat intake to no more than 30 per cent of total calories. So modern trends towards convenience foods have resulted in consumer demand for pre-cooked and shelf-stable meat based snack products. So there is strong need to develop nutritionally superior meat products having longer shelf life at room temperature conditions especially for those places which are either energy deficient or do not have refrigeration facilities.

Snack foods are convenient food items typically designed to be portable, quick and satisfying for one person, house hold, working women, school age children and during travelling or to satisfy short term hunger (Lusas and Rhee, 1987). Today majority of world's population is enjoying snack foods which are prepared from natural ingredients or components to yield products with some specific functional properties. Snack is a "titbit" which is a small meal in broadest sense. Snacking can be defined as problem free consumption of easy to handle, miniature portioned, hot or cold products in solid or liquid form which need little or no preparation and are intended to satisfy the short term hunger (Macrae *et al.*, 1993). These snack foods are available in the market in different forms and shapes (Tettweiler, 1991). In 2004-05, the value of total market of world snack foods was Rs 82.9 billion including semi-processed/cooked and ready to eat foods, and is increasing with a growth rate of 20 per cent (FICCI, 2006). But according to Global Industry Analysts, the market of world snack foods is predicted to reach almost US$ 335 billion by 2015 (Anonymous, 2012). The present value of Indian snacks market is worth around US$ 3 billion, having half of the market share from organized sector with an annual growth rate of 15-20 per cent whereas unorganized snack market is valued at US$ 1.56 billion with an annual growth rate of 7-8 per cent (Parmar, 2011). Moreover with the proliferation of high protein diets such as the Atkins Diet, snacking on meat products is nowadays just as common as munching on potato chips or cookies (White, 2004).

Characteristic Features of Meat Snacks

1. Snacks may be classified as first generation (conventional potato chips and baked crackers), second generation (directly expanded snacks) and third generation (semi-products or pellets, half-products or intermediate-products) (Pansawat, 2007).
2. Meat snacks are less perishable, more durable, more appealing and shelf stable in nature.

3. The snacks can be prepared by extrusion technology in the form of various shapes such as cookies, pretzels, corn chips, tortilla chips, snack nuts, meat snacks (Lusas and Rhee, 1987).
4. Cereal based snacks are usually deficient in some essential amino acids such as threonine, lysine and tryptophan (Kumar *et al.*, 2012). So incorporation of meat greatly enhances its nutritional value especially with respect to essential amino acids content and high biological value protein along with some sensory attributes like flavor and taste (Raja *et al.*, 2014).
5. Meat snacks are superior to other meat products by virtue of their good nutritional profile, less fat, sodium and calorie content.
6. Many meat based snack products are now marketed with "fat free" slogans, (97- to 98-percent fat-free) and in fact, most have been low in fat since their early development (Kale, 2009).
7. The meat snacks are highly crispy and crunchy, attractive, (Zeuthen, 1984), nutritionally sound and shelf-stable in nature (Bhattarcharya *et al.*, 1988).

Extrusion Technology for Meat Snacks

Food extrusion is a technique which is used to form snacks of different shapes by forcing the raw food material through a special die which is designed to form and/or puff dry extrudates (Rhee *et al.*, 1999a). In snack food industry, the use of extruders was started in 1930, when corn curls were first extruded. Later on, their use for production of second and third generation snacks was increased (Moore, 1993). Today a large number of single screw, multiple and twin screw extruders are available for preparation of a wide variety of snacks such as pastas, multigrain snacks, jelly beans, breakfast cereals, cookies, pretzels, French Fries, Idiappam, corn chips, tortilla chips, snack nuts, fish crackers, turkey meat papads, extruded snacks, meat snacks etc.

Extrusion cooking is a highly versatile and efficient technology for food processing as it denatures anti-nutritional factors and improves the quality and digestibility of proteins (Harper, 1978). During extrusion, raw materials undergo various structural and chemical changes such as starch gelatinization, protein denaturation, complex formation between amylase and lipids, inactivation of raw food enzymes, destruction of naturally occurring toxins and reduction of microbial load in the final product. Some of the scientific studies related to extrusion cooking have been reported for starch conversion (Zheng and Wang, 1994), degradation of starch (Cai *et al.*, 1995) and destruction of some vitamins or pigments (Ilo and Berghofer, 1998). The final quality of the product depends upon structural and chemical alterations occurring due to extensive shear forces during extrusion (Ilo and Berghofer, 1999). For meat snacks, extrusion technology is more important due to its technological advantages over the traditional food processing techniques. Extruded snack foods were produced with meats in combination with other ingredients (Mittal and Lawrie, 1984;Megard *et al.*, 1985; Alvarez *et al.*, 1990; Alvarez *et al.*, 1992). Application of extrusion technology helps in nutrient retention and inactivation of both contaminating and disease producing micro-organisms. The detail of various meat snacks developed is shown in Table 7.1.

Table 7.1: Some of the Meat Snacks Developed from different Meat Species

Sl.No.	*Description of Meat Snacks Developed*	*References*
1.	Beef snacks- using beef, defatted soy flour and corn starch	Park *et al.* (1993)
2.	Chile-flavored, jerky-type extruded products frompotato flour and either partially defatted chopped beef, mechanically separated chicken or chicken thigh meat	McKee *et al.* (1995)
3.	Turkey meat *papads*- using turkey meat and rice flour in ratio of 50:50	Berwal *et al.* (1996)
4.	Snacks using extruded potato and chicken thigh meat (0-40 per cent) and twin screw extruded	Jean *et al.* (1996)
5.	Jerky-type extruded products- using potato flour with beef/chicken and chile powder	Ray *et al.* (1996)
6.	Lamb blends- using vacuum packaged lamb and corn starch	Rhee *et al.* (1999a)
7.	Blends of corn starch (81.72-84.86 per cent) and ground meat (goat meat, lamb, mutton, spent hen meat, beef 15.14-18.28 per cent)	Rhee *et al.* (1999b)
8.	Tapioca-fish and Tapioca-peanut snacks	Suknark *et al.* (1999)
9.	Chickpea snacks- containing 0, 5 and 10 per cent of bovine lung meat with extrusion at 130°C and 13 per cent moisture content of the feed	Cardoso-Santiago and Arêas (2001)
10.	Chicken chips- using four different levels of spent hen meat *i.e.* 75 per cent, 80 per cent, 85 per cent and 95 per cent	Sharma and Nanda (2002)
11.	Keropok-like snacks using wheat flour, cassava flour and pulverized meat offal	Subba (2002)
12.	Snacks, one containing chickpea and either 0, 5 or 10 per cent bovine lung and another with pure amaranth	ChávezJáuregui *et al.* (2003)
13.	Fermented cabrito snack sticks using goat meat and different proportions of soy protein concentrate	Cosenza *et al.* (2003)
14.	Popped cereal snacks having various blends of spent hen meat and grains (potato starch, corn starch and rice flour)	Lee *et al.* (2003)
15.	Extruded blends of catfish flesh using a single-screw extruder	Rhee *et al.* (2004)
16.	Extruded tripe snacks by using buffalo rumen meat (50 per cent) and corn flour (40 per cent, 50 per cent and 60 per cent)	Anna Anandh *et al.* (2005)
17.	Snack dips- using combination of 50 per cent ham, 26 per cent bacon or 28 per cent pepperoni with added sour cream, unflavored yogurt and tofu	Defreitas and Molins (2006)

Contd...

Table 7.1–Contd...

Sl.No.	*Description of Meat Snacks Developed*	*References*
18.	Extruded snacks using chick pea, corn and bovine lungs	Moreira-Araújo *et al.* (2008)
19.	Value-added jerky-style fish meat snacks from salmon flesh	Kong *et al.* (2008)
20.	Chicken snack sticks using spent hen meat with 4.2 per cent oat meal and 8.4 per cent ragi flour	Kale (2009)
21.	Fish crackers- using different ratios of fish and tapioca flour- 1:1, 1.5:1, 2:1 and 2.5:1	Nurul *et al.* (2009)
22.	Snack foods fortified with fish flesh/protein	Maga and Reddy (1985); Clayton and Miscourides (1993); Choudhury (1994); Mathews *et al.* (2003); Shankar and Bandyopadhyay (2005); Kong *et al.* (2008); Pansawat *et al.* (2008); Dileep *et al.* (2010)
23.	Chicken snacks using spent hen meat, rice flour and sodium caseinate	Singh *et al.* (2002); Singh *et al.* (2011)
24.	Kilishi- dried and grilled meat snacks	Shamsuddeen (2009), Mgbemere *et al.* (2011)
25.	Corn-fish snacks - using 3 per cent, 5 per cent, 7 per cent and 9 per cent fish protein powder	Shaviklo *et al.* (2011)
26.	Chicken meat caruncles- using 65 per cent spent hen meat, rice flour, tapioca starch, potato starch, baking powder in different proportions	Singh *et al.* (2012a); (2012b); (2013a); (2013b)
27.	Chicken meat biscuits by using different levels *viz.* 5 per cent, 10 per cent, 15 per cent, 20 per cent and 25 per cent of defatted chicken and refined wheat flour	Bukya *et al.* (2013)
28.	Chicken meat noodles	Khare *et al.* (2014); Verma *et al.* (2014)
29.	Fish curls using different levels of flours *viz.* 20 per cent cornflour, 10 per cent black gram flour and 10 per cent peanut flour	Raja *et al.* (2014)

Cooking of Meat Snacks

Cooking is the process of preparing food by application of heat. The ultimate satiety or enjoyment that comes from eating meat is largely dependent on cooking method. Cooking not only ensures palatability and microbial safety but also increases the shelf life of product and enhances flavor and taste, improves digestibility, softens the connective tissue and thus improves texture, coagulates and denatures meat proteins thereby changing their solubility and color, inactivates proteolytic enzymes and thus preventing the development of off flavors, lowers the water activity of meat and improve the peelability of frankfurters and lastly stabilizes the red color of cured meats.

Microwave treatment is a convenient way for thawing, heating, cooking, drying, frying, pasteurization and blanching of foods (Datta and Davidson, 2000). Microwave ovens make use of a portion of the electromagnetic spectrum. The two commonly used frequencies for microwave heating are 915 and 2450 MHz with wavelengths of 32.8 and 12.25 cm respectively. Once microwave energy is absorbed, polar molecules (such as water molecules) and ions (dissolved salts) inside the food will rotate or collide according to the alternating electromagnetic field and heat is subsequently generated for cooking. Microwave cooking has many advantages over conventional cooking such as rapid and thorough cooking of product, ease of control, wide degree of selectivity, reduction in cooking losses, less energy usage and retaining the initial characteristics of the product. Microwave ovens are about 40 per cent efficient as compared to 14 per cent for standard electric ovens and 7 per cent for gas ovens. The use of microwave oven in the development of snacks, fried without oil, envisages a very interesting field in product development and innovation.

Oven cooking is widely used in commercially processed poultry meats, particularly in the food service systems. The air forced convection method is the most representative cooking system, and it results in desirable foods quality traits. Air convection is frequently mixed with steam injection in the oven chamber in order to determine an improvement of meat tenderness and also to reduce cooking losses. Some workers have observed that microwave oven cooked meat products had lower moisture content than conventional oven cooking (Hoda *et al.*, 2002) but Nath *et al.* (1996) and Mendiratta *et al.* (1998) reported no moisture difference in microwave oven and conventional oven cooked chicken patties.

Use of Non-Meat Ingredients in Meat Snacks

Incorporation of functional ingredients (flours and starches) into snack products by using extrusion cooking can improve their processing and nutritional quality (Veronica *et al.*, 2006). Various non-meat ingredients such as rice flour, tapioca starch, potato starch, oat meal, ragi flour etc can be explored for the production of meat based snacks. Tapioca starch can be used to improve flavour and flavour release, increase moisture retention as well as reduce cooking losses (McAuley and Mawson, 1994). Potato starch is also a rich source of carbohydrates and it provides crispiness to meat products and improves suitability to extrusion. Potatostarches are recommended as water binders in comminuted meat products to increase yields, reduce losses from cooking, improve texture and extend shelf life (Murphy, 2000). Amylopectin potato starch can be used as a thickener and or stabilizer in a wide range of sectors of the food industry (Vries, 1995). The pregelatinized waxy starch and potato starch are hydrated prior to cooking which allows bubbles to be produced and retained and thus controls the crisp and crunchy texture of the snack foods (Carey *et al.*, 1998). When starch was used in large quantity, water absorption index of extrudates increased due to increased starch gelatinization (Cheftel, 1986). Expansion in extrudates with 20 per cent chicken thigh meat was more due to hydrophilic nature of chicken proteins which facilitated starch gelatinization as compared to 40 per cent chicken meat (Chandrashekhar and Kirleis, 1988). Lower value of pH in snacks due to increase in level of meat and decrease in starch content in the mix was reported by Mittal and Usborne (1986).

Rice flour is also a good source of carbohydrates, proteins and a few amino acids like leucine, isoleucine, phenylalanine etc (Traitilevich, 1984), yet it lacks some other essential amino acids which are required for good health. So incorporation of essential amino acids can be done by addition of nuts, fish, meat etc to rice flour (Wu *et al.*, 2002). Since rice like other cereals can be puffed or popped easily, so it is commonly used in snack foods to increase cooking yield, flavor, colour, taste, crispiness etc. Maga and Reddy (1985) prepared extrudates of varying amounts (10–35 per cent) of deboned minced carp. They blended these extrudates with non-extruded rice flour and prepared *pakodas* (a fried Indian snack food). Clayton and Miscourides (1992) prepared shelf stable, texturised and puffed snack food products from underutilized fish tissue alone and in combination with rice flour blends using a single screw extruder. Singh *et al.* (2002) developed chicken snacks using spent hen meat, rice flour, sodium caseinate, spice mix, condiments, salt, phosphate and baking powder. They observed that product with 50 per cent spent hen meat got highest score for colour, appearance, texture, crispiness and overall acceptability.

Physico-chemical and Rheological Characteristics of Meat Snacks

In extruded snack foods, the quality evaluation seems to be correlated with sensory, instrumental and microstructure characteristics which all together will account for a product with high acceptability (Anton and Luciano, 2007). The acceptability of meat snacks depends on certain physical characteristics of the product along with sensory attributes. It has been observed that some properties especially expansion, density, hardness, crispiness etc are more important for such types of products (Peri *et al.*, 1983). Shear force value is considered as the most critical factor in determining the texture of extruded meat based snack products and generally lower shear force value indicates lower bulk density, higher expansion ratio and higher water absorption by products (Park *et al.*, 1993). Also it was observed that >5 per cent fat level interfered with extrusion process and produced irregular extrudates and die head blockage. They also observed that high corn starch and low fat level resulted in higher water absorption index of extrudates. Siriburi and Hill (2000) reported that water absorption index and water solubility index can be used to characterize extruded products. They also concluded that increasing WSI was due to increasing starch conversion.

Mohamed (1990) observed that by increasing the meat protein resulted in decreased expansion of the extruded corn starch. Extrudates containing meat with more moisture expanded more. But moisture content of the meat extrudates decreased as the concentration of non-meat binders increased. Extrudates should have a final moisture content of less than 5 per cent to make the products brittle (Jean *et al.*, 1996).Snack meat products are dried/low- moisture meat products with lower water activity (a_W) level and commonly stored at room temperature conditions which makes them more prone to the fungal spoilage. Cereal based snacks and thermally processed meat products are shelf stable.

Packaging of Meat Snacks

Packaging is an important tool for preservation of meat as it avoids contamination, delays spoilage; permit some enzymatic activity to improve tenderness, reduce weight loss and sometimes to ensure an oxymyoglobin or cherry-red colour in red meats at retail or customer level (Brody, 1989). The type of packaging *i.e.* aerobic, vacuum or modified atmosphere to be used for a particular product depends on the nature of product and their shelf life.

Vacuum packaging is most widely used for extruded products as this type of packaging can keep the product safe especially beef products for about 45 days at ambient temperature (32±2°C) (Edward *et al.*, 1987). Packaging can lower the weight loss, cost of transportation and increase the shelf life of food products (Rozbehet al., 1993). Extruded type products are mostly packaged in laminated pouches for convenience and to maintain their quality. Vacuum packaging can delay the growth of aerobic spoilage micro-organisms and retard oxidation of lipids in fresh meats (Genigeorgis, 1985).

Modified Atmosphere Packaging (MAP) provides too many advantages of cost, shelf life, product uniformity, label information, and supply chain integration for most industrialized countries to return to in-store cutting and packaging to supply self-service meat cases. During MAP of sliced-pastrima (dry meat product), the storage temperature, storage period and the storage period x the storage temperature interaction showed significant ($P<0.01$) effects on the amount of the residual nitrite (Aksuet al.,2005). Modified atmosphere packaged pastrimas also give higher sensory ratings than those packaged by Vacuum Packaging (VP) or Aerobic Packaging (AP) and stored at 4°C for 120 days (Gok *et al.*, 2008).

Popular Meat Based Snacks

1. Popped Pork Rinds

This is also called as bacon skin. The skin of pig is processed in such a way that it gets puffed off to many times its original volume thereby giving a crispy and friable texture to the product. The flavor from such type of product depends upon the type of fat which is used for its processing. Acceptable pellets are processed as needed in 400-425°F fat. The puffs are sprinkled with flake flour salt or a seasoning mixture. Some authors also suggest dipping of rinds in acid solution (acetic acid solution for 15 seconds) prior to puffing so as ensure the complete puffing of the product. The reticular tissue gets weakened by acetic acid treatment and does not interefere with puffing of the snacks. Immediately after popping, the total solids content will be over 99 per cent. Other typical analytical values are protein - 57 per cent, fat - 34 per cent, carbohydrates - 5 per cent and ash - 4 per cent. The shelf-life of this product is variable from region to region and is in range of 3-8 weeks(Matz, 1976).

2. Kilishi

For a successful industrial production of Kilishi a sundried, roasted meat product containing 46.4 per cent beef and 53.6 per cent non-meat ingredients (spices and condiments), an understanding of the rheological properties of the nonmeatingredient portion used in its production is very important (Nkama and Badau, 2000). It is popular especially in Northern Nigeria, Cameroun, Chad, Niger Republic and other countries in the Sahelian region of Africa. Kilishialso has been an important export commodity to Saudi Arabia and other Asian countries where it is also widely consumed as it is brown in colour, light in weight and rich in protein and other vital nutrients (Igene *et al.*, 1990). Addition of spices is crucial stage during preparation of Kilishi, as it is a critical control point (Shamsuddeen and Ameh, 2008). The entire fat is removed before meat is cut into thin slices using sharp knife. These slices of meat are then spread on a wooden sheet and are then sun dried to reduce the moisture in it. The moisture content of Kilishi varies from 10 to 20 per cent (wet basis). It is then coated with other ingredients of the formulation and again dried and then roasted. Traditionally, it is stored without special packaging and is eaten as a snack. One of the drawbacks of the current processing method is the high losses that occur during coating. Optimizing Kilishi processing is essential to improve yields and ensure that the end product is stable (Kalilou *et al.*, 1998). The resultant product is wrapped in a cardboard paper and is sold at retail shops, restaurants, hotels, markets and airports.

3. Jerky

Jerky is a popular meat product item in the world due to its stability and nutritional value. The word jerky has been derived from a Spanish word - "charqui" which means to burn (meat). Jerky type products are characterized by a considerable diversity in the, with different kind of the applied raw materials (beef, pork, poultry, venison, and fish), spices and other functional additives such as *e.g.* antioxidants and stabilizers, individual technological procedures (comminution, curing, smoking, drying, and packaging) and applied equipment (Konieczny *et al.*, 2007). Jerky is one of the oldest types of meat products that is preserved by salting and drying to reduce water activity, and its easy preparation, light weight, rich nutrient content, and stability without refrigeration make it a popular item for sports enthusiasts, travelers, and mountaineers (Choi *et al.*, 2008). Jerky has traditionally been made from sliced whole muscle of large animal which have been marinated and dried. But it has also been prepared with by-products (beef heart and tongue) (Miller *et al.*, 1988).Processing of restructured pork jerky with added meat emulsion to improve the binding ability has also been done (Choi *et al.*, 1993). Jerky can also be formed from comminuted meat. Generally, because restructured jerky can be made with muscles of poorer quality and trimmings including meats of small size relative to sliced jerky, manufacturers are saving production expenses, making possible the mass production of standardized products due to control of the product size and shape. Traditionally meat jerky was prepared American Indians who salted strips of muscle tissue from deer, buffalo and other game animals, and cured them in the sun or over smoky fires for long periods of time. Currently it is also being produced from Alligator, Bison, Emu, Elk, Duck, Goat, Tuna, Trout, Yak and even Kangaroo meat. Whether it is prepared from extruded ground meat or from whole muscle, typically it will have moisture content of less than 15 per cent and pH is less than 4. Curing is done mostly with sodium nitrite which helps in the development of color and flavor in the product. Beef jerky is also prepared from ground meat which has been given saline treatment for about 12 hours and freezed for 1-3 weeks. Such meat is then sliced into strips and dried. Once packageded, shelf stable jerky meat has a shelf-life of one year or more. Jerky does not need refrigeration due to its low water activity (0.70–0.85) (Quinton *et al.*, 1997). Also, jerky is nutritious (high in protein and low in fat), shelf-stable (0.75:1.00 moisture protein ratio), it is still in high demand as a snack food in many countries (Yang *et al.*, 2009). The commercial fraud of jerky products, particularly for yak jerky which is much more expensive than cattle jerky, may cause many problems (Chen *et al.*, 2010).

4. Pepperettes

These are also called Pepperoni sticks. This is a low pH fermented meat based snack which is usually prepared from ground meat. There are several categories and varieties of allbeef or beef/pork (55 per cent/45 per cent) fermented sausage that have a moisture/protein (M/Pr) ratio of 1.6.1 and, as such are called pepperoni (Faith *et al.*, 1997). Though it is a fat rich product, its toughness can be best controlled by the fat content and amount of drying. The product is stuffed into small diameter casings (mostly collagen, although sheep casings are still used), after which a number of production methods can be applied. Some pepperettes are simple smoked and dried, some smoked and cooked, some smoked, cooked and dried, all with various levels of fermentation. Fermentation at 41°C to pH 4.6 or pH 5and post-fermentation heating of sticks to an internal temperature of 54°C resulted in a > 5 $\log_{10}$ decrease in counts of *E. coli* 0157:H7 in summer sausage. As with Jerky, it is becoming common to find not-so-common flavor profiles in pepperoni sticks.

5. Pickled Pig's Feet

Pickled pig's feet is a type of pork associated with Cuisine of the Southern United States, Scandinavian cuisine and Korean cuisine. Pig's feet that are pickled are usually consumed as something of a snack or a delicacy rather than as the primary focus of a meal as its meat course. However, pig's feet are not always pickled and in the aforementioned cultures, may be cooked as a part of a meal, often with vinegar and water to preserve their natural flavor. They have a high fat content, with almost an equal portion of saturated fat to protein. For preparation of this product forefeet are mainly used as hind feet are usually not so meaty in case of pigs. Moreover, the hind feet also get deformed due to shackles from which the carcass is suspended. If hind feet are used for preparation of this product, it is then known as "tid-bits". The feet are pickled in a solution of brine containing 175 ppm of sodium nitrite at 4.4°C. These are then cooked in boiling water and then immediately chilled to an internal temperature of 12.8-15.6°C. These are then sliced and packed in jars for retail sale.

Microbiological Quality of Meat Snacks

As far as microbiological quality of meat snacks is concerned, it is very important because several outbreaks of foodborne disease have been attributed to consumption of both home dried and commercial jerky due to contamination with *Staphylococcus aureus* and *Escherichia coli* O157:H7 (Nummer *et al.*, 2004). In addition, beef jerky recalls have been issued due to contamination with *Salmonella* spp. and *Listeria monocytogenes* (Porto-Fettetal.,2009). The most common pathogen associated with beef jerkyoutbreaks has been Salmonella, with the largest outbreak occurring in 1994, affecting 93 persons who consumed beef jerky made inNew Mexico. A more recent outbreak in 2012 involved4 cases of salmonellosis from individuals who consumed turkeyjerky made in Minnesota (Scheinberg *et al.*, 2014). Additionally, consumption of venison jerky has resulted in infections with Escherichia coliO157:H7 in Oregon in 1994 and Connecticut in 2002 (Rabatsky-Her *et al.*, 2002). Some of the pathogens such as *L. monocytogenes* can tolerate low pH, water activity (a_w) and high salt concentrations (Ryser and Marth, 2007). Traditional pepperoni processing involves aging of meat (with sodium chloride), fermentation (with added spices, sugar and sodium nitrate) with starter culture bacteria, and drying of the product. These manufacturing stepsmay reduce levels of *L. monocytogenes* in sausages but not necessarily eliminate it from the final product. Therefore, microbial cells that survive processing may acquire salt and acid resistance or cross-protection to various stresses. In addition to potential survivors, pepperoni may become re-contaminated with *L. monocytogenes* originating from the processing environment or food contact surfaces, during peeling, slicing or packaging. Hence microbiological quality of meat based snacks must be monitored from production till the final product is consumed.

Conclusions

At present the sales of meat snacks rising day by day. While meat snacks have traditionally been popular with men and teens, their increase in popularity can partially be traced to the fact that more and more women are buying them - for themselves, and as nutritious snacks for their children. Three key consumer trends in meat snacks have been identified: convenience, healthy choice and meal replacement. As a result, the brands and producers who cater to these trends will be the ones to watch as sales continue to climb. In addition, available flavours are changing, and the range is expanding, with the newest flavours ranging from hot and spicy to smoky and chipotle, and incorporating a number of new meat options, like wild game, in addition to the more traditional beef, pork, chicken and turkey. The rise in meat based snacks industry is continuing giving more opportunities to employers, industry professionals and common people.

References

Aksu MI, Kaya M, Ockerman HW 2005. Effect of modified atmosphere packaging, storage period, and storage temperature on the residual nitrate of sliced-pastrima, dry meat product, produced from fresh meat and frozen/thawed meat. Food Chem 93: 237-242.

Alvarez VB, Ofoli RY, Smith DM 1992. Protein in solubilization and starch gelatinization of mechanically deboned chicken meat and cornstarch during twin-screw extrusion. Poultry Sci71: 1087-1095.

Alvarez VB, Smith DM, Morgan RG, Booren AM 1990. Restructuring of mechanically deboned chicken and nonmeat binders in a twin-screw extruder.J Food Sci55(4): 942-946.

Anna Anandh M, Lakshmanan V, Mendiratta SK, Anjaneyulu ASR, Bist GS 2005. Development and quality characteristics of extrude tripe snack food from buffalo rumen meat and corn flour. J Food SciTechnol 42(3): 263-267.

Anonymous 2012.Global Snacks and Sweets Industry. Snack and Sweet Industry: Market Research Reports, Statistics and Analysis. http: //reportlinker.com/. Website visited onSeptember 15, 2012.

Anton AA, Luciano FB 2007. Instrumental texture evaluation of extruded snack foods: a review. Ciencia y Technol Aliment 5: 245-251.

Berwal JS, Dhanda JS, Berwal RK 1996. Studies on rice and turkey meat blend papads. J Food SciTechnol 33: 237-239.

Bhattarcharya S, Das H, Bose AN 1988. Effect of extrusion process variables on *in vitro* protein digestibility of fish-wheat flour blends. Food Chem 28: 225.

Brody AL 1989. Modified atmosphere/vacuum packaging of meat. In: Controlled/ modified atmosphere/vacuum packaging of foods. Brody AL (ed). Food and Nutrition Press, Trumbull, Connecticut, pp. 17-37.

Bukya A, Sunooj KV, SurendraBabuA 2013.Standardization and evaluation of physical, textural and organoleptic properties of chicken biscuits.Int J Adv Res 1: 163-168.

Cai W, Diosady LL, Rubin LJ 1995. Degradation of wheat starch in a twin-screw extruder.J Food Engg **26(3):** 289-300.

Cardoso-Santiago RA, Arêas JAG 2001. Nutritional evaluation of snacks obtained from chickpea and bovine lung blends. Food Chem 74(1): 35-40.

Carey JM, Moisey MJ, Levine H, Slade L 1998. Production of crispy wheat-based snacks having surface bubbles. United States Patent: US5747092.

Chandrashekar A, Kirleis AW 1988. Influence of protein on starch gelatinization in sorghum. Cereal Chem 65(6): 457-462.

ChávezJáuregui RN, CardosoSantiago RA, Silva ME, Arêas JA 2003. Acceptability of snacks produced by the extrusion of amaranth and blends of chickpea and bovine lung. Int J Food SciTechnol 38(7): 795-798.

Cheftel JC 1986. Nutritional effects of extrusion cooking.Food Chem 20: 263-283.

Chen SY, Liu YP, Yao YG 2010.Species authentication of commercial beef jerky based on PCR-RFLP analysis of the mitochondrial 12S rRNA gene. J GenetGenom 37(11): 763-769.

Choi JH, Jeong JY, Han DJ, Choi YS, Kim HY, Lee MA, Lee ES, Paik HD, Kim CJ 2008.Effects of pork/beef levels and various casings on quality properties of semi-dried jerky. Meat Sci 80(2): 278-286.

Choi YI, An YS, Hong SK 1993. Effect of emulsion addition on binding ability and storage characteristics of restructured pork jerky. Korean J AnimSciTechnol 35: 223-229.

Choudhury GS 1994.Application of extrusion technology to process fish muscle. In: Nutrition and Utilization Technology in Aquaculture. Sessa DJ, Lim C (eds). American Oil Chemists Society, Champaign, pp 61821–5930, pp 233-245.

Clayton JT, Miscourides DN 1993. Extruder texturized foods from underutilized fish tissue. J Aquatic Food Product Technol 1(3-4): 65-89.

Cosenza GH, Williams SK, Johnson DD, Sims C, McGowan CH 2002. Development and evaluation of a fermented cabrito snack stick products. Meat Sci 64: 51-57.

Datta AK, Davidson PM 2000. Microwave and radio frequency processing.Kinetics of microbial inactivation for alternative processing technologies.J Food Sci 65(S8): 32-41.

De-Freitas Z, Molins RA 1988. Development of meat snack dips: chemical, physical, microbiological and sensory attributes characteristics. J Food Sci 53(6): 1645-1649.

Department of Health 1994. Nutritional Aspects of Cardiovascular Disease, Report on Health and Social Subject No. 46. London: Her majesty's stationary office.

DGCIS 2009.Directorate general of statistics and commercial intelligence (DGCIS) - Export of agricultural commodities.Kolkata.

Dileep AO, Shamasundar BA, Binsi PK, Howell NK 2010. Composition and quality of rice flour-fish mince based extruded products with emphasis on thermal properties of rice flour. J Texture Stud 2(41): 190-207.

Edwards RA, Dainty RH, Hibbard CM, Ramantanis SV 1987. Amines of fresh beef of normal pH and the role of bacteria in changes in concentration observed during storage in vacuum packs at chill temperature. J ApplBacteriol 63: 427-434.

Faith NG, Parniere N, Larson T, Lorang TD, Kaspar CW, Luchansky JB 1997.Viability of *Escherichia coli* 0157: H7 in pepperoni during the manufacture of sticks and the subsequent storage of slices at 21, 4 and - 20°C under air, vacuum and CO_2. International J Food Microbiol 61(4): 377-382.

FICCI 2006.FICCI Food and Beverages Survey. New Delhi: Federation House, 1, Tansen Marg.

Genigeorgis CA 1985. Microbiological and safety implicationa of the use of modified atmosphere to extend the storage life of fresh meat and like fish.J Food Microbiol1: 237.

Gok V, Obuz E, Akkaya L 2008. Effects of packaging method and storage time on the chemical, microbiological and sensory properties of Turkish pastrima-a dry cured beef product. Meat Sci80: 335-344.

Harper JM 1978.Food Extrusion.Crit Rev Food SciNutr 11(2): 155-215.

Hoda I, Ahmad S, Srivastava AK 2002.Effect of microwave oven processing, hot air oven cooking, curing and polyphosphate treatment on physico-chemical, sensory and textural characteristics of buffalo meat products.J Food SciTechnol 39(3): 240-245.

Igene JO, Farouk MM, Akanbi CT 1990.Preliminary studies on the traditional processing of "Kilishi" J Sci Food Agric50: 89-98.

Ilo S, Berghofer E 1998. Kinetics of thermo mechanical destruction of thiamine during extrusion cooking.J Food Sci**63**(2): 312-316.

IloS, Berghofer E 1999. Kinetics of colour changes during extrusion cooking of maize grits. J Food Engg 39(1): 73-80.

Jean IJ, Work R, Camire ME, Briggs J, Barrett AH, Bushway AA 1996. Selected properties of extruded patato and chicken meat.J Food Sci 61(4): 783-789.

Khare AK, Biswas AK, Sahoo J 2014.Comparison study of chitosan, EDTA, eugenol and peppermint oil forantioxidant and antimicrobial potentials in chicken noodles and theireffect on colour and oxidative stability at ambient temperaturestorage.LWT- Food SciTechnol 55: 286-293.

Kale JS 2009.Processing technology and extension of shelf life of chicken snack sticks incorporated with oat meal and ragi flour. M.V.Sc. Thesis, Guru AngadDev Veterinary and Animal Sciences University, Ludhiana, India.

Kalilou S, Collignan A, Zakhia N 1998. Optimizing the traditional processing of beef into Kilishi. Meat Sci 50(1): 21-32.

Kong J, Dougherty M, Perkins L, Camire M 2008. Composition and consumer acceptability of a novel extrusion-cooked salmon snack.J Food Sci73(3): S118-S123.

Konieczny P, Stangierski J, Kijowski J 2007. Physical and chemical characteristics and acceptability of home style beef jerky. Meat Sci 76(2): 253-257.

Kumar G, Goswami M, Pathak V, Singh VP 2012. Development of chicken sticks by replacement of rice flour with different levels of minced chicken meat. *J Food ProcessTechnol* 3: 10.

Lee SO, Min JS, Kim IS, Lee M 2003. Physical evaluation of popped cereal snacks with spent hen meat. Meat Sci 64: 383-390.

Lusas E W and Rhee K C. 1987. Extrusion processing as applied to snack foods and breakfast cereals. Ch.16. In: Cereals and Legumes in the Food Supply.Dupont J, Osman EM (eds). Iowa State University Press, Ames, I.A. pp. 201.

Macrae R, Robinson RK, Sadler MJ 1993.Encyclopaedia of Food Science, Food Technology and Nutrition, Academic press, London, pp. 4167.

Maga JA, Reddy T 1985. Coextrusion of carp (cyprinuscarpio) and rice flour.J Food Process Pres 9(2): 121-128.

Mathews K, Ahmedna M, Goktepe I 2003. Value-added snacks from defatted peanut flour and fish mince: optimizing formulation and consumer acceptability. In: Institute of Food Technologists Annual Meeting, Chicago, IL, July 12-16, pp. 38.

Matz SA 1976. Snack Food Technology. The AVI publishing company, Inc. Westport, Connecticut, pp.109-115.

McAuley C, Mawson R 1994. Low fat and low salt meat product ingredients.Food Aust 46(6): 283-286.

McKee LH, Ray EE, Remmenga M, Christopher J 1995. Quality evaluation of chile-flavoured, jerky-type extruded products from meat and potato flour.J Food Sci 60(3): 587-591.

Mendiratta SK, Kumar S, Keshri RC, Sharma BD 1998 Comparative efficacy of microwave oven for cooking of chicken meat. FleischwirtschaftInt 78(7): 827-829.

Megard D, Kitabatake N, Cheftel JC 1985. Continuous restructuring of mechanically deboned chiken meat by HTST extrusion-cooking.J Food Sci50: 1364-1369.

Mgbemere VN, Akpapunam MA, Igene JO 2011. Effect of groundnut flour substitution on yield, quality and storage stability of kilishi i–a Nigerian indigenous dried meat product. African JFood AgricNutrDev 11(2).

Miller MF, Keeton JT, Cross HR, Leu R, Gomez F, Wilson JJ 1988. Evaluation of the physical and sensory properties of jerky processed from beef heart and tongue. J Food Qual 11: 63–65.

Mittal GS, Usborne WR 1986. Meat emulsion functionality related to fat-protein ratio and selected dairy and cereal products. Meat Sci 18(1): 1-21.

Mittal P, Lawrie RA 1984. Extrusion studies of mixtures containing certain meat offals: part 1 - objective properties. MeatSci10(2): 101-116.

Mohamed S 1990. Factors affecting extrusion characteristics of expanded starch-based products. J Food Process Pres14: 437-452.

Moreira-Araújo RS, Araújo MA, Arêas JA 2008. Fortified food made by the extrusion of a mixture of chickpea, corn and bovine lung controls iron-deficiency anaemia in preschool children. Food Chem 107(1): 158-164.

Moore G 1993. Snack food extrusion. In: Technology of Extrusion Cooking. Frame ND (ed).An ASPEN Publishers, Inc., Gaithersburg, Maryland, pp 110.

Murphy P 2000.Starch. In: Handbook of Food Hydrocolloids. Phillips GO, Williams PA (eds).Woodhead Publishing House, Cambridge, pp 41-65.

Nath RL, Mahapatra CM, Kondaiah N, Singh JN 1996. Quality of chicken patties as influenced by microwave and conventional oven cooking.J Food SciTechnol33(2): 162-164.

Nkama I, Badau MH 2000. Rheological properties of reconstituted Kilishi ingredient mix powder. J Food Engg 44(1): 1-4.

Nummer BA, Harrison JA, Harrison MA, Kendall P, Sofos JN 2004. Effects of preparation methods on the microbiological safety of home-dried meat jerky. J. Food Prot. 67: 2337-2341.

Nurul H, Boni I, Noryati I 2009. The effect of different ratios of Dory fish to tapioca flour on the linearexpansion, oil absorption, colour and hardness of fish crackers.Int Food Res J 16: 159-165.

Pansawat N 2007. Development of an extruded snack supplemented with fish protein and N-3 fatty acids. Ph.D Thesis, Graduate School, Kasetsart University, Thailand.

Pansawat N, Jangehud K, Jangchud A, Wuttijumnong P, Saalia FK, Eitenmiller RR, Phillips RD 2008.Effects of extrusion conditions on secondary extrusion variables and physical properties of fish, rice-based snacks. LWT-Food SciTechnol 4(41): 632-641.

Park J, Rhee KS, Kim BK, Rhee KC 1993. High-protein texturized products of defatted soy flour, corn starch and beef: shelf-life, physical and sensory properties. J Food Sci 58(1): 21-27.

Parmar N 2011.Indian Snacks Industry.http: //scribd.com. Website visited onSeptember 15, 2012.

Peri C, Barbieri R, Casiraghi EM 1983. Physical, chemical and nutritional quality of extruded corn germ flour and milk protein blends. J Food Technol 18: 43-52.

Porto-Fett ACS, Call JE, Hwang CA, Juneja V, Ingham S, Ingham B, Luchansky JB 2009. Validation of commercial processes for inactivation of *Escherichia coli* O157: H7, *Salmonella typhimurium* and *Listeria monocytogenes* on the surface of whole-muscle turkey jerky. Poultry Sci 88: 1275-1281.

Quinton RD, Cornforth DP, Hendricks DG, Brennand CP, Su YK 1997. Acceptability and composition of some acidified meat and vegetable stick products. J Food Sci 62(6): 1250-1254.

Rabatsky-Her T, Dingman D, Marcus R, Howard R, Kinney A, Mshar P 2002. Deer meat as the source for a sporadic case of *Escherichia coli* O157: H7 infectionConnecticut. Emerging Infect Dis 8: 525-527.

Raja WH, Kumar S, Bhat ZF, Kumar P 2014. Effect of ambient storage on the quality characteristics of aerobically packaged fish curls incorporated with different flours. Springer Plus. 3: 106.

Ray EE, MckeeLH, Gardner BJ, Smith DW1996.Properties of an extruded jerky - type meat snack containing potato flour. J Muscle Foods 7(2): 199-212.

Rhee KS, Cho SH, Pradahn AM 1999a. Expanded extrudates from corn starch–lamb blends: process optimization using response surface methodology. Meat Sci 52: 127-134.

Rhee KS, Cho SH, Pradahn AM 1999b. Composition, storage stability and sensory properties of expanded extrudates from blends of corn starch and goat meat, lamb, mutton, spent fowl meat, or beef. Meat Sci 52: 135-141.

Rhee KS, Kim ES, Kim BK, Jung BM, RheeKC2004.Extrusion of minced catfish with corn and defatted soy flours for snack foods. J Food Process Pres28(4): 288-301.

Riaz MN 2004. Snack Foods, Processing. In: Encyclopedia of Grain Science.Corke H, Walker CE, Wrigley C (eds). 3 Volume set, Academic press, pp. 98-108.

Rozbeh M, Kalchyanand N, Field RA, Johnson MC 1993. The influence of bio-preservative on the bacterial level of refrigerated vacuum packaged beef. J Food Safety 13: 99-111.

Ryser ET, Marth EH 2007.Listeria, Listeriosis and Food Safety.CRC Press, Boca Raton, FL.

Scheinberg JA, Svoboda AL, Cutter CN 2014. High-pressure processing and boiling water treatments for reducing *Listeria monocytogenes, Escherichia coli* O157: H7, *Salmonella* spp. and *Staphylococcus aureus* during beef jerky processing. Food Cont 39: 105-110.

Shamsuddeen U 2009. Microbiological quality of spice used in the production of kilishi a traditionally dried and grilled meat product. Bayero J Pure ApplSci 2(2): 66 -69.

Shamsuddeen U, Ameh JB 2008. Survey on the possible critical control points in kilishi (a traditional dried and grilled meat snack) produced in kano. Int J Biosci 3(2): 34-38.

Shankar TJ, Bandyopadhyay S 2005.Process variables during single screw extrusion of fish and rice-flour blends. J Food Process Pres 29: 151–164.

Sharma BD, Nanda PK 2002.Studies on the development and storage stability of chicken chips.Indian J Poultry Sci37: 155-158.

Shaviklo GR, Olafsdottir A, Sveinsdottir K, Thorkelsson G, Rafipour F 2011. Quality characteristics and consumer acceptance of a high fish protein puffed corn-fish snack. J Food SciTechnol 48(6): 668-676.

Singh P, Chatli MK, Sahoo J, Biswas AK 2012b. Effect of different cooking methods on the physico-chemical and sensory attributes of chicken meat caruncles. Ind J PoultSci 47(3): 363-367.

Singh P, Sahoo J, Chatli MK, Biswas AK 2013a. Effect of different levels of baking powder on the physico-chemical and sensory attributes of chicken meat caruncles. Haryana Vet47: 363-367.

Singh P, Sahoo J, Talwar G, Chatli MK, Biswas AK 2012a. Effects of processing conditions on physico-chemical parameters of chicken meat caruncles. In: Proceedings of the International Conference on Sustainable Agriculture for Food and Livelihood Security. Sandhu SK, Sidhu N, Rang A (eds).November 27-29, 2012, Ludhiana, India: Crop Improvement Vol 39 (Spl. Issue), pp 1495-1496.

Singh P, Sahoo J, Talwar G, Chatli MK, Biswas AK 2013b. Development of chicken meat caruncles on the basis of sensory attributes: process optimization using response surface methodology. J Food SciTechnolDOI 10.1007/s13197-013-1160-2.

Singh VP, Sanyal MK, Dubey PC 2002. Quality of chicken snack containing broiler spent hen meat, rice flour and sodium caseinate. J Food SciTechnol 39(4): 442-444.

Singh VP, Sanyal MK, Dubey PC, Sachan N, Kumar V 2011. Chicken snacks as affected by storage conditions under aerobic and vacuum packaging at 30±2°C. *African J Food Sci* 5(11): 620-625.

Siriburi R, Hill SE 2000. Extrusion of cassava starch with either variations in ascorbic acid concentration or pH.Int J Food SciTechnol 35: 141-154.

Subba D 2002.Acceptability and nutritive value of keropok-like snack containing meat offal.Int J Food SciTechnol 37: 681-685.

Suknark K, Phillips RD, Huang YW 1999.Tapioca-fish and tapioca-peanut snacks by twin-screw extrusion and deep-fat frying. J Food Sci 64: 303-308.

Tettweiler P 1991. Snack foods worldwide. Food Technol 45(2): 58.

Traitilevich M 1984. Amino acid composition and biological value of rice proteins.Mukomolnoelevatornaya I KombikarmovayaPromyshleunost 2: 31.

Verma AK, Pathak V, Singh VP 2014. Quality characteristics of value added chicken meat noodles. J Nutr Food Sci 4: 255.

Veronica AO, Olusola OO, Adebowale EA 2006.Qualities of extruded puffed snacks from maize/soybean mixture. J Food Process Engg *29*(2): 149-161.

Vries J A 1995. New possibilities with amylopectin potato starch. *Voedingsmiddelentechnologies* 28(23): 26–27.

White L 2004. Ahead of the game. Meat MarkTechnol 4: 44-48.

Wu JG, Shi C, Zhang X 2002. Estimating the amino acid composition in milled rice by near-infrared reflectance spectroscopy.Field Crops Res 75(1): 1-7.

Yang HS, Hwang YH, Joo ST, Park GB 2009. The physico-chemical and microbiological characteristics of pork jerky in comparison to beef jerky. Meat Sci *82*(3): 289-294.

Zeuthen P 1984. Industrial applications of extrusion cooking (present or potential) In : Thermal Processing and Quality of Foods.ZeuthenP (ed). Elsevier Applied Science Publishers, London. pp. 282.

Zheng X, Wang SS 1994. Shear induced starch conversion during extrusion. *J Food Sci*59(5): 1137-1143.

2016, Dairy and Food Product Technology *Pages 105–129*
Editors: **Birendra Kumar Mishra and Subrota Hati**
Published by: **BIOTECH BOOKS, NEW DELHI**

Chapter 8

Fluidized Bed Dryers for Food Processing

A. Chaudhari and J.B. Upadhyay

Department of Dairy Engineering
S.M.C. College of Dairy Science,
A.A.U., Anand, Gujarat

Introduction

Drying is one of the oldest technique for drying of food products. It is a thermal processing technique. It is mainly used for increasing the shelf life of the food products. It is the most economic processing technique for industries. In this article we have reviewed different types of dryers and also we have discussed in detail about fluidized bed drying. This review will be helpful to use fluidized bed drying for different types of food products economically.

Drying commonly describes the process of thermally removing volatile substances (moisture) to yield a solid product. Moisture held in loose chemical combination, present as a liquid solution within the solid or even trapped in the microstructure of the solid, which exerts a vapour pressure less than that of pure liquid is called bound moisture. Moisture in excess of bound moisture is called unbound moisture. When a wet solid is subjected to thermal drying, two processes occur simultaneously:

1. Transfer of energy (mostly as heat) from the surrounding environment to evaporate the surface moisture.
2. Transfer of internal moisture to the surface of the solid and its subsequent evaporation due to process.

Drying is perhaps the oldest, most common and most diverse of chemical engineering unit operations. Over 400 types of dryers have been reported whereas over 100 distinct types are commonly available. It competes with distillation as the most energy-intensive unit operation due to the high latent heat of vapourization and the inherent inefficiency of using hot air as the (most common) drying medium.

Types of Dryer

Batch Tray Dryer

In this basic type of dryers the material to be dried is placed in pans or trays on the hollow shelves, which are heated by the heating medium, which can range from high pressure steam for moderate to high temperature operation to sub-atmospheric steam for low-temperature operation to hot oil, or even by an electric heater in the case of smaller units. The trays are generally metal to ensure good heat transfer between the trays and the shelves. The number of shelves ranges from 1 to more than 20 shelves in a large unit (Oakley, 1997).

This type of dryers, along with the use of vacuum, is used extensively for drying heat-sensitive or easily oxidized materials. They are particularly well suited for materials, which require gentle handling and where material loss must be minimized. Hygroscopic materials may also be dried completely at temperatures lower than the maximum permissible temperatures of those materials. The major disadvantages of this type of dryers are the high labor cost involved during the loading and unloading of the drying materials and the low capacities of the units.

Indirect-Contact Rotary Dryers

The most common type of indirect-contact rotary dryers is the steamtube dryer. This type of dryer basically consists of a cylindrical shell, typically inclined slightly (1 to 5°C) to the horizontal to facilitate the transportation of the wet feedstock through its body, and in which a number of steam-heated tubes are placed symmetrically (in one, two, or three concentric rows) around its perimeter and rotate with it. The length to diameter ratio of the rotating shell can vary from 4:1 to 10:1.

This type of dryer is used for continuous drying or heating of granular or powdery solids, which cannot be exposed to ordinary atmospheric or combustion gases. It is especially suitable for fine dusty particles as only low gas velocities are needed to purge the drying chamber.

The thermal efficiency of steam-tube dryers is in the range of 70–90 per cent, if well insulated. This number does not allow for boiler efficiency, so it cannot be directly compared with that of direct-heat units, however. Heat transfer coefficients in steamtube dryers may range from 30 to 85 W/(m^2.K) and these values increase with increasing steam temperature due to an increased effect of radiation. The heat flux of these dryers, carrying saturated steam at 140–170C, may range from 6300 W/m^2 for difficult-to-dry and organic products to between 1900 and 3800 W/m^2 for fine inorganic materials (Moyers and Baldwin, 1997).

Rotating Batch Vacuum Dryers

The major disadvantages of this type of dryers are the limited heat transfer areas, especially in larger units. One possible solution to this problem is the addition of internal heating panels; however, this alternative is prone to fouling. The stickiness problem can be alleviated by rotating the vessel slowly or intermittently when the drying material is at high moisture content. However, truly sticky materials may not still be processed in this type of dryers.

The overall heat transfer coefficients of this type of dryer, which may vary near 35 W/(m^2 K), depend largely on the resistance between the inner jacket wall and the drying solids, which depends to a large extent on the solid characteristics. The overall heat transfer coefficients may drop considerably if the dryer walls are fouled; the values in the range of 5–10 W/(m^2 K) are not uncommon.

Agitated Dryers

The first type of agitated dryers is the vertically agitated dryer. Batch cone dryer (or conical mixer dryer) is a conically shaped vessel, with capacities ranging from 50 l to 25 m^3, and is commonly used for drying solvent or water-wet, free-flowing pharmaceuticals, and fine chemical products, so the dryer is normally operated under vacuum. Agitating action is provided by an internally mounted screw, which can be heated to provide additional heating (possibly in the range of 10–30 per cent) to that available through the vessel wall, which rotates about its own axis and also moves around the dryer on an orbiting arm.

The heating medium for this type of dryer is hot water, steam, or hot oil in the temperature range of 50–150C and pressure in the range of 3–30 kPa absolute. The vapour generated during drying is evacuated by a vacuum pump and, if necessary, passed through a condenser for recovery of solvent. For design purposes the values of the heat transfer coefficients can be assumed to be around 60 W/(m^2 K) (Moyers and Baldwin, 1997).

For design purposes the overall heat transfer coefficients of this type of dryer, which depend largely on the film coefficient between the inner jacket wall and the solids, which in turn depend largely on the solid characteristics, can be assumed to be around 50 W/(m^{2}K) (Moyers and Baldwin, 1997). However, higher values are expected for products that have surface moisture (Van't Land, 1991).

Rotary Dryers

Rotary drying is one of the many drying methods existing in unit operations of chemical engineering. The drying takes place in rotary dryers, which consist of a cylindrical shell rotated upon bearings and usually slightly inclined to the horizontal. Wet feed is introduced into the upper end of the dryer and the feed progresses through it by virtue of rotation, head effect, and slope of the shell and dried product withdrawn at the lower end. The direction of gas flow through the cylinder relative to the solids is dictated mainly by the properties of the processed material. Co-current flow is used for heatsensitive materials even for high inlet gas temperature due to the rapid cooling of the gas during initial evaporation of surface moisture, whereas for other materials countercurrent flow is desirable in order to take advantage of the higher thermal efficiency that can be achieved in this way. In the first case, gas flow increases the rate of solids flow, whereas it retards it in the second case (Poersch,1971).

Types of Rotary Dryers

The main types of rotary dryers include the following:

1. **Direct rotary dryer:** It consists of a bare metal cylinder with or without flights, and it is suitable for low and medium temperature operations, which are limited by the strength characteristics of the metal.
2. **Direct rotary kiln:** It consists of a metal cylinder lined in the interior with insulating block or refractory brick, in order to be suitable for operation at high temperatures.
3. **Indirect steam-tube dryer**: It consists of a bare metal cylindrical shell with one or more rows of metal tubes installed longitudinally in its interior. It is suitable for operation up to the available steam temperature or in processes requiring water-cooling of the tubes.
4. **Indirect rotary calciner:** It consists of a bare metal cylinder surrounded by a fired or electrically heated furnace and it is suitable for operation at temperatures up to the maximum that can be tolerated by the metal of the cylinder, usually 800–1025 K for stainless steel and 650–700 K for carbon steel.
5. **Direct Roto-Louvre dryer:** It is, perhaps, the most important of the special types, as the solids progress in a crosscurrent motion to the gas, and it is suitable for low- and medium temperature operations.

Drum Dryers

The drum dryer is commonly used to dry viscous, concentrated solutions, slurries or pastes on rotating steam-heated drums(Moore 1987 and Moore 1995). It can also be used to dry concentrated solutions or slurries that become more viscous or pasty because of flashing or boiling off of moisture or of irreversible thermochemical transformations of their content that occur on their first contact with the hot drum surface (Okos *et al.*, 1993 and Daud and Armstrong, 1987).The viscous slurry or paste is mechanically spread by the spreading action of two counter-rotating drums into a thin sheet that adheres on the hotter drum in single drum dryers or split sheets on both hot cylinders in double drum dryers. The adhering thin sheet of paste is then rapidly dried conductively by the high heat flux of the condensing steam inside the drum. For very wet slurries that produce wet sheets, the drying of the wet thin sheet can be further enhanced by blowing hot dry air on the sheet surface. The thin sheet containing heat-sensitive materials, such as vitamins, can also be dried at a lower temperature in a vacuum.

Types of Drum Dryers

The drum dryer was first patented for use in the manufacture of pre-gelatinized starch in Germany in 1921. Since then a host of other patents have appeared especially in the United States where extensive variations of feeding methods, number and configuration of drums, heating system and product removal were considered. The diameter of the drum varies from 0.45 to 1.5 m and its length varies from 1 to 3m. The thickness of the drum wall is between 2 and 4 cm. The drum dryer is classified according to the number and configuration of the steam-heated drums and the pressure of the atmosphere around the drying sheet.

Atmospheric Double Drum Dryer

This type of dryer has a higher production rate, can handle a wider range of products, and is more efficient (Moore,1987; Moore, 1995; Okos *et al.*, 1993 and Tang *et al.*, 2003). The slurry or paste is fed through a pendulum nozzle or through a header with multiple nozzles on to the nip of two steam-heated drums counterrotating toward each other, forming a boiling pool at the nip. The feed can also be fed at the nips of applicator rollers and the drums. Starch slurries gelatinize in the boiling pool forming pastes that become more viscous. The counter-rotation of the drums spread the slurry or paste into two thin sheets on both drums that consequently dry conductively.

Atmospheric Single Drum Dryer

The slurry or paste is fed through a pendulum nozzle or through a header with multiple nozzles similar to those of the double drum dryers, on to the nip of a steam-heated drum and a much cooler applicator roller counter rotating toward each other, forming a boiling pool at the nip (Moore, 1987; Moore,1995; Okos *et al.*, 1993 and Tang *et al.*, 2003).

Atmospheric Twin Drum Dryer

The slurry is applied by direct dip coating of the twin drums in the feed tray at the bottom of the dryer or by splash or spray feeders from a feed reservoir at the bottom of the dryer onto the surface of the two steam heated drums that are counter rotating away from each other (Moore 1987; Moore 1995; Okos *et al.*, 1993 and Tang *et al.*, 2003). The sheet is formed by adhesion on to the drum surface and is held up against gravity by its surface tension. The sheets consequently dry conductively. This type of dryer is suitable for solutions that produce a dusty product.

Enclosed Drum Dryer

If solvent vapour other than water released during drum drying needs to be recovered or if the dried products generate a lot of dust, atmospheric double or twin drum dryers can be enclosed in vapour or dusttight enclosures (Moore 1987; Moore 1995; Okos *et al.*, 1993 and Tang *et al.*, 2003). The vapour can be recovered by using a suitable condenser and the dust can be removed by using a wet scrubber.

Vacuum Double Drum Dryer

Heat sensitive materials can be dried in a vacuum double drum dryer where the dryer is enclosed in an air tight enclosure under vacuum (Moore1987; Moore, 1995; Okos *et al.*,1993 and Tang *et al.*, 2003). This type of dryer is also fitted with a condenser, a scrubber, and a vacuum pump. The operation of the dryer is similar to its atmospheric version except that there are two product troughs, namely one for breaking the vacuum and the other for product discharge.

Industrial Spray Dryer

Spray drying is a suspended particle processing (SPP) technique that utilizes liquid atomization to create droplets that are dried to individual particles when moved in a hot gaseous drying medium, usually air. It is a one-step continuous unit processing operation. It has become one of the most important methods for drying the fluid foods in the Western world. The development of the process has been intimately associated with the dairy industry and the demand for drying of milk powders.

Spray drying used in dairy industry dates back to around 1800, but it was not until 1850 that it became possible in industrial scale to dry the milk. However, this technology has been developed and expanded to cover a large food group that is now successfully spray-dried. Over 25,000 spray dryers are now estimated to be commercially in use to dry products from agrochemical, biotechnology products, fine and heavy chemicals, dairy products, dyestuffs, mineral concentrates to pharmaceuticals in capacities ranging from a few kg/h to over 50 tons/h evaporation capacity.

Wheel Atomizers

Liquid is fed into the center of a rotating wheel, moves to the edge of the wheel under the centrifugal force, and is disintegrated at the wheel edge into droplets. The spray angle is about 180°C and forms a broad cloud. Because of the horizontal trajectory these atomizers require large diameter chambers. The most common design of the wheel atomizer has radial vanes. The linear peripheral speed ranges from 100 to 200 m/s. For the usual wheel diameter, angular speeds between 10,000 and 30,000 rpm are necessary(Kessler, 1981).

Pressure Nozzle

A pressure nozzle, sometimes called a single-fluid nozzle, creates spray as a consequence of pressure to velocity energy conversion as the liquid passes through the nozzle under pressure within the usual range of 5–7MPa.

Pneumatic Nozzle

Pneumatic nozzles are also known as two-fluid nozzles as they use compressed air or steam to atomize the fluid. In this case the feed is mixed with the air outside the body of the nozzle. Less frequently, the mixing occurs inside the nozzle. The spray angle ranges from 20 to 60° and depends on the nozzle design. Approximately, 0.5m^3 of compressed air is needed to atomize 1 kg of fluid. The capacity of a single nozzle usually does not exceed 1000 kg/h of feed.

Freeze Dryers

Certain biological materials, pharmaceuticals, and foodstuffs, which may not be heated even to moderate temperatures in ordinary drying, may be freeze dried. The substance to be dried is usually frozen. In freeze drying, the water or another solvent is removed as a vapour by sublimation from the frozen material in a vacuum chamber. After the solvent sublimes to a vapour, it is removed from the drying chamber where the drying process occurs.

Tray and Pharmaceutical Freeze Dryer

Pharmaceutical freeze dryers are very often used to produce raw materials like ampicillin, cloxacillin, and cefazolin (usually as sodium salt), or other specialty materials like collagen. In these systems, the product is usually charged on stainless steel or polyethylene film trays and the plant is usually a medium or a large unit with a loading surface varying from 15 to 60 m^2 (Trappler, 1989).

Multi-batch Freeze Dryer

The freeze drying process in a batch plant is normally program controlled to minimize the drying time and to maximize the production of the plant. With a single batch plant the load on the various systems will be very variable throughout the drying cycle.

Tunnel Freeze Dryer

In the tunnel type of freeze dryer, the process takes place in a large vacuum cabinet into which the tray carrying trolleys are loaded at intervals through a large vacuum lock at one end of the tunnel and discharged similarly at the other end (Mellor,1978).

Vacuum Spray Freeze Dryer

The vacuum-spray freeze dryer has been developed for coffee extract, tea infusion, or milk. The product is sprayed from a single jet upward or downward in a cylindrical tower of 3.7-m diameter by 5.5-m high (Mellor, 1978; Thuse *et al.*, 1968).Generally, sprayed freeze-dried coffee has less flavor than normal freeze-dried coffee and the product from this plant is no exception. However, it is hoped retention can be improved in the dried product by concentration before spraying into the tower.

Continuous Freeze Dryer

Recent years have shown a growing interest in freeze drying plants operating with a continuous flow of material through the process. Particularly in industries working with a single standardized product and the preparation of the product is by a continuous process, such plants are really profitable. They give continuity in processing throughout and constant operating conditions that are easily controlled, and they require less manual operation and supervision. A particular incentive comes from the prospect of balancing the load imposed on the water vapour condensation system and the vacuum system. In a batch process, the water vapour evolution rate from the foodstuff is quite high at the start of drying and becomes less as drying proceeds. The condenser system must be designed to handle the maximum water vapour removal requirement.

Continuous freeze dryers are used for freeze drying of product in trays and for freeze drying of agitated bulk materials. When handling the product in trays, the most delicate treatment of the product is achieved. The product is stationary in the tray and therefore is not exposed to abrasion, and it comes in contact only with surfaces that fully meet standards of hygiene.

Microwave and Dielectric Dryers

The terms "dielectric" and "microwave" are somewhat confusing and must be defined as best we can. The term "dielectric heating" can be applied logically to all electromagnetic frequencies up to and including at least the infrared spectrum. The lower frequency systems operate at frequencies through at least two bands: high frequency (HF) (3–30 MHz) and very high frequency (VHF) (30–300 MHz). Thus the names HF, dielectric, radio frequency (RF), and RF heating can often be used interchangeably. However, it is generally accepted that dielectric heating is done at frequencies between 1 and 100 MHz, whereas microwave heating occurs between 300 MHz and 300 GHz. This makes the wavelengths in dielectric heating extend to many meters.

Solar Dryers

Open-air sun drying has been used since time immemorial to dry plants, seeds, fruits, meat, fish, wood, and other agricultural or forest products as a means of preservation. However, for large-scale production the limitations of open-air drying are well known. Among these are high labor costs, large area requirement, lack of ability to control the drying process, possible degradation due to biochemical or microbiological reactions, insect infestation, and so on. In order to benefit from the free and renewable energy source provided by the sun several attempts have been made in recent years to develop solar drying mainly for preserving agricultural and forest products.

Impingement Dryers

Impinging jets of various configurations are commonly used in numerous industrial drying operations involving rapid drying of materials in the form of continuous sheets (*e.g.*, tissue paper, photographic film, coated paper, nonwovens, and textiles) or relatively large, thin sheets (*e.g.*, veneer, lumber, and carpets), or even beds of coarse granules (*e.g.*, cat or dog food). Specific, large-scale applications of impingement drying, such as drying of tissue paper on Yankee dryers, combined impingement and through-drying of newsprint, and drying of wood.

Pneumatic and Flash Dryers

One of the most widely used drying systems is flash drying and is also known as pneumatic drying. Flash dryers are most commonly direct drying units and are also known as convective dryers. Pneumatic or flash dryers may be classified as gas–solid transport systems that are characterized by continuous convective heat and mass transfer processes. Hot air produced by indirect heating or direct firing is the most common drying medium in these systems. In direct flash dryers, the gas stream transports the solid particles through the system, and makes direct contact with the material to be dried. This gas stream (drying medium) also supplies the heat required for drying and carries away the evapourated moisture. Superheated steam can also be used as drying medium yielding sometimes to higher efficiencies and often to higher product quality.

Conveyor Dryers

The conveyor dryer is conceptually very simple. Product is carried through the dryer on conveyors and hot air is forced through the bed of product. It is often described as simply a conveyor in a box with hot air. The reality, however, is that the conveyor dryer is one of the most versatile dryers available. Few drying technologies can match the conveyor dryer's ability to handle such a wide range of products. Products as varied in composition, shape, and size as coated breakfast cereals, nuts, animal feed, charcoal briquettes, and rubber can be dried in a conveyor dryer. Although it is simple in concept, an improper understanding of the heat and mass transfer processes in the conveyor dryer will surely lead to poor product handling, wasted energy, and nonuniform product quality. It also further classified as:

1. Single pass/single stage dryer
2. Single pass/Multistage dryer and
3. Multi-pass dryer.

Infrared Dryers

The most common current applications of IR drying are in dehydration of coated films and webs and to correct moisture profiles in drying of paper and board. Theoretical work and laboratory-scale experimental results on IR drying of paints, coatings, adhesives, ink, paper, board, textiles, etc. can be found in the literature (Navarri *et al.*, 1992; Kuang *et al.*, 1992; Therien *et al.*, 1991; Cote *et al.*, 1990). On the other hand, reports on IR drying applied to other products like foodstuffs, wood or sand are not very common as yet.Most published data on IR drying of foods comes from (the former) USSR, the United States, and the East European countries (Hallstrom *et al.*, 1988). Ginzburg (1969) described IR drying of grains, flour, vegetables, pasta, meat, fish, etc. and showed that IR drying can be successfully applied to foodstuffs. There are many current industrial applications of drying agricultural produce by IR. Sandu (1986) pointed out as advantages of IR drying in foods, the versatility of IR heating, simplicity of the required equipment, easy accommodation of the IR heating with convective, conductive, and microwave heating, fast transient response, and also significant energy savings. Experimental and theoretical works on IR drying, of opaque and semitransparent materials (silica sand, brick, brown coal, graphite suspensions and slurry of surplus activated sludge) have been performed by Hasatani *et al.* (1983, 1988).

Basics of Fluidization

Fluidization occurs when a flow of fluid upwards through a bed of particles reaches sufficient velocity to support the particles without carrying them away in the fluid stream. The bed of particles then has the characteristics of a boiling liquid, hence the term fluidization is given (Gupta and Mujumdar,1983). Various sized foods ranging from fine powders to particulate foods can be fluidized. The fluid responsible for fluidization may be a gas or a liquid. The choice of which confers different properties on the fluidizing system. This, in turn, affects the choice of processes that may be used. At low fluid velocities the particles will simply remain in the loosely packed state in the bed. The minimum fluid velocity required to support the bed is known as the minimum fluidization velocity and at this point the bed can be described as being incipiently fluidized (Gardner, 1971). At high fluid flow velocities, which are greater than the terminal settling velocity, particles will be conveyed out of the column and hydraulic or pneumatic transport will occur depending on whether the fluidizing medium is a liquid or a gas, respectively. This process can be used to transport particulate materials around the processing area, thus saving on complex conveying equipment. In practice, therefore, a fluidized bed operates with the fluid velocity lying between the minimum fluidization velocity and the terminal settling velocity. Further, between these two velocities a wide variety of fluidization types are observed.

Different Types of Fluidized Beds

- ☆ *Slugging bed* is a fluid bed in which air bubbles occupy entire cross sections of the vessel and divide the bed into layers.
- ☆ *Boiling bed* is a fluid bed in which the air or gas bubbles are approximately the same size as the solid particles.
- ☆ *Channeling bed* is a fluid bed in which the air (or gas) forms channels in the bed through which most of the air passes.
- ☆ *Spouting bed* is a fluid bed in which the air forms a single opening through which some particles flow and fall to the outside.
- ☆ **Aggregative** or **bubbling** is a fluid bed at higher airflow rates, agitation becomes more violent and the movement of solids becomes more vigorous. Additionally, the bed does not expand much beyond its volume at minimum fluidization. At sufficient airflow rates, the terminal velocity of the solids is exceeded, the upper surface of the bed disappears, entrainment becomes appreciable, and the solids are carried out of the bed with the air stream (Gardner, 1971).

Fluidized Bed Drying

In fluidized bed dryer, hot air or hot gas is passed via a plenum chamber and through a diffuser plate into the fluidized bed of material. Wet feed can be fed continually into the bed and dry material extracted via a down-comer. This is made possible by the rapid mixing in the bed and the high rates of mass transfer. For fluid-bed dryers, the evaporation rate, determining the drying intensity, ranges from 1×10^{-2} to 4×10^{-2} kg/m^3 s; for spray dryers, this value does not exceed 3×10^{-3} kg/m^{3}s (Gupta and Mujumdar, 1983).Fluidized beds have been applied for mixing, granulation, coating, and drying in the pharmaceutical industry. The fluid bed is an excellent gas-solid contactor that can be applied for the drying of granular and liquid feeds, allowing for high convective heat and mass transfer rates that avoid excessive thermal damage to the products (Hovmand, 1995). The main advantages of this equipment are the narrow particle size distribution obtained for the dried products and the high productivity. However, fluidized beds cannot be applied to particle sizes greater than 1mm due to the tendency to develop undesirable regimes, like slugging and channeling. Fluidized bed drying technique has been used for long, to dry food materials like grains, nuts, potato pieces, diced carrots, chopped fruits, peas, beans, diced vegetables, fruit pieces, onion flakes and fruit juice powders etc. (Bourquin 1978; Jayaraman and Gupta 1995; Somchart 1999; Chalida *et al.*, 2005). Marella and Shah (1997) reviewed application of fluidized bed technology in dairy and food industry. Fluidized beds offer a very high efficiency of heat and mass transfer, as well as a high degree of mixing within the bed. Other possible advantages include the cushioning effect of the air, which will help to prevent the dried material getting damaged, or broken (Devahastin, *et al.*, 2001).

Classification of Fluidized Bed Dryers

Basic

1. The bed is formed of the material being dried
2. The bed is formed of inert material that does not take part in the drying process

Classification: Based on following criteria

1. Stages- Single, double or multistage
2. Nature of particle movement (fluidized, spouted, vortex, Centrifugal)
3. Geometrical shape of the drying chamber (cylindrical, rectan-gular, conical, conical-cylindrical)
4. With and without recirculation of the drying agent (open or closed cycle)
5. Drying with simultaneous grinding or without it.
   (Gupta and Mujumdar, 1983, Strumillo *et al.*, 1983)

Mechanism of Drying on Fluidized Bed

Drying of shreddes in fluidized bed with hot air, has been presented as an alternative to tray drying, in an attempt to obtain powdered products with the same quality, at low cost. In **fluid bed drying**, heat is supplied by the fluidization air, but the air flow need not be the only source. Heat may be effectively introduced by heating surfaces (panels or tubes) immersed in the fluidized layer.

The drying mechanism depends on initial moisture and quantity. Because of the very large surface area of the particles and the intensive fluidization the moisture is removed in a very short time and dried particles remaining on the bed.

Conventional Fluidized Bed Drying

Spouted Bed Dryers

The applicability of the spouted bed technique (Gishler and Mathur 1957; Vukovich *et al.*, 1975; Peterson, 1974; Mathur and Epstein1974; and Mathur and Gishler 1955) to drying of granular products that are too coarse to be readily fluidized (*e.g.*, grains) was recognized in the early 1950s. Interest in this area received appreciable impetus two decades later as the energy-intensive drying processes were reexamined with renewed vigor. Spouted bed dryers (SBDs) display numerous advantages and some limitations over competing conventional dryers. Because of the short dwell time in the spout, SBDs can be used to dry heat-sensitive solids, such as foods, pharmaceuticals, and plastics.

With simple modification the so-called modified spouted beds can be designed to ensure good mixing, controlled residence time, minimum attrition, and other desirable features. Also, the operations of coating, granulation agglomeration, and cooling, among others, can be carried out by the same apparatus by varying the operating parameters. SBDs can be used for solids with constant as well as falling rate drying periods. Using inert solids as the bed material, SBDs have been used successfully to dry pastes, slurries, and heat-sensitive materials.

Centrifugal Fluid Bed Dryers

The low density materials transit from fixed bed condition to slugging bed condition almost instantaneously in conventional fluidized beds. But it is not desired for efficient drying conditions. So in centrifugal fluidized bed dryers the body force becomes an adjustable parameter, which can be controlled with the basket radius and the speed of rotation.

Centrifugal fluidized bed (CFB) drying is a technology in which the wet material undergoes a highly enhanced heat and mass transfer process in a centrifugal field through rotation of the bed. The bed essentially is a cylindrical basket rotating around its axis with a porous cylindrical wall (Brown and Farkas, 1972). The drying material is introduced into the basket and forced to form an annular layer at the circumference due to the rotation. Typical application areas are drying of bio sludge, fish and meat processing, protein and starch, paper and pulp, mineral sludges, sliced foods as potato, carrots, radish *etc* (Rebeca *et al.*, 2004).

Cannon (1975) described a centrifugal fluidized bed drier intended for the dehydration of vegetables and other particulate solids. Hot air was introduced at a velocity in the range of 1500-3000 ft/min with a typical speed of rotation of 100-120 rev/min for a drier with a 3 ft diameter cylinder. Advantages of the equipment claimed were: high evaporation rate; the use of higher velocities during the initial stages of drying when particle surfaces are wet and the evaporation rate is almost directly proportional to air velocity; it could perform the same evaporative duty as a first-stage band drier and would occupy less space and also less cost. Because of the high rates of heat transfer it can be used as a continuous puffing device.

Shi *et al.* (2000) studied the drying process of wet sand, wet glass balls and sliced potato, in a centrifugal fluid bed dryer. The study showed that the drying rate increases with increasing superficial gas velocity and particle diameter and decreasing bed rotation speed and initial bed thickness. But it was found that the speed of rotation should be kept the minimum so as to maintain the efficient fluidizing conditions. Sliced food products were dried very well and efficiently.

Pulsed Fluid Bed Dryer

Low-frequency (2-8 Hz) pulsations in the gas flow to an fluidized bed, brought about by means of a rotary or solenoid valve have been found to improve the fluidization quality and bed stability, and also to eliminate channeling. The amplitude of these pulsations depends on the air velocity, bed height, particle size,and distributor design. Improvement in the heat transfer is more for larger particles and for a horizontal configuration of heaters. Nitz and Taranto (2004) reported a pulsed fluidized bed dryer. The bed is pneumatically vibrated by a rotary valve that sequentially directs the gas stream to different sections of the gas plenum in the dryer.

Mechanically Agitated Fluidized Bed Drying

Ambrosio and Taranto (2002) reported a mechanically stirred fluid bed dryer for chemical drying. They studied drying of 2-hydroxybenzoic acid, with moisture contents 10 per cent and 26 per cent. It was concluded that agitation allowed fluidization of solids with cohesive characteristics, which is not likely to occur in the conventional fluidized bed. Reyes and Castro (2004) studied effect of operating conditions on the residence time distribution (RTD) in a mechanically stirred fluidized bed of $CaCO_3$ particles with lateral air flows. The results showed that the solids with 5 per cent moisture contents present zigzag shaped RTD curves, which are characteristic for recycling beds, and which disappeared when the feed was dry. The values of the solids mean residence time ranged between 5.7 and 7.5 minutes for all operating conditions.

Vibrated Fluidized Bed Drying (VFB)

High frequency (50-200/s) and low-amplitude (1-10 x 10^{-3} m) vibrations imparted to a bed of solids have been found to help in fluidizing sticky and fine particles. In addition, vibration of the bed results in reduced air flow requirements, closer control of the residence time, and simultaneous drying, cooling, and conveying actions. Commercial applications of VFBs include "instantizing" of spray-dried milk powder, increasing surface activity of detergents, and drying of granulated fertilizers, polymer chips, inorganic and organic chemicals, pharmaceuticals, and food products (Pallai *et al.*, 1995). A state of vigorous solid mixing, a phenomenon termed *pseudo-fluidization,* can be attained without aeration with VFB.

Fluidized Bed Dryer with Indirect Heating

Indirect heating of a fluidized bed dryer means an energy supply by tubes placed within the bed. This kind of heating can be used for two reasons: First, if the capacity of an existing fluid bed dryer has to be increased. Second, if limitations concerning the maximal drying temperature are imposed. Both types of application tear profit from the two main advantages of indirect heating: The high heat transfer coefficients allowing compact apparatus design and the heat supply at the point of lowest temperature–in the bed, where evaporation takes place. Three physically different mechanisms of heat transfer contribute to indirect heating:

1. Gas convection from the wall of the immersed tube to the fluidization gas.
2. Radiation.
3. Particle convection, the most important and the most difficult to describe.

Mortier *et al.* (2012) worked on the drying behaviour of a single wet granule before tabletting, using a six-segmented fluidized bed drying system, which is a part of a fully continuous from-powder-to-tablet manufacturing line, and consists of two sub-models. In the first drying phase (submodel 1), the surface water evapourates, while in the second drying phase (submodel 2), the water inside the granule evapourates.

Jet-Zone Drying

A novel method of fluidized bed drying has been developed by Jet-Zone Inc., U.S.A. (Anonymous, 2013b) drying of plastic pellets, food pellets, sand, metal particles etc. The process essentially consists of impinging hot air in the form of multiple jets over a thin bed of granular solids, thereby agitating the bed so as to create an appearance of fluidization. Advantages claimed by the manufacturer are modular construction, ease of operation and automa-tion, and plug flow operation.

Super Heated Steam Fluidized Bed Dryer

Superheated steam as the fluidizing medium offers a number of advantages, *e.g.*, no fire or explosion hazards, no oxidative damage, better operation performance (higher drying rate) and product quality, environmental friendliness, high energy efficiency, suitability for drying product containing toxic or expensive organic liquids, ability to permit pasteurization, sterilization, and deodorization of food product.Wathanyoo and co-worker (2005) studied the effect of superheated steam in fluidized bed dryer over hot air drying. The experimental results showed that, for the same duration of drying, drying rates of paddy dried by superheated steam were lower than those dried by hot air due to an initial steam condensation during the first few minutes of superheated steam drying.

Heat Pump Fluidized Bed Dryer

An ordinary fluidized bed drying system consist of blower, heater, dehumidifier (optional), fluidized bed chamber, and cyclone (filter), whereas an ordinary heat pump drying system consists of evaporator, compressor, condenser, and an expansion valve. By combining fluidized bed and heat pump drying system, where the evaporator acts as a dehumidifier and the condenser as a heater, a heat pump fluidized bed dryer is formed.The 95–90 per cent of this energy may be recovered by means of a heat pump assisted drying, leading to substantial energy savings. Vázquez *et al.* (1997) suggested that the use of driers with heat pump as an alternative method to dry fruits and vegetables with lower values in consumption of energy, relative humidity and temperatures.

Drying of Different Material in Fluidized Bed Dryer

Fluidized Bed Drying for Grains/Granular/Agricultural Products

In 2009, Solis and *et al.*, used fluidized bed technology to attrition reduction and quality improvement by drying puffed wheat coated with sweet covering. Coated puffed wheat is ready-to-eat (RTE) breakfast cereal of popular consumption worldwide.Placebo granules consisting of lactose monohydrate, corn starch, and poly-vinyl-pyrolidone were prepared using de-ionized water in a high-shear mixer and dried in a conical fluidized bed dryer at various superficial gas velocities. Acoustic, vibration, and pressure data obtained over the course of drying was analyzed using various statistical, frequency, fractal, and chaos techniques. Traditional monitoring methods were also used for reference. Analysis of the vibration data showed that the acceleration levels decreased during drying and reached a plateau once the granules had reached final moisture content of 1–2 per cent; this plateau did not differ significantly between superficial gas velocities, indicating a potential criterion to support drying endpoint identification. Acoustic emissions could not reliably identify the drying endpoint. However, high kurtosis values of acoustic emissions measured in the filtered air exhaust corresponded to high entrainment rates. This could be used for process control to adjust the fluidization gas velocity to allow drying to continue rapidly while minimizing entrainment and possible product losses (Briens and Bojarra 2010).

Optimization of drying and inactivation of heat-labile inhibitors conditions of soybean by using a fluidized bed dryer was done in order to shorten treatment time and to reduce losses in end-product quality such as soy flour color and soy protein solubility. The independent variables were initial moisture of soybeans, heating time and temperature of air entering the fluidization chamber. The response variables studied were final moisture of soybeans, inactivation of urease, soy flour color and soy protein solubility. Response surface methodology was able to model the response of the different studied variables. For each response group, relevant terms were included into an equation; the behavior of response was predicted within the experimental area and was presented as a response surface. The results suggested that a combination of soybean initial moisture of 0.14 g/g (w.b.), treatment time of 3.4 min and hot-air temperature of 136.5 °C could be a good processing combination of parameters for heating soybean using hot-air in order to reduce treatment time and quality losses in soybean flour. Thus, fluidized bed drying technology may be used as an alternative industrial method to eliminate the anti-nutritional factors (Martinez *et al.*, 2011).Somchart (1999) studied continuous fluidized bed drying of the paddy. Higher initial moisture in paddy was desirable for efficiency. Izadifar and Mowla (2003) studied a cross-flow fluid bed dryer for drying of paddy and developed mathematical model for the same.Chalida and *et al.* (2005) studied the fluidized bed drying of coconut. An industrial-scale batch (50 kg) fluidized bed dryer was used to dry finely chopped coconut pieces. Drying was fastest when using an inlet air velocity of 5.94 m/s. This indicated that most drying was in the constant-rate drying period where external resistances to heat and mass transfer dominate the drying process. It was found that the highest inlet air velocity (5.94 m/s) led to a high drying rate but a lot of oil on the surface of the dried coconut was observed.

Somkiat *et al.* (2005) compared the performance of pulsed and conventional fluidized bed dryers. Their experimental results have shown that the variation of moisture content exists in both dryer types. Heat utilization was more effective when such dryers were used to dry paddy at moisture contents above 24 per cent dry basis and up to 50 per cent of the thermal energy was saved by recycling 70-80 per cent of the air. Paddy qualities *i.e.* head-rice yield and colour of the dried white rice were similar with both dryers and almost the same as the original undried values, or slightly higher in the case of head-rice yield, depending upon the drying conditions. Below 28 per cent dry basis, it is recommended that inlet-air temperature should be lower than 145°C in order to maintain white colour. The cooked rice obtained from paddy dried at a temperature of 145°C was harder than naturally dried control samples.

The fluidized bed drying of castor oil seeds (*R. communis*) was carried out in falling-rate period, this fact indicates that moisture transfer during drying is controlled mainly by internal diffusion. The effect of drying air temperature in the range of 80°-110° C and 7 m/s of air velocity for seeds was characterized. The DR and effective diffusivity coefficient increased with air temperature and consequently the drying time decreased. The effective moisture diffusivity values were ranged from 8.21×10^{-10} m^2/s to 2.61×10^{-9} m^2/s. The temperature dependence of effective diffusivity coefficients was described adequately by Arrhenius-type equation. The activation energy for moisture diffusion was found as 41.41 kJ/mol (Perea-Flores *et al.*,2012).

Hot air thermal treatment's effect on volume and density of whole soybean product during drying in fluidized bed drying checked by Irigoyen and Giner in 2011. To produce a whole soybean snack, four type of samples were processed by first increasing the water content and then reducing it by hot-air fluidization: moistened dried at 60° C (MD), soaked–dried at 60° C (SD), soaked and dried toasted at 140° C (SDT) and soaked–cooked and dried–toasted at 140° C (SCDT). A semi-theoretical model was proposed to describe volume and density as a function of moisture during fluidization. An equilibrium shrinkage coefficient was determined. Volume expansion achieved by increasing the moisture content was not totally lost during fluidization, allowing for lighter products, whose density decreased with the reduction in moisture. As the overall treatment was more severe (SCDT > SDT > SD > MD), shrinkage coefficients increased, up to 0.75. The SCDT sample became crispy and glassy after cooling.

Effect of fluidized bed drying temperature and tempering time on quality of waxy rice measured by Patcharat *et al.*,(2009). The effects of fluidized bed drying temperature (90°, 110°, 130 °C) and tempering time (30–120 min) on the quality of waxy rice, *i.e.*, head rice yield, thermal properties, color, and microstructure, were investigated. The results showed that head rice yield of waxy rice after drying was significantly lower than that of the reference sample even when tempering was performed. Higher drying temperatures led to higher head rice yield while the tempering time did not have any effect on the head rice yield except when the drying temperature of 90°C was used. Drying at higher temperatures also affected the starch granule morphology and the pasting properties. Waxy rice changed its appearance from opaque white to translucent when being dried at 130°C.

The investigation done by Somkiat *et al.* (2006) on drying characteristics and inactivation of urease in soybeen dried by superheated steam and hot air fluidized bed dryer. The results showed that the value of effective diffusion coefficient was increased with increased in drying temperature and increased moisture content. Effective diffusion coefficient also depends on drying medium, with higher moisture diffusion for soybean dried by hot air. The enzyme activation was faster in superheated steam than in hot air. Temperature of hot air for inactivation of urease enzyme was 135°C to 150 °C and for super heated steam temperature was 135°C.

In 2005, Nattapol *et al.*,analyzed the influence of temperature in the fluidized bed dryer, moisture content and tempering period on head rice yield, and operating time. And they found that the moisture content after first-stage drying and tempering have a dominant effect on head-rice yield and operating time in reducing high-moisture contents to a safe level. And they recommended the allowable temperature should be not higher than 250°C for the first stage and the moisture content after first stage drying should be not lower than 22.5 per cent dry basis, with subsequent tempering for at least 25 min.Somkiat and *et al.*, in 2004 studied the moisture removal rate and quality of maize with temperature range of 90° to 170°C and air velocity 0.075 to 0.375 m/s. The initial moisture of maize was between 37 per cent and 43 per cent dry basis. They found that the two stage drying improve the efficiency of operation as well as increase the moisture removal rate. Two stage drying darken the color of maize but also reduce the cracks. They concluded that less then 150°C temperature increased the efficiency and ambient air velocity of 0.075 m/s to 0.375 m/s for cooling gave good color and less stress cracks.

Somchart *et al.* (2001) studied the fluidized bed drying characteristics of soyabeans at temperature of 110 to 140°C and moisture contents, 31 to 49 per cent. They had used air speeds of 2.4-4.1 m/s and bed depths of 10 to 15 cm. The mininumfluidized bed velocity was 1.9 m/s. From a quality point of view, fluidized bed drying was found to reduce the level of urease acitvity which is an indirect measure of trypsin inhibitor, with 120°C being the minimum required to reduce the urease activity to an acceptable level. Increasing air temperature caused increased cracking and breakage, with temperature below 140°C giving an acceptable level. The protein level was not significantly reduced in this temperature range. The results also showed that from 33.3 per cent dry basis, soybeen should not be dried below 23.5 per cent dry basis in the fluidized bed dryer, to avoid excessive grain cracking. They concluded the optimum parameters based on quality criteria, drying capacity, energy consumption and drying cost. They were 140 °C drying temperature, 18 cm bed depth and 2.9 m/s air velocity.

Fluidized Bed Drying for Food Products (Vegetables and Fruits)

Niamnuy and Devahstin(2005) used fluidized bed dryer for checking drying kinetics and quality of coconut drying. They conclude from experiments that the inlet air velocity has more effect on the quantity of oil on the surface of dried coconut than the inlet air temperature. However the inlet air temperature was found to be the most significant parameter affecting the color of the dried product.

Optimization of drying parameters of bird's eye chili in a fluidized bed dryer performed and concluded the optimum condition for humidity contents at constant 2 cm bed height while varying the air velocity, temperature and drying time are at 1.09 m/s of air velocity, 70° C of temperature and 120 min of drying time by Tasirin *et al.*,2007.Optimum conditions of 79.4°C drying air temperature, 35.8 min tempering time, pretreatment of the once pricked peas with chemical blanching in a solution of

2.5 per cent NaCl and 0.1 per cent $NaHCO_3$, and mass per unit area of 6.8 g/cm^2 were recommended for the fluidized bed drying of green peas. At those conditions the rehydration ratio was 3.49 (Burande *et al.*, 2008).

Drying kinetics of olive pomace was investigated in a fluidized bed dryer. The drying experiments were performed at different temperatures of the drying air (50, 60, 70 and 80 °C) and bed heights of the sample (41, 52, and 63 mm), using a constant air velocity of 1.0 m/s. A constant rate period was not observed in the drying of olive pomace; all the drying process occurred in falling rate period. Ten thin layer drying models were evaluated and fitting to the experimental moisture data. For all drying experiments, the effective diffusivity values were found to be varying between 0.68 and 2.15×10^{-7} m^2/s depending on air temperature and bed height of the sample. The activation energies were 34.05, 36.84 and 38.10 kJ/mol for 41, 52 and 63 mm bed height of the sample, respectively (Meziane 2011).

Rossi and Clementi (1985) described drying of washed cells of *Lactobacillus bulgaricus*and *Streptococcus thermophilus* in a fluidized bed at 30°C for 15, 22, 28 or 40 min. The residual moisture content was 37 per cent after 15 min drying but live cells had decreased by a power of 10. The survival during storage after drying for 28 or 40 min was satisfactory, particularly with storage at 12°C.Wijitha and *et al.* (2003) studied the influence of shapes of selected vegetable materials on drying kinetics during fluidized bed drying. Three different particular geometrical shapes of parallelepiped, cylinder and sphere were taken from cut green beans (length:diameter 1:1, 2:1 and 3:1) and potatoes (aspect ratio 1:1, 2:1 and 3:1) and peas, respectively. Their drying behaviour in a fluidized bed was studied at three different drying temperatures of 30, 40 and 50 ºC (RH 15 per cent). Drying curves were constructed using non-dimensional moisture ratio (MR) and time and their behaviour was modeled. The diffusion coefficient was least affected by the size when the moisture movement was considered three dimensional, whereas the drying temperature had a significant effect on diffusivity as expected. The drying constant and diffusivity coefficients were on the descending order for potato, beans and peas. The Arrhenius activation energy for the peas was also highest, indicating a strong barrier to moisture movement in peas as compared to beans and skinless cut potato pieces.Pan *et al.* (1999a) studied intermittent drying of carrot cubes in a vibro-fluidized bed. It was concluded that with intermittent regulation of the airflow significantly improves β-carotene content. Pan *et al.* (1999b) conducted drying experiments for squash slices in a vibrated fluidized bed. From their experimental observations, a reduction in drying time of up to 40 per cent was observed based on a drying time of 96 min for the squash slice to reach 14.75 per cent moisture content.Gheorghita (2004) studied the effect of oscillation in a fluidized bed. It was found that the fluidization begins at the bottom of the bed and afterwards in the whole bed, contrary to the non-vibrating fluidization, when the fluidization begins from the top. Raut *et al.* (2004) reported that designing of a VFB dryer, is governed by the selection and percentage opening of the sieve plate or distributor plate. By selecting the optimum size we can enhance the rate of drying.

Fluidized Bed Drying for Milk

Upadhyay (1998) designed a spouted fluid bed dryer for drying of milk on inert bed. A mathematical model for designing of inert fluid bed dryer was also reported. Various inert materials *viz.* polypropylene bids, ABS and glass balls were tried and optimum dimensions were determined. The optimum velocity of air at the base of the drying chamber was found to be in the range of 2-3.5 m/s. The temperature above 120 °C was found to be unsatisfactory as it gave burnt particles. The optimum air temperature for inlet and outlet were found to be 100 and 70 ° C. The particle size obtained in this kind of dryer was found to be smaller than the conventional process.Very fine powders (<100 µm) are found to agglomerate in Fluidized beds, due to inter-particle forces of attraction. Typical examples for application of these kinds of dryers are: starch, wheat flour, zinc oxide, milk powder, and carbon. Attempts to fluidize such materials result in "rat-holing" and channeling through the bed. The agitation speed obtained for sodium bi-chromate crystals was 4 r/min, and for milk powders, 36 r/min (Hayashi and Wada, 1980).Mukundrao (2006) developed and evaluated performance of inert spouted bed dryer. Energy analysis of the drier was carried out and observed following parameters like thermal efficiency (48.56 – 81.09 per cent), overall volumetric heat transfer coefficient (3.37- 8.34 kW/m^{3}K), specific energy consumption (5.839-12.217 MJ/kg), surface heat transfer coefficient (0.0027 – 0.0067 kW/m^{2}K).

Fluidized Bed Drying for Milk Products

Swanson (1971) reported that as compared to the conventional tray method, the fluidized bed method shortens the time required for drying grated Italian cheese. Makarov *et al.* (1972) discussed the use of ultrasonics acoustic vibrations to accelerate the rate of lactose drying in a fluidized bed dryer. Ultrasonic waves with 5000-15000 hz, produced by sirens operating at less than or equal to 7 kg/cm^2 air pressure and less than or equal to 160 dB used in an experimental fluidized bed drier, reduced the time required for drying lactose of 18 per cent moisture content from nearly 80 to 30 min with air at 50 °C temperature and 1.4 m/s flow.Bentzien (1981) suggested improvements in the rotating disk fluidized-bed drier to facilitate salvage of the product particles as cheese. Tesch and Lavallee (1982) described particle injection system for a fluid bed drier and observed that the particles are immediately surrounded by and suspended in the drying air and do not tend to stick together. This is particularly important for such products as cheese and yeast, which readily agglomerate into lumps.

Advancement in Fluidized Bed Drying System

In this direction Correa *et al.* (1999) studied improving operability of spouted beds using a simple optimizing control structure. The air flow was regulated by a frequency inverter, at the speed of blower rotation. A PI controller was used and the set-point for the air flow rate was calculated on-line by a simple and well-known minimization method called Golden Section Search (GSS). It was concluded that the optimizing control strategy, using the simple Golden Section Search method of minimization, is fast and works fine in the control of the fluid dynamics of a spouted bed. Costa *et al.* (2001) investigated modeling and simulating the drying of suspensions, such as organic and biological pastes, in conical spout-fluid beds of inert particles. A computer program had been developed combining the air flow and particle circulation models with the mass and energy balances and the drying kinetic equations in order to describe this drying process. The effect of cohesive forces was also incorporated to the fluid flow model.

Wachiraphansakul and Devahastin (2005) investigated experimentally the effects of various drying parameters. Those were inlet air velocity, inlet air temperature, initial bed height and heating duration. The study was on both the drying kinetics and various quality attributes of dried okara *viz.* percentage changes of the total protein content, color, urease index, as well as the specific energy consumption during drying in a jet spouted bed dryer. Based on the quality and energy consumption the selected drying conditions were those using all drying parameters at their highest levels (air velocity -1.5 m/s, bed height -18 cm, and inlet air temperature - 130°C); these conditions yielded the specific energy consumption of 3.69 MJ/kg evaporated water.

Conclusion

Fluidized bed drying has many practical applications in drying of granular solids in the food, ceramic, pharmaceutical and agriculture industries. They listed some advantages like,

1. High drying rate due to good gas particle contact, leading to optimal heat and mass transfer rates;
2. Smaller flow area;
3. High thermal efficiency;
4. Lower capital and maintenance costs and
5. Ease of control.

References

Ambrosio, M.C.B. and Taranto, O.P. (2002).The drying of solids in a modified fluidized bed.*Brazilian Journal of Chemical Engineering*, 19, 3.

Anonymous (2013b).Cited from http: //www.jetzone.com

Bentzien, L. (1981). Product reclamation in a fluid bed dryer. Universal Foods Corp. United States Patent., 4272 895 (Cited from CAB abstracts, Acc. no. 820473725).

Bourquin, O. (1978). Use of fluidization in the food industry.*IndustriaLechera*, 57(660), 12-16.

Briens, L. and Bojarra, M. (2010).Monitoring fluidized bed drying of Pharmaceutical granules, *AAPS Pharmscitech*, 11, 4.

Brown, G. E. and Farkas, D. F. (1972). Centrifugal fluidized bed. *Food Technology*, 26(12), 23–30.

Burande, R., Kumbhar, B., Ghosh, P. and Jayas, D. (2008).Optimization of fluidized bed drying process of greens using response surface methodology.*Dairy Technology*, 26, 7.

Cannon, M. W. (1975). Centrifugal fluidized bed dryer. *Food Trade Review.*, 45(12), 9-12.

Chalida, Niamnuy, Sakamon andDevahastin (2005). Drying kinetics and quality of coconut dried in a fluidized bed dryer. *Journal of Food Engineering*, 66, 267–271.

Chua, K. J. and Chou, S. K. (2003). Intermittent drying of bioproducts: An overview. Trends in food Science and Technology, 14, 519-528.

Correa, N. A., Freire, J. T. and Correa, R. G. (1999).Improving operability of spouted beds using a simple optimizing control structure.*Brazilian Journal of Chemical Engineering*, 16(4).

Costa Jr., E.F., Cardoso, M. and Passos, M. L. (2001). Simulation of drying suspension in spout fluid beds of inert particles. *Drying Technology*, 19(8), 1975 - 2001.

Cote, B., Broadbent, A.D. and Therien, N. (1990).Modelisationet simulation du sechage en continudescouches minces par rayonnementinfrarouge. *Canadian Journal of Chemical Engineering*, 68, 786–94.

Daud, W.R.W. and Armstrong, W.D. (1987).Pilot plant study of the drum dryer.In Mujumdar, A. S. (Eds), *Industrial Drying*. (pp. 101) Hemisphere Publishing Corporation, New York.

Devahastin, Sakamon and Mujumdar, A. S. (2001).Some hydrodynamic and mixing characteristics of a pulsed spouted bed dryer.*PowderTechnology*, 117, 189–97.

Gardner, W. A. (1971). Industrial Drying.*Chemical and Process Engineering Series, I.L. Hepne*r (series ed.), Leonard Hill, London.

Gheorghita, J. (2004). The Romanian school contributions on the oscillations Influence in the intensification of process transfer in gas fluidized bed. *Drying 2004* – Proceedings of the 14th International Drying Symposium (IDS 2004) São Paulo, Brazil, 22-25 August 2004, A, 272-279.

Ginzburg, A.S. (1969). Application of Infrared Radiation in Food Processing.*Chemical and Process Engineering Series*, Leonard Hill, London.

Gishler, P.E. and Mathur, K.B. (1957). U.S. Patent 2, 736, 280 to National Research Council of Canada, filed 1954.

Gupta, R. and Mujumdar, A.S. (1983).Recent developments in fluidized bed drying.In Mujumdar, A. S. (Eds).*Advances in Drying*.(pp. 155–92). Vol. 2, Hemisphere Publishing, New York.

Hallstrom, B., Skjoldebrand, C. and Tragardh, C. (1988).Heat Transfer and Food Products.*Elsevier Applied Science*, London.

Hasatani, M., Harai, N., Itaya, Y. and Onoda, N. (1983).Drying of Ceramics, *Drying Technology* 1(2), 193–214.

Hasatani, M., Itaya, Y. and Miura, K. (1988). Sorption drying of soybean seeds with silica gel, *Drying Technology* 6(1), 43–68.

Hayashi, H. and Wada, T. (1980).Drying.In Mujumdar, A.S. (Eds), *Drying*, Hemisphere, Washington D. C.

Hovmand, S. (1995).Fluidized bed drying.In Mujumdar, A.S. (Eds), *Handbook of Industrial Drying*. Marcel Dekker, New York, , 195–248.

Irigoyen, R. M. T. and Giner, S. A. (2011).Volume and density of whole soybean products during hot air thermal treatment in fluidised bed.*Journal of food engineering*, 102, 224-32.

Izadifar, M. and Mowla, D. (2003).Simulation of a cross-flow continuous bed dryer for paddy rice.*Journal of food engineering*, 58, 325-29.

Jayaraman, K. S. and Gupta Das, D. K. (1995).In Mujumdar A. S. (Ed.), Handbook of industrial drying.(pp. 643–90). Marcel Dekker, New York.

Kessler, H.G. (1981).Food Engineering and Dairy Technology, Verlag A. Kessler, Freising, Germany.

Kuang, H., Chen R., Thibault, J. and Grandjean, B.P.A. (1992).Theoretical and experimental investigation of paper drying using gas-fired IR dryer.In Mujumdar, A. S. (Ed.), *Drying*, (pp 941). Elsevier, Amsterdam.

Makarov, A. P., Polyanskii, K. K. and Karnaukh, V. I. (1972). Use of acoustic vibrations to accelerate the rate of lactose drying. *Izvestiya Vysshikh Uchebnykh Zavedenii Pishchevaya Tekhnologiya*, 4, 181-83

Marella, C. and Shah, U.S. (1997).Application of fluidized bed technology in dairy and food industry - a review.*Journal of Dairying, Foods and Home Sciences*, 16(1), 1-7.

Mathur, K.B. and Epstein, N. (1974).*Spouted Beds*, Academic Press, New York.

Mathur, K.B. and Gishler, P.E. (1955).*Journal of Applied Chemistry*, 5, 624.

Mellor, J.D. (1978).*Fundamentals of Freeze Drying*. Academic Press, London.

Meziane, S. (2011).Drying kinetics of olive pomace in a fluidized bed dryer.*Energy conversion and Management*, 52, 3.

Moore, J.G. (1987). Drum dryers.InMujumdar, A. S. (Eds), *Handbook of Industrial Drying, 1st ed.* Marcel Dekker, New York, 227

Moore, J.G. (1995). Drum dryers.In Mujumdar, A. S. (Eds), *Handbook of Industrial Drying, 2nd ed.*, Marcel Dekker, New York, 249

Mortier, S. T. F. C., Beer, T. D., Gernaey, K. V., Vercruysse, J., Fonteyne, M., Remon, J. P., Vervaet, C. and Nopens, I. (2012).Mechanistic modelling of the drying behaviour of single pharmaceutical granules.*European journal of pharmaceutics and biopharmaceutics*, 80, 682-89.

Moyers C. G.and Baldwin G. W. (1997).Psychrometry, In Perry, R. H., and DW (Eds), *Evaporative cooling, and solids drying*.

Mukundrao, N. (2006). Development and performance evaluation of inert spouted bed dryer for drying of milk.A PhD thesis submitted to Anand Agricultural University, Anand.

Nattapol, P., Apichit, T., Somkiat, P. and Somchart, S. (2005). Investigation on head rice yield and operating time in the fluidised bed drying process: experiment and simulation. *Journal of stored products research*, 41, 387-400.

Navarri, P., Gevaudan, A. and Andrieu, J. (1992). Preliminary study of drying of coated film heated by infrared radiation. In Mujumdar, A. S. (Eds), *Drying*, (pp 722), Elsevier, Amsterdam.

Niamnuy, C. and Devahastin, S. (2005). Drying kinetics and quality of coconut dried in a fluidized bed dryer. *Journal of food Engineering*, 66, 267-271.

Nitz, M. and Taranto, O. P. (2004). Drying of beans in a pulsed-fluid bed dryer – fluid-dynamics and the influence of temperature, airflow rate and frequency of pulsation on the drying rate, *Drying 2004* – Proceedings of the 14th International Drying Symposium (IDS 2004) São Paulo, Brazil, 22-25 August 2004, vol. B, 836-843.

Oakley, D. (1997). Contact dryers. In Baker, C. G. J. (Eds), *Industrial Drying of Foods*.(pp 115-33).Blackie Academic and professional, London.

Okos, M.R., Narsimhan, G., Singh, R.K. and Weitnauer, A.C. (1993).Food dehydration. In Heldman, D.R. and Lund, D.B. (Eds), *Handbook of Food Engineering*, Marcel Dekker, New York, 437.

Pallai, E., Szentmarjay, T. and Mujumdar, A. S. (1995).In Mujumdar, A. S. (Eds), *Handbook of industrial drying*, Marcel Dekker, New York, 453-88.

Pan, Y. K., Zhao, L. J. and Hu, W. B. (1999b).The effect of tempering intermittent drying on quality and energy of plant material.*Drying Technology*, 17, 1795-1812.

Pan, Y. K., Zhao, L. J., Dong, Z. X., Mujumdar, A. S., and Kudra, T. (1999a). Intermittent drying of carrot in a vibrated fluid bed: effect on product quality. *Drying Technology*, 17, 2323-40.

Patcharat, J., Somkiat, P., Sakamon, D. and Somchart, S. (2009). Effect of fluidised bed drying temperature and tempering time on quality of waxy rice.*Journal of food engineering*, 95, 517-24.

Perea-Flores, M. J., Garibay-Febles, V., Chanona-Perez, J. J., Calderon-Dominguez, G., Mendez-Mendez, J. V., Palcios-Gonzalez, E. and Gutierrez-Lopez, G. F.(2012). Mathematical modelling of castor oil seeds (*Ricinuscommunis*) drying kinetics in fluidized bed at high temperature. *Industrial Crops and Products*, 38, 64-71.

Peterson, W.S. (1974). *Canadian Journal of Chemical Engineering*, 9, 11.

Poersch, W. (1971).Verfahrenstechnik, 5(5), 186.

Raut, N. B., Pangarkar, B. L. and Parjane, S. B. (2004).Critical approach for design of vibrofluidized bed dryer.*Chemical Engineering World*, 62-3.

Rebeca, Y. C. P., Limaverde, J. R. and Finzer, J. R. D. (2004).Drying of banana paste in rotary dryer with inert bed.*Drying 2004* – Proceedings of the 14th International Drying Symposium (IDS 2004) São Paulo, Brazil, 22-25 August 2004, vol. B, pp. 876-83.

Rossi, J. and Clementi, F. (1985).Fluid bed drying of lactic starters.*Latte*, 10(6), 553-54.

Shi, M. H., Wang.H. and Hao Y.L. (2000).Experimental investigation of the heat and mass transfer in a centrifugal fluidized bed dryer.*Chemical Engineering Journal*, 78, 107–13.

Somchart, S. (1999).Fluidised Bed Paddy Drying.*Science Asia*, 25, 51-56.

Somchart, S., Thanit, S., Somboon, W. and Wivat, W. (2001).Fluidised bed drying of soybeens.*Journal of stored products research*, 37, 133-51.

Somkiat, P., Paveena, P. and Somchart, S.(2006). Heating process of soybean using hot air and superheated steam fluidised bed dryers. LWT, 39, 770-778.

Somkiat, P., Somchar, S., Somboon, W. and Kongsak, C. (2004).*Journal of stored products research*, 40, 379-93.

Somkiat, P., Warunee, T., Korakot, P. andSomchart, S. (2005). Comparison of performances of pulsed and conventional fluidised bed dryers.*Journal of stored products research*, 41, 479-97.

Strumillo, C., Markowski, A. and Kaminski, W. (1983).Modern developments in drying of pastelike materials. InMujumdar A. S. (Eds), *Advances in Drying*. 2nd ed.(pp 193-231). Hemisphere publishing, Washington.

Swanson, A. M. (1971). New technique for Italian cheese drying.*Modern Dairy*, 50(9), 20-22.

Tang, J., Feng, H. and Shen, G.Q. (2003). Drum drying. In *Encyclopedia of Agricultural, Food, and Biological Engineering*, (p. 211). Marcel Dekker, New York.

Tasirin, S. M., Kamarudin, S. K., Ghani, J. A. and Lee, K. F. (2007).Optimization of drying parameters of bird's eye chilli in a fluidized bed dryer.*Journal of food engineering*, 80, 695.

Tesch, W. and Lavallee, M. H. (1982).Particle injection system for a fluid bed drier. United-States-Patent., US 4 309 829

Thuse, E., Ginnette, L.F. and Derby, R. (1968).U.S. Patent 3, 362, 835.

Trappler, E.H. (1989).*Pharmaceutical Technology*, pp. 56–60.

Upadhyay J. B. (1998). Design and development of a spouted fluidized bed dryer for drying of milk. A PhD thesis submitted to Gujrat Agricultural University, Anand Campus, Anand.

Van't Land CM. (1991).*Industrial Drying Equipment: Selection and Application*. Marcel Dekker, New York.

Vázquez, G., Chenlo, F., Moreira, R. and Cruz, E. (1997).Grape drying in a pilot plant with a heat pump.*Drying Technology*, 15 (3-4), 899–920.

Vukovich, D.V., Zdanski, F.K. and Littman, H. (1975). Present status of the theory and application of spouted bed technique, International Congress on Chemical Engineering (CHISA), Prague, Czechoslovakia.

Wachiraphansakul, S. and Devahastin, S. (2005). Drying kinetics and quality of soy residue (Okara) dried in a jet spouted-bed dryer. *Drying Technology*, 23, 1229–42.

Wathanyoo, R., Adisak, N., Warunee, T. and Somchart, S. (2005).Comparative study of fluidized bed paddy drying using hot air and superheated steam.*Journal of Food Engineering*.71, 28-36.

Wijitha, S., Bhes, R. B., Gordon, Y. and Bandu, W. (2003). Influence of shapes of selected vegetable materials on drying kinetics during fluidized bed drying. *Journal of Food Engineering*, 58, 277-83.

2016, Dairy and Food Product Technology
Editors: Birendra Kumar Mishra and Subrota Hati
Published by: BIOTECH BOOKS, NEW DELHI

Pages 131–143

Chapter 9

The Next Frontier: Genetically Modified Animal

Tarun Pal Singh, Parminder Singh and Pavan Kumar

*Department of Livestock Products Technology,
College of Veterinary Science,
Guru Angad Dev Veterinary and Animal Sciences University,
Ludhiana – 141 004, Punjab*

Historical Background

Developments in biotechnology over the past 25 years have allowed scientists to engineer genetically modified (GM) animals. An increasing set of evidence has been reported on how consumers could potentially react to the introduction of genetically modifiedanimal. Genetically modified animals are derived from gene modification technology which involves deliberate modification of the genetic material of animals. A wide range of genetically modified (GM) animal is being developed with a great variety of purposessuch as growth-rate, quality of meat, milk composition, disease resistance and survival.The great majority of GM animals and fish are currently only at the research stage. The first GM livestockincluding sheep, rabbit and pig were produced in 1985 via microinjection of DNA with human GH-mMT promoter gene construct into pro-nuclei of fertilized eggs (Hammer *et al.*, 1985) and since then the number of GM animal and fish species has increased several folds (Niemann and Kues, 2007; Kues and Niemann, 2011; FAO, 2012).

Introduction

Food products derived from genetically modified (GM) animals have not yet entered the marketand there are unlikely to be for some years to come.There is rapid development of modern biotechnology challenges governments to develop the GM animals that will maximize the economics and social benefits to consumers.The first methods for genetic modification of animals were developed in the 1980s and many of the ideas for applications of the technology to livestock and fish were suggested at that time. Genetic modification of animals from the main livestock and fish species is now possible. Several different technologies have been developed, with different uses depending on the specific modifications required, but there are still significant advances required to enable efficient and sophisticated modifications to be generated. Transgenic technology could offer alternative and allowed a generation of GM animals and fish to be used for consumption. The success of trans-genesis for animal production has been limited for 15 years by the difficulty and the cost to generate transgenic animals and fish. The term trans-genesis refers to creation of an animal in which there has been a deliberate modification of the genome *i.e.* the material responsible for the inherited characteristics of the animal. The production of transgenic animals takes place within an existing set of regulations with a history of national and international political conflict.

GM animals and fish are genetically modified in vitro for many purposes by introduction of several traits.However, it can be roughly divided into two categories: food products and non-food products. The first category includes the animals that have been modified to enhance economically important traits (*e.g.* growth, feed efficiency, health and longevity). The second includes animals that have been transformed to produce specific substances in their milk, eggs, bloodetc. The first GM mammals resulted from the early finding that transgenic mice could be generated following injection of viral (SV40) DNA into pre-implantation mouse blastocysts (Jaenisch *et al.*, 1975). Today, groups of GM animals that contain the same recombinant genetic material are often identified via their commercial names (*e.g.*Enviropig™, Onco-Mouse[1], NeonMice™, GloFish[1], and ATryn[1]) (Flavio *et al.*, 2013). These names are, in many cases, well known even among non-specialists. Many of them are trademarked and for this reason they can be used only with the explicit permission of the owner. This protects the name from fraudulent and illegal usage, but at the same time it limits its usefulness. The Aqua Advantage® salmon is currently being considered by the Food and Drug Administration (FDA) and may become the first GM animal to enter the food chain. The increase in global food demand arising from the growth of the human population, the need of cheap food in large sectors of the population and the need for more environment-friendly animals might motivate food production from GM animals in the future. GM animals play a significant role in enhancing food production to satisfy the needs of an increasing global food demand. Moreover, improving living standards globally will increase the demand for meat and other products from livestock origin, thereby generating a powerful drive for livestock production. Genetic engineering of mammals offers opportunities to produce more meat products with superior quality and improved nutritional properties. It has been proposed that GM techniques focused on the improvement of economically important traits could fill gaps missed by conventional genetic and genomic selection and other non-GM techniques (such as embryo transfer, cloning, MOET)(Niemann *et al.*, 2011). The development of genetically modified (GM) animals has been a matter of considerable interest and worldwide public controversy. Research and development in this field is enormously dynamic and with recent findings and innovations beginning to anticipate a significant role for GM animals in addressing some of the fundamental challenges facing humanity. However, progress in this area is very slow and has a long way to go before having an impact at commercial usage level.

The Potential Risks and Benefits of GM Animals

Over the past years, it has become evident that reactions of the public to GM food animals vary widely. Both potential positive attributes and negative attributes (*e.g.*unnatural) have been identified. Realistic scenarios representing technological applications of GM animals that may enter the market in the future were developed through a combination of literature review, data mining activities and expert consultation with industry and academic specialists (Houdebine and Jolivet, 2011). The results suggested that the techniques available to generate GM animals have been improved considerably and so development costs no longer represent a major barrier to the development of transgenic animals. In particular, the cost of genetically modifying larger animals is no longer prohibitive, as was the case in the past. Advances in research into GM farm animals have resulted in foods derived from these animals having enhanced quality or production yields (Laible, 2009), or improved nutritional value (*e.g.* (Lyall *et al.*, 2011; Lock and Bauman, 2004). Foreign genes might alter nutritional value of foods in unpredictable ways by decreasing levels of some nutrients while increasing levels of others. This will cause a difference between thetraditional strain and the GM-counterpart.Foods may be changed to meet the needs of individuals with specific dietary requirements, such as the modification of milk fat composition to enhance fatty acid content (Muysson and Verrinder-Gibbins, 1989). Developments are frequently intended to improve food security or human health, although the benefits and long-term impacts on agricultural sustainability are difficult to predict (NRC,1989). The most technologically advanced projects are related to the expression of bioactive compounds such as human lactoferrin in bovine milk (Yang *et al.*, 2008; Brundige *et al.*, 2010), and the production of meat enhanced with omega-3 fatty acids through the expression of roundworm desaturase gene in transgenic pigs (Lai *et al.*, 2006). Improved food security through increased production efficiency may be facilitated by the development of GM animals, assuming this is not compromised and reduced availability of genetic resources. Religious concerns are also voiced as some of the reasons for opposing genetic engineering of animal foods, while some people object to bio-engineered foods for personal, ethical, cultural and aestheticreasons, as well as infringement on consumer choice and inability to distinguish GM animal foods fromnon-GM counterparts (Robinson, 1997; Thompson, 1997).A further concern relates to the possible privatization of genetic resources due to the application of intellectual property rights. A further issue relates to environmental impact (*e.g.* unintended environmental release of animals) which may be less controllable for high profligacy species such as fish (SOSC, 2008). In addition, the implementation of effective traceability systems is required to ensure that GM animals and their products can be identified within supply chains. The main risk assessment concern, which can be identified when considering the use of any GM organism for food or feed production is food safety, is the potential for introduction of allergens or toxins in food chain (Kleter and Kok, 2010).

Genetically Modified Animals

A number of genes have been inserted into animals to improve their growth, health and reproduction and to alter the animal's meat composition in ways that are better for the environment and safer; and provide leaner and healthier meat for consumers (Table 9.1). Many of the GM animals for food purposes have been transformed with growth-related genes, such as those for growth hormone (somatotropin), growth hormone releasing factor and insulin-like growth factor (Pinkert and Murray, 1999), with an increased growth-rate, a greater volume of muscle and a higher percentage of lean meat, while at the same time reducing the feed intake per kg of meat produced. In addition, GM animals have been created with novel enzymes in their intestinal epithelium to increase the efficiency of feed utilization as an alternative to the use of feed additives. Examples include animals expressing phytase to increase the uptake of organically bound phosphorus (Golovan *et al.*, 2001) and animals expressing bacterial enzymes catalysing the synthesis of the essential amino acid cysteine (Ward *et al.*, 1999).

Table 9.1: Overview of the Genes in GM Animals and Fish Used in Food Production

Species	*Trait*	*Transgenic Gene*	*Gene Origin*	*Reference*
Cattle	Health	Lysozyme	Human	Yang *et al.* (2011)
	Udder	Omega-3	Nematode	Wu *et al.* (2012)
		α and k-Casein	Bovine	Brophy *et al.* (2003)
		Antimicrobial	Bacteria	Wall *et al.* (2005)
Sheep	Growth	IGF-1	Ovine	Damak *et al.* (1996)
Goat	Health	Lysozyme	Bovine	Scharfen *et al.* (2007)
	Udder	Mono sat. fatty acid	Bovine	Reh *et al.* (2004)
Pig	Growth	GH	Human	Nottle *et al.* (1999)
		cSKI	Chicken	Pursel *et al.* (1992)
	Milk	α-Lactalbumin	Bovine	Wheeler *et al.* (2001)
		Bacteria resistance	Mouse	Lo *et al.* (1991)
	Health	Lysozyme	Human	Tong *et al.* (2011)
		Unsat. Fattyacid	Spinach	Saeki *et al.* (2004)
		Omega-3	Nematode	Lai *et al.* (2006)
	Feed efficiency	Phytase	Bacterium	Phillips *et al.* (2006)
Chicken	Health	lacZ	Bacteria	Mozdziak *et al.* (2003)
	Growth	GH	Bovine	Chen *et al.* (1990)
Fish	Growth	GH	Fish	Du *et al.* (1992)
		Follistatin	Fish	Medeiros *et al.* (2009)
	Health	Cecropin	Insect	Dunham *et al.* (2002)
		Lactoferrin	Human	Mao *et al.* (2004)
		Lysozyme	Fish	Fletcher *et al.* (2011)
	Envir. tolerance	Antifreeze	Fish	Hew *et al.* (1999)
		GH	Fish	Saunders *et al.* (1998)

Enhancement in Carcass Composition and Growth

Using transgenic technology, it is possible to manipulate growth factors, growth factor receptors and growth modulators. Transgenic livestock and fish have been produced which contain an exogenous GH gene. The first hypertrophic pig was produced with the microinjection of chicken cSKI (Sloan–Kettering protein), resulting in individuals, showed evident hypertrophy and enlargement of the hams and shoulders (Pursel *et al.*, 1992). A transgene consisting of the zinc-inducible metallothionein promoter driving ovine growth hormone expression, showed significant decrease in fat and increase in muscle tissue (Pursel *et al.*, 1997). Saeki *et al.* (2004) developed transgenic pigs carrying unsaturated fatty acid genes, originally from linoleic and linolenic acids that are essential for mammalian nutrition and regulate many biological processes, including the growth of piglets. Transgenic pigs showed 3 fold higher levels of n-3 fatty acids and 23 per cent lower level n-6 fatty acids then non transgenic pigs. Humans can't produce these fatty acids, nor can livestock animals. The tissues of livestock animals contain high level of omega 6 fatty acid but livestock lack the enzymes (n-3 fatty acid de-saturase) to convert omega 6 fatty acids to omega 3 fatty acids. The fat-1 gene desaturase found in the roundworm *C. elegans* convert omega 6 fatty acids to omega 3 fatty acids by introducing double bond into the gene (Saeki *et al.*, 2004). Phytase transgenic pigs have been developed to address the problem of manure-related environmental pollution. These pigs carry a bacterial phytase gene under transcriptional control of a salivary gland specific promoter, which allows the pigs to digest plant phytate. Without the bacterial enzyme, phytate passes through the animal undigested and pollutes the environment with phosphorus. With the bacterial enzyme, fecal phosphorus output was reduced up to 75 per cent (Golovan *et al.*, 2001).In chicken, more investigations focus on fat deposition, such as the percentage of hypodermal fat, abdominal fat and intramuscular fat in breast and legs. The gene for extracted extracellular fatty acid binding protein (EXFABP) was considered as a candidate locus or linked to a major gene that significantly affected abdominal fat traits in chicken. The intramuscular fat (IMF) was in positive correlation with meat flavour, quality and succulency, especially tenderness of meat (Yu *et al.*, 2007). The growth factor, myostatin, has been shown to be a negative regulator of muscle growth and is highly conserved in mammals and birds. So myostatin manipulation via transgenesis leads to increase in muscles growth and muscles differentiation and may have a beneficial phenotypic effect in terms of increase in lean meat yield. In fact, spontaneous mutations in the myostatin gene in cattle have been found to be responsible for the double muscling phenotype in the Belgian Blue breed (Grobet *et al.*, 1997). In a recent strategy to increase meat production directly aimed at "double muscling" (resulting phenotype similar to that of Belgian Blue cattle) in rainbow trout using transgenes expressing follistatin, which inhibits myostatin, resulting in two muscle layers (Medeiros *et al.*, 2009).

Out of all the transgenic, domesticated animals that have been produced so far, fish are considered to be the safest for human consumption and are expected to be the first transgenic animal to be approved as a food item (Niiler, 2000). The growth-enhanced trait has attracted great interest as it could be an answer for increasing the yield of fish farms in addressing the need for greater availability of high quality protein, whilst at the same time overcoming environmental concerns. In fish, dramatic increases have been shown in growth rate of transgenic Atlantic salmon using the gene promoter and growth hormone gene derived from fish species. A salmon modified to grow four to six times faster than conventional salmon. It contains an additional salmon growth hormone gene and an anti-freeze gene from an ocean pout (*Macrozoarces americanus*) fish which will allow it to produce growth hormone all year-round rather than just in the warm months like conventional salmon. (Du *et al.*, 1992; Devlin *et al.*, 1994). In addition, the production of these growth-enhanced salmon will have vast positive environmental benefits.

Enhancement in Milk and Milk Products

Milk has been described as an almost perfect food. It is a rich source of vitamins, calcium and essential amino acidsbut lactose-intolerance limits the consumption of dairy products by many people. A DNA construct containing the cow intestinal lactase-phlorizin hydrolase cDNA under the control of the mammary specific α-lactalbumin promoter was introduced into cow with the possibility of reducing the lactose content of milk by transgenic expression of lactase in the mammary gland (Jost *et al.*, 1999). Transgenic alteration of milk composition has the potential to enhance the production of certain proteins and/or growth factors that are deficient in milk (Wall *et al.*, 1991). The increased expression of a number of these proteins in milk may improve growth, development, health and survivability of the developing offspring. Some of these factors are insulin-like growth factor 1 (IGF-I), epidermal growth factor (EGF), transforming growth factor (TGF-β) and lactoferrin (Konakci *et al.*, 2005; Malcarney *et al.*, 2005; Zhang *et al.*, 2008). GM goats and cattle expressing high levels of the human lysozyme enzyme in their milk were engineered to improve udder health, food safety and consumer health (Scharfen *et al.*, 2007). Scharfen *et al.* (2007) demonstrated that lysozyme has positive effects on the development of lactic acid bacteria and particularly the *lactococcal* species, throughout the cheese-making process.

Caseins represent about 80 per cent of the total milk protein and have high nutritional value and functional property. The amount of caseins in milk is an important factor for cheese manufacturing, since greater casein content results in greater cheese yield and improved nutritional quality. The genetic variants of α1-casein are positively related to curd firmness and tend to give more consistent curd; the 8-casein is negatively related to clotting time and β-casein is positively related to proteolysis resistance and to curd consistency (Mariani and Summer, 1995). Increased 8-casein content (100 per cent) has been linked to decreases the size of the micelle in milk and to improve the heat stability and cheese yield (Brophy *et al.*, 2003). Moreover, Jimenez *et al.* (1988) demonstrated that increased β-casein (8-20 per cent) content has been correlated with improved processing properties, including reduced rennet clotting time and increased whey expulsion. The protein β-lactoglobulin is present in cow's milk but not in human milk and approximately 10 per cent of consumers develop an allergic response to the protein. Transgenic removal of the protein or modification of the cow's BLG gene in its genome would benefit these consumers. We added silencing sequences of DNA to the cow's genome. These sequences are called miRNAs (micro-RNAs) and when expressed in the cell (by processing of their sequences), bind to the BLG RNA. This binding prevents the BLG protein from being produced. Moreover, these 'silencing sequences' have been inserted in such a fashion that these miRNAs are only produced when the cow is lactating. (Anower *et al.*, 2012).

In pigs, the increased transgenic expression of a bovine α-lactalbumin construct in the mammary gland resulted in increased lactose content and increased milk production which resulted in improved survival and development of the piglets (Wheeler *et al.*, 2001). This is truly a "value-added" opportunity for animal agriculture by increasing the concentrations of existing proteins or producing entirely new proteins in milk. The end result would be more saleable product for the dairy producer and a reduced environmental footprint.

Enhancement in Hair and Fiber

The manipulation of the quality, length, fibreelasticity, strength fineness and crimp of the wool and hair fibre from sheep and goats has been examined using transgenic methods (Hollis *et al.*, 1983; Powell *et al.*, 1994; Bawden *et al.*, 1998). In the future transgenic manipulation of wool will focus on the surface of the fibres.The limited investigation has been made into GM sheep for agriculture purposes such as the wool production via the microinjection of an IGF-1 gene construct (Damak *et al.*, 1996). Moreover this field is yet to be documented convincingly.

Labelling

People now a dayshave raised concerns about the long-term safety of genetically modified (GM) animal's products with regards to human health and the environment. Frequently critics of biotechnology make the statement that labelling for GM is costless – all that is required is to put a label on the products. The cost of labelling involves far more than the paper and ink to print the actual label. Accurate labelling requires an extensive identity preservation system from farmer to elevator to grain processor to food manufacturer to retailer (Maltsbarger and Kalaitzandonakes, 2000). However, others feel that labelling will benefit both the consumer and manufacturers for the following reasons:

1. It will enable consumers who prefer specially engineered GM animal foods (*e.g.* those with health enhancing properties) to get them while enabling others to avoid certain foods for ethical, cultural or religious reasons.
2. Labelling would enable manufacturers to emphasize the improved quality of their product, for example, improved taste and longer shelf-life and these would be good selling points that could appeal to consumers.
3. Lack of labelling would deny producers a chance to build brand identity.

Pro-labelling Arguments

Consumers have a right to know what's in their food, especially concerning products for which health and environmental concerns have been raised (Raab and Grobe, 2003). Mandatory labelling will allow consumers to identify and steer clear of food products that cause them problems. Surveys indicate that a majority of Americans support mandatory labelling. (However, such surveys often do not specify the effect on food prices.) At least 21 countries and the European Union have established some form of mandatory labelling. (Gruere and Rao, 2007; Phillips and McNeill, 2000)

Anti-labelling Arguments

Labels on GM food imply a warning about health effects, whereas no significant differences between GM and conventional foods have been detected. If a nutritional or allergenic difference were found in a GM food, current FDA regulations require a label to that effect. Labeling of GM foods to fulfil the desires of some consumers would impose a cost on all consumers. Experience with mandatory labelling in the European Union, Japan, and New Zealand has not resulted in consumer choice. Rather, retailers have eliminated GM products from their shelves due to perceived consumer aversion to GM products (Carter and Gruere, 2003). Consumers who want to buy non-GM food already have an option: to purchase certified organic foods, which by definition cannot be produced with GM ingredients. The food system infrastructure (storage, processing, and transportation facilities) in this country could not currently accommodate the need for segregation of GM and non-GM products. Consumers who want to avoid animal products need not worry about GM food. No GM products currently on the market or under review contain animal genes. However, there is no guarantee that this will not happen in the future.

Conclusion

There is still a lot of research required to identify useful targets for genetic modification and to increase overall efficiency of genetic modification methods, from frequency of production of transgenic animals to effective transgene expression. There is little support from research funding agencies for development of the technology in relation to genetically modified food products production or from the relevant industries, as costs are too high. There are no food products derived from GM livestock and poultry near to production for the market. Knowledge gained from these approaches can be beneficial in defining and optimizing the assurance of food quality. The human health benefits will now be realized based on the science-based regulatory framework for governing how these animals will provide benefits.We have embraced the human health aspects of genetically modified animals as well as the food aspects. The challenges ahead are not simple but if we follow the lead of the science, rigorous regulatory approval will portend compelling consumer benefits.We now need to adopt a more holistic approach to understand such Nobel biotechnological techniques, in the delivery of consistent quality products.

References

Anower, J., Stefan, W., Judi, M., David, N. W. and Goetz, L. (2012). Targeted microRNA expression in dairy cattle directs production of β-lactoglobulin-free, high-casein milk. ProcNatlAcadSci, U S A, 109(42): 16811–16816.

Bawden, C.S., Powell, B.C., Walker, S.K. and Rogers, G.E. (1998). Expression of a wool intermediate filament keratin transgene in sheep fibre alters structure. Transgenic Res, 7: 273-287.

Brophy, B, Smolenski, G., Wheeler, T., Wells, D., L'Huillier, P. and Laible, G. (2003). Cloned transgenic cattle produce milk with higher levels of β-casein and κ-casein. Nature Biotechnology, 21: 157-162.

Brundige, D.R., Maga, E.A., Klasing, K.C., Murray, J.D. (2010). Consumption of pasteurized human lysozyme transgenic goats' milk alters serum metabolite profile in young pigs. Transgenic Research, 19(4): 563–74.

Carter, C.A., and Gruere, G.P. (2003). Mandatory labeling of genetically modified foods: Does it really provide consumer choice? AgBio Forum, vol. 6, no. 18, www.agbioforum.org/.

Chen, W.Y., Wight, D.C., Wagner, T.E. and Kopchick, J.J. (1990). Expression of a mutated bovine growth hormone gene suppresses growth of trans- genic mice. Proc. Natl. Acad. Sci., USA, 87: 5061–5065.

Damak, S., Su, H., Jay, N.P., Bullock, D.W. and Su, H.Y. (1996). Improved wool production in transgenic sheep expressing insulin-like growth factor 1. Biotechnology, 14: 185–188.

Devlin, R.H., Yesaki, T.Y., Biagi, C.A., Donaldson, E.M., Swanson, P. and Chan, W. (1994).Extraordinary salmon growth. Nature, 371: 209-210.

Du, S.J., Gong, Z., Fletcher, G.L., Shears, M.A., King, M.J., Idler, D.R. and Hew, C.L. (1992). Growth enhancement in transgenic Atlantic salmon by the use of an "all fish" chimeric growth hormone gene construct. Nat. Biotechnol., 10: 176–181.

Dunham, R.A., Warr, G.W., Nichols, A., Duncan, P.L., Argue, B., Middleton, D. and Kucuktas, H. (2002). Enhanced bacterial disease resistance of transgenic channel catfish Ictaluruspunctatus possessing cecropin genes. Mar. Biotechnol., 4: 338–344.

FAO, 2012. GMOs in the pipeline: looking to the next five years in the crop, forestry, livestock, aquaculture and agro-industry sectors in developing countries. In: Proceedings of the 18th FAO Biotechnology Forum Conference, 128.

Flavio, F., Fredrik, L., Sundstrom and Lotta, R. (2013).An algorithm for the identification of genetically modified animals. Trends in Biotechnology, 31(5): 272-274.

Fletcher, G.L., Hobbs, R.S., Evans, R.P., Shears, M.A., Hahn, A.L. and Hew, C.L. (2011). Lysozyme transgenic Atlantic salmon(Salmosalar L.). Aquac. Res., 42: 427–440.

Golovan, S.P., Meidinger, R.G., Ajakaiye, A., Cottrill, M., Wiederkehr, M.Z., Barney, D.J., Plante, C., Pollard, J.W., Fan, M.Z., Hayes, M.A., Laursen, J., Hjorth, J.P., Hacker, R.R., Phillips, J.P. and Forsberg, C.W. (2001). Pigs expressing salivary phytase produce low-phosphorus manure. Nat. Biotechnol., 19: 741–745.

Grobet, L., Poncelet, D., Royo, L. J., Brouwers, B., Pirottin, D. and Michaux, C. (1998).Molecular definition of an allelic series of mutations disrupting the myostatin function and causing double-muscling in cattle. Mammal Genome, 9(3): 210–213.

Gruere, G.P., and Rao, S.R. (2007). A review of international labeling policies of genetically modified food to evaluate India's propose rule. AgBioForum, vol. 10, no. 1, www.agbioforum.org/.

Hammer, R.E., Pursel, V.G., RexroadJr., C.E., Wall, R.J., Bolt, D.J., Ebert, K.M., Palmiter, R.D. and Brinster, R.L. (1985). Production of transgenic rabbits, sheep and pigs by microinjection. Nature (UK), 315: 680–683.

Hew, C., Poon, R., Xiong, F., Gauthier, S., Shears, M., King, M., Davies, P. and Fletcher, G. (1999). Liver-specific and seasonal expression of trans- genic Atlantic salmon harboring the winter flounder antifreeze protein gene. Transgenic Res., 8: 405–414.

Hollis, D.E., Chapman, R.E., Panaretto, B.A. and Moore, G.P. (1983). Morphological changes in the skin and wool fibers of Merino sheep infused with mouse epidermal growth factor. Aust J BiolSci, 36(4): 419-434.

Houdebine, J.M. and Jolivet, G. (2011). Generation and use of genetically modified animals: a technical overview (PEGASUS Project Deliverable 2.1). Jouy-en-Josas: National Institute for Agricultural Research (INRA).

Jaenisch, R., Fan, H. and Croker, B. (1975). Infection of preimplantationmouse embryos and of newborn mice with leukemia virus: Tissue distribution of viral DNA and RNA and leukemogenesis in the adult animal. Proceedings of the National Academy of Sciences USA 72: 4008-4012.

Jimenez-Flores, R. and Richardson, T. (1998). Genetic engineering of the caseins to modify the behavior of milk during processing: a review. J Dairy Sci, 71: 2640-2654.

Jost, B., Vilotte, J. L., Duluc, I., Rodeau, J. L. and Freund, J. N. (1999).Production of low-lactose milk by ectopic expression of intestinal lactase in the mouse mammary gland. Nature Biotechnology, 17, 135–136.

Kleter, G.A. and Kok, E. (2010). Safety assessment of biotechnology used in animal production, including genetically modified (GM) feed and GM animals – a review. Animal Science Papers and Reports, 28(2): 105–14.

Konakci, K.Z., Bohle, B., Blumer, R., Hoetzenecker, W., Roth, G., Moser, B., Boltz-Nitulescu, G., Gorlitzer, M., Klepetko, W., Wolner, E. and Ankersmit, H.J. (2005). Alpha-gal on bioprostheses: xenograft immune response in cardiac surgery. Eur. J. of Clin. Invest. 35: 17 33.

Kues, W.A. and Niemann, H., 2011. Advances in farm animal transgenesis. Prev. Vet. Med., 102: 146–156.

Lai, L., Kang, J.X., Li, R., Wang, J., Witt, W.T., Yong, H.Y., Hao, Y., Wax, D.M., Murphy, C.N., Rieke, A., Samuel, M., Linville, M.L., Korte, S.W., Evans, R.W., Starzl, T.E., Prather, R.S. and Dai, Y. (2006). Generation of cloned transgenic pigs rich in omega-3 fatty acids. Nat. Biotechnol., 24 (4): 435–436.

Lo, D., Pursel, V., Linton, P.J., Sandgren, E., Behringer, R., Rexroad, C., Palmiter, R.D. and Brinster, R.L. (1991). Expression of mouse IgA by transgenic mice, pigs and sheep. Eur. J. Immunol., 21: 1001–1006.

Lock, A.L. and Bauman, D.E. (2004). Modifying milk fat composition of dairy cows to enhance fatty acids beneficial to human health. Lipids, 39(12): 1197–206.

Lyall, J., Irvine, R.M., Sherman, A., McKinley, T.J., Nunez, A. and Purdie A. (2011). Suppression of avian influenza transmission in genetically modified chickens. Science, 331(6014): 223–226.

Malcarney, H.L., Bonar, F. and Murrell, G.A. (2005). Early inflammatory reactions after rotator cuff repair with porcine small intestine submucosal implant: a report of 4 cases. Am. J. of Sports Med. 33: 907.

Maltsbarger, R., and N. Kalaitzandonakes. (2000). Direct and hidden costs in identity preserved supply chains. AgBio Forum, Vol. 3, no. 4, www.agbioforum.org/.

Mariani, P., Summer, A., 1995. La caratterizzazionedellaqualitatecnologica del latte. In: Atti S.I.S. Vet., SalsomaggioreTerme 27–30 SettembreXLIX(I), 135–143.

Medeiros, E., Phelps, M., Fuentes, F. and Bradley, T. (2009).Overexpressionoffollistatin in trout stimulates increased muscling. Am.J.Physiol.Regul., Integrative Comp. Physiol. 297: 235–242.

Mozdziak, P.E., Pophal, S., Borwornpinyo, S. and Petitte, J.N. (2003). Trans- genic chickens expressing beta-galactosidase hydrolyze lactose in the intestine. J. Nutr., 133: 3076–3079.

Muysson, D.J. and Verrinder-Gibbins, A.M. (1989). The alteration of milk content by genetic engineering and recombinant DNA-mediated selection techniques. Canadian Journal of Animal Science, 69(3): 517–27.

National Research Council. Animal biotechnology: science based concerns 1989. Washington, DC: National Academies Press, 2002, http: //books.nap. edu/ openbook.php? record_id=10418 per cent 26page=1 (accessed 10.04.14).

Niemann, H. and Kues, W.A., 2007. Transgenic farm animals: an update. Reprod., Fertil. Dev., 19: 762–770.

Niemann, H., Kuhla, B. and Flachowsky, G. (2011). Perspectives for feed- efficient animal production. J. Anim. Sci., 89: 4344–4363.

Niiler E. 2000. FDA researchers consider first transgenic fish. Nature Biotechnology 18: 143.

Nottle, M.B., Nagashima, H. and Verma, P.J. (1999). Meeting on the Application of Transgenic Animals for Use in Production Agricul- ture Location: TAHOE CITY, CA. In: Murray, J.D., Anderson, G.B., Oberbauer, A.M. (Eds.), Transgenic Animals in Agriculture, 145–156.

Phillips, P.W.B., and McNeill, H. (2000).A survey of national labeling policies for GM foods.AgBio Forum, Vol. 3, no. 4, www.agbioforum.org/.

Phillips, J.P., Golovan, S.P., Meidinger, R.G. and Forsberg, C.W. (2006). Trans- genic enhancement of nutrient cycling: moving toward an envir- onmentally sustainable animal agriculture. In: Proceedings of the 8th World Congress on Genetics Applied to Livestock Production. Belo Horizonte, Minas Gerais, Brazil, 13–18 August, 2006. InstitutoProciencia, Minas Gerais, 19–02.

Pinkert, C.A. and Murray, J.D. (1999).Transgenic farm animals. pp. 1–18. In: Transgenic animals in agriculture. Eds. Murray, J.D., Anderson, G.B., Oberbauer, A.M., McGloughlin, M.M. Wallingford, CAB International.

Powell, B.C., Walker, S.K., Bawden, C.S., Sivaprasad, A.V. and Rogers, G.E. (1994). Transgenic sheep and wool growth: possibilities and current status. ReprodFertilDev, 6(5): 615-623.

Pursel, V.G., Sutrave, P., Wall, R.J., Kelly, A.M. and Hughes, S.H. (1992). Transfer of C-Ski gene into swine to enhance muscle development. Theriogenology, 37: 278.

Pursel, V. G., Wall, R. J., Soloman, M. B., Bolt, D. J., Murray, J. D. and Ward, K. A. (1997). Transfer of an ovine metallothionein-oveine growth hormone fusion gene into swine. Journal of Animal Science, 75: 2208–2214.

Raab, C., and D. Grobe. (2003). Labeling genetically engineered food: The consumer's right to know?, AgBio Forum, vol. 6, no. 4, www.agbioforum.org/.

Reh, W.A., Maga, E.A., Collette, N.M., Moyer, A., Conrad-Brink, J.S., Taylor, S.J., DePeters, E.J., Oppenheim, S., Rowe, J.D., Bon-Durant, R.H., Anderson, G.B. and Murray, J.D. (2004). Hot topic: using a stearoyl-CoA desaturase transgene to alter milk fatty acid composition. J. Dairy Sci., 87: 3510–3514.

Robinson, C. (1997). Genetically modified foods and consumer choice. Trends in Food Science and Technol, 8: 84–8.

Saeki, K., Matsumoto, K., Kinoshita, M., Suzuki, I., Tasaka, Y., Kano, K., Taguchi, Y., Mikami, K., Hirabayashi, M., Kashiwazaki, N., Hosoi, Y., Murata, N. and Iritani, A. (2004).Functional expression of a Delta12 fatty acid desaturase gene from spinach in transgenic pigs. Proc. Natl. Acad. Sci. USA 101: 6361–6366.

Saunders, R.L., Fletcher, G.L. and Hew, C.L. (1998). Smolt development in growth hormone transgenic Atlantic salmon. Aquaculture, 168: 177–193.

Scharfen, E.C., Mills, D.A. and Maga, E.A. (2007). Use of human lysozyme transgenic goat milk in cheese making: effects on lactic acid bacteria performance. J. Dairy Sci., 90: 4084–4091.

Scientific opinion of the Scientific Committee on food safety, animal health and welfare and environmental impact of animals derived from cloning by somatic cell nucleus transfer (SCNT) and their offspring and products obtained from those animals. EFSA, Journal, 2008, 767: 1–49, http: //www.efsa.europa.eu/en/efsajournal/doc/767.pdf (accessed 15.04.14).

Thompson, P.B. (1997). Food biotechnology's challenge to cultural integrity and individual consent. Hastings Center Report, 27(4): 34–8.

Tong, J., Wei, H., Liu, X., Hu, W., Bi, M., Wang, Y., Li, Q., Li, N., Tong, J., Wei, H.X., Liu, X.F., Hu, W.P., Bi, M.J., Wang, Y.Y., Li, Q.Y. and Li, N. (2011). Production of recombinant human lysozyme in the milk of trans- genic pigs. Transgenic Res., 20: 417–419.

Wall, R.J., Powell, A.M., Paape, M.J., Kerr, D.E., Bannerman, D.D., Pursel, V.G., Wells, K.D., Talbot, N. and Hawk, H.W. (2005). Genetically enhanced cows resist intramammary*Staphylococcus aureus* infection. Nat. Biotechnol., 23: 445–451.

Wall, R.J., Pursel, V.G., Shamay, A., McKnight, R.A., Pittius, C.W. and Henninhausen, L. (1991).High-level synthesis of a heterologous milk protein in the mammary glands of transgenic swine.ProcNatlAcadSci USA 88: 1696-1700.

Ward, K.A., Leish, Z., Brownlee, A.G., Bonsing, J., Nancarrow, C.D. and Brown, B.W. (1999). The utilization of bacterial genes to modify domestic animal biochemistry. In: Murray, J.D., Anderson, G.B., Oberbauer, A.M., McGloughlin, M.M. (Eds.), Science 287: 303–305.

Wheeler, M.B., Bleck, G.T. and Donovan, S.M. (2001).Transgenic alteration of sow milk to improve piglet growth and health. Reproduction (58): 313–324.

Wu, X., Ouyang, H., Duan, B., Pang, D., Zhang, L., Yuan, T., Xue, L., Ni, D., Cheng, L., Dong, S., Wei, Z., Li, L., Yu, M., Sun, Q., Chen, D., Lai, L., Dai, Y., Li, G., Wu, X., Ouyang, H.S., Duan, B., Pang, D.X., Zhang, L., Yuan, T., Xue, L., Ni, D.B., Cheng, L., Dong, S.H., Wei, Z.Y., Li, L., Yu, M., Sun, Q.Y., Chen, D.Y., Lai, L.X., Dai, Y.F. and Li, G.P. (2012). Production of cloned transgenic cow expressing omega-3 fatty acids. Transgenic Res., 21: 537–543.

Yang, B., Wang, J., Tang, B., Liu, Y., Guo, C., Yang, P., Yu, T., Li, R., Zhao, J., Zhang, L., Dai, Y., Li, N., Yang, B., Wang, J.W., Tang, B., Liu, Y.F., Guo, C.D., Yang, P.H., Yu, T., Li, R., Zhao, J.M., Zhang, L., Dai, Y.P. and Li, N. (2011). Characterization of bioactive recombinant human lysozyme expressed in milk of cloned transgenic cattle. PLoS One, 6 (3): 17593.

Yang, P., Wang, J., Gong, G., Sun, X., Zhang, R. and Du, Z. (2008).Cattle mammary bioreactor generated by a novel procedure of transgenic cloning for large-scale production of functional human lactoferrin.PloS One, 3(10): 3453.

Yu, G., Ran, Z., Xiaoxiang. H. and Ning, L. (2007).Application of genomic technologies to the improvement of meat quality of farm animals. Meat Science, 77: 36–45

Zhang, J., Li, L., Cai, Y., Xu, X., Chen, J., Wu, Y., Yu, H., Yu, G., Liu, S., Zhang, A., Chen, J. and Cheng, G. (2008) Expression of active recombinant human lactoferrin in the milk of transgenic goats. Protein ExprPurif 57: 127-135.

2016, Dairy and Food Product Technology *Pages 145–169*
Editors: Birendra Kumar Mishra and Subrota Hati
Published by: BIOTECH BOOKS, NEW DELHI

Chapter 10

High Pressure Processing: A Novel Non-thermal Approach for Microbial Safety of Foods

*Manpreet Kaur and Sunil Kumar**

Division of Livestock Products Technology,
Faculty of Veterinary Science and Animal Husbandry,
Sher-e-Kashmir University of Agricultural Sciences and Technology,
R.S. Pura, Jammu – 181 102, J&K

Introduction

High pressure processing (HPP) is an emerging non-thermal pasteurization technique in the production of various food products which involves the static treatment of pre-packed food at or above 100 MPa by means of a liquid transmitter.HPP is also known as High hydrostatic pressure processing (HHP) or ultrahigh pressure processing (UHP). The technology has been cited as one of the best innovations in food processing in 50 years (Dunne, 2005). The beneficial effect of this technique in food preservation isknown since 19th century. High pressure can kill microorganisms by interrupting their cellular function without the use of heat that can damage the taste, texture, and nutrition of the food. In 1980's Blaise Pascal, who first explored this technique began testing HPP on bacteria. It was in 1899 that an agricultural scientist, Bert Hite designed and constructed a high pressure unit to pasteurize milk and other products at Agricultural University, West VirginiaUSA (Hite, 1899). Hite's high pressure unit was so constructed that it could reach pressures of about 6800 atmospheres and along with his coworkers he found that HPP could destroy harmful

* Corresponding Author

microorganisms in foods like milk, meat, fruitsand vegetables. In 1899, Hite reported that treatment of milk at pressure of 450 MPa or greater (600 MPa) at room temperature for 60 minutes could improve the keeping quality of milk (Hite, 1899) and can extend its shelf life by 4 days.

As reported by Patterson *et al.* (1995) Hite in 1914 showed that yeasts and lactic acid bacteria associated with sweet, ripe fruit were more susceptible to pressure than other organisms, especially spore forming bacteria associated with vegetables. Although thetechnique is known for long but its application has been recently recognized widely. The first high pressure treated product was a high acid jam and it reached Japanese market in April, 1990 (Knorr, 1993). It was followed by installation of two semi-continuous citrus fruit juice processors in Japan. In 1991 high pressure processed yogurts, fruit jellies, salad dressings, and fruit sauces were also introduced.

Principle

HPP is based on two scientific principles, one is Le Chatelier principle and the other is isostatic rule.Le Chatelier principle states "If a chemical system at equilibrium experiences a change in concentration, temperature, volume or pressure then the equilibrium shifts to counteract the imposed change and a new equilibrium is established". According to this principle the application of pressure shifts equilibrium to the state that occupies the smallest volume and this forms the basis of HPP.Also, the isostatic rule states that pressure is instantaneously and uniformly transmitted throughout a sample under pressure, whether the sample is in direct contact with the pressure medium or hermetically sealed in a flexible package that transmits pressure (Olsson, 1995). Hence the time period for HPP of a product is independent of its size, shape and composition unlike thermal processing. Compression will uniformly increase the temperature of foods approximately 3°C per 100 MPa. The temperature of homogeneous food will increase uniformly due to compression. An increase in food temperature above room temperature and to a lesser extent a decrease below room temperature increases the inactivation rate of microorganisms during HPP treatment. Temperatures in the range of 45°C to 50°C appear to increase the rate of inactivation of food pathogens and spoilage microbes. Temperatures ranging from 90°C-110°C in conjunction with pressure 500-700 MPa have been used to inactivate spore forming bacteria such as *Clostridium botulinum* (Kadam *et al.*, 2012).

Technique and Equipment

Despite the HPP being known to be in use as a means of cold pasteurization since late 1800s it is only recently that developments in mechanical engineering have permitted large high-pressure vessels to be constructed at reasonable cost with sufficient durability to withstand thousands of pressure cycles without failure.High-pressure processing systems were initially developed in the chemical and material process industries for applications such as making artificial diamonds and sintered materials from powders. It is only during the past two decades that the food industry has begun using pressure treatment for food preservation (Balasubramaniam,2011). Commercial-scale, high-pressure processing systems cost approximately $500,000 to$2.5 million, depending on equipment capacity and extent of automation.

Commercialbatch vessels have internal volumes ranging from 30 to more than 600 liters.Currently, HPP treatment costs are quoted as ranging from 4–10 cents/lb, including operating cost and depreciation, and are not "orders of magnitude" higher than thermal processing(Saiz *et al.*, 2008).With the growth in demand for HPP equipment, innovation is expected to further reduce capital and operating costs of HPP.

In HPP treatment the product is submerged in a liquid (which acts as pressure transmitting medium) in a pressure vessel capable of sustaining the required pressure. Selection of suitable liquid medium depends upon four main factors *i.e.* the ability of the liquid medium to protect the inner vessel wall from corrosion, the specific HP system being used, and the temperature range of process and the viscosity of liquid medium. Various liquids that can be used as medium are water, castor oil, silicone oil, sodium benzoate, ethanol, glycol etc.

Industrial HPP involves two types of processes, batch process suitable for solid food products and semi-continuous process which is suitable for liquids, slurries and other pumpable products (Ting and Marshall, 2002). Both horizontal and vertical pressure vessel configurations are available.

A typicalhigh pressure system consists of a pressure vessel and pressure generating device. Food packed in high-barrier, flexible pouch or a plastic container are loaded into the high pressure vessel and to top closed. The pressure medium, usually water containing a small amount of soluble oil, is pumped into the vessel from the bottom. The desired pressure can be maintained by pressurizing either by pumping medium into the vessel or by reducing volume of the pressure chamber by using piston. Once the desired pressure is achieved the pump or piston is stopped, all valves are closed and the pressure is maintained without further energy output(Ting and Marshall, 2002). When the required holding time is passed the vessel is opened and the system is depressurized followed by unloading of the product. Thetime taken for pressurization, holding and depressurization forms one 'cycle time'. For any HP system, the working pressure is an important parameter because not only the initial prices of equipment increases significantly with its maximum working pressure but also because a decrease in working pressure can reduce significantly the number of failures, increasing the working life of the equipment (Otereo *et al.*, 2000). The process being isostatic, allows quick and uniformtransmission of pressure throughout the liquid medium and the food with little or no heating. It is equal from all sides so there is no squashing effect and product is not affected *i.e.* the size and geometry of product remains unaltered. High pressure modifies non-covalent bonds and does not affect small molecules like flavour compounds and vitamins (Simonin,2012). Pressure in the range of 300-700 MPa are generally used for foods (500 MPa isequivalent to 20 family cars bearing down on an area the size of postagestamp). These pressures, maintained for approximately 15 minutes, cancause more than a 10,000 fold reduction in numbers of food poisoningbacteria such as *Salmonella, Campylobacter* and *Listeria* in milk andpoultry meat (Kadam *et al.*, 2012).

Market Trends in HPP of Foods

HPP market includes both HPP equipment market and HPP products market.

Avure Technologies Inc. (U.S.) and Hyperbaric Espuna (Spain) are main manufacturers of HPP equipments. Some other manufacturers are Multivac (Germany), Kobelco (Japan) and Chic (China). Market value of HPP equipment is based on type of orientation, application, end users and geography. HPP equipment market value is projected to grow at a CAGR of 24 per cent.Hormel food corporation (U.S.) and Esteban Espuna SA (Spain) are among leading companies that are commercializing their product range with this technology.U.S. is the largest consumer of HPP products including meat and seafood in North America. The market in Spain is highly developed and fastest growing European market in HPP equipment for meat and seafood.Asia dominates the meat market for HPP equipment in the Asia Pacific region which is also a fastest growing market. According to Mark de Boevere (2010) the contribution of U.S., Europe, Asia and Oceania to the production of HPP machines is 55 per cent, 25 per cent, 15 per cent and 5 per cent, respectively.

Percentage of industrial high pressure process equipmentdedicated to various food manufacturing segments in decreasing order arevegetable products (35 per cent) > meat (29 per cent)> seafood including fish (15 per cent) >juice and beverages (15 per cent)> other products like dairy etc (6 per cent)(Mark de Boevere, 2010).

Table 9.1: Segmentation of HPP Equipments and Products Market

Type	*Basis*				
	Orientation	*Vessel Size*	*Food Application*	*End-Users*	*Geography*
1	Vertical	<100 L	Meat	SMEs	North America
2	Horizontal	100–250 L	Fruit/ vegetable	Large production plants	Europe
3		250–500 L	Juice/ beverage	Groups (MNCs and Conglomerates	Asia
4		>500 L	Seafood		Pacific
5			Others (dairy etc.)		Rest of the world

Effect of HPP on Microorganisms

The microorganisms are inactivated by HPP through various mechanisms that include changes in cell wall, cell membranes, proteins and enzyme mediated cellular functions. According to recent studies Cell membrane alterations are probably the main cause of cell death (Patterson, 2005; Moussa *et al.*, 2009). Disruption of ribosome is also a key mechanism that leads to inactivation of microbial cell (Abe, 2007). Different organisms have different degree of resistance towards high pressure treatment. Pressure resistance of microorganisms depends upon various factors like species, strain, stage of growth and food composition. However factors like pH, water activity, temperature, pressure and holding time that affect the response of microorganisms to pressure treatment can be optimized to ensure food safety. As a thumb rule the order of susceptibility of microorganisms is as follows, Gram negatives > Yeast/Mould > Gram positives > Spores(only in combination with heat also)> Prions.

Bacteria and Spores

The bacterial cells of prime concern in food industry are *Staphylococcus aureus, Listeria monocytogenes, Salmonella, Bacillus cereus, Escherichia coli, Campylobacter* spp., *Clostridium botulinum* and *Vibrio*. *Staphylococcus aureus* and *E.coli* O157:H7 drives special attention for processing as they are highly barotolerant pathogens while as Yersinia enterocolitica is one of the most sensitive pathogenic bacteria.Barotolerance of a pathogen in NaCl solution is known to increase by pre incubation at temperature below 0ºC (Noma and Hayakawa, 2003). Most of the rod shaped bacteria are sensitive whereas cocci are mostly resistant for HPP (Cheftel,1995;Ludwig and Schreck,1997). As reported by Yuste *et al.* (2003) the sensitivity of pathogens is in the following order *Y.enterocolitica* > *S. typhimurium* > *E. coli* >*L. monocytogenes*>*S. aureus*.The stage of growth of bacteria also determines the susceptibility of microorganism. Bacterial cells in stationary phase are more resistant than those in the exponential phase (Mc Clements *et al.*, 2001). Psychrotrophic bacteria like *Pseudomonas* spp a potential pathogen in meat and meat products is pressure sensitive and are more susceptible to high pressure than are mesophiles. There was a reduction of 4.74log CFU/g for psychrotrophs and of 3.7log CFU/g for mesophiles in mechanically recovered poultry meat treated at 450 MPa for 15 min at 20°C (Yuste *et al.*, 2001). Garriga *et al.* (2004) reported that reduction in the psychrotrophic bacteria was quite more than mesophilic bacteria in high-pressure treated (600 MPa and 16°C, for 6 min) sliced dry cured ham and sliced cooked ham during 60th day of subsequent storage. One possible explanation is that when most psychrotrophs are subjected to high pressure, they lose their ability to grow at low temperatures, preventing their recovery during subsequent chilled storage. Processing meat at pressures between 300 and 450 MPa appears to be sufficient to completely inactivate indigenous *Pseudomonas* (Carlez *et al.*, 1994; L´opez-Caballero *et al.*, 2002). In the study by Diez *et al.* (2008) on blood sausages treated between 300 and 600 MPa at 15°C, *Pseudomonas* was the most pressure-sensitive microbial group (along with *Enterobacteria*) and the bacterial count remained below the level of detection during 28 d of cold storage.

A high-pressure treatment of about 400 to 600 MPa at 7 to 18°C for 5 to 10 minutes is generally effective for reducing the number of *enterobacteria* below the level of detection in high a_w products, such as cooked ham, marinated beef loin, or blood sausages (Lopez- Caballero *et al.*, 1999; Garriga *et al.*, 2004; Diez *et al.*, 2008). In cooked ham, dry-cured ham, and marinated beef loin inoculated at 3.5 log CFU/g, and in marinated beef with an endogenous load of 1.18 log CFU/g, *E. coli* was reduced below the level of detection during 120th day of chilled storage by a high-pressure treatment at 600 MPa for 6 min at 31°C (Garriga *et al.*, 2004; Jofr´e *et al.*, 2009). In the study of Garriga *et al.* (2002), a mixture of 2 strains of *E. coli* displayed a 4.5 log CFU decline after 24 hrs of high pressure treatment (400 MPa, 10 min, 17 °C). As reported by Morales *et al*(2008) a high-pressure treatment of 400 MPa at 12°C for 20 min was sufficient to give a reduction of 2.45 log CFU/g (of 7 log initially present) of a pressure-resistant strain of the serotype O157:H7 in ground beef. Porto-Fett *et al.* (2010) showed a total reduction of the initial 5 log CFU/g of *E. coli* O157:H7 inoculated into dry-fermented salami after a high-pressure treatment at 483 MPa at 19°C for 5 min. Gola *et al.*(2000) also reported a total reduction in 8 strains of *E.coli* O157:H7 inoculated in raw minced meat treated at 700 MPa at ambient temperature for 5 minutes.

According to various studiesLactic acid bacteria (LAB)is reported to have baroresistance,although the resistance and ability of LAB to recover from ahigh-pressure treatment can be positive or negative dependingon whether the strain is used for its technological properties or isa spoilage strain. Different behaviors of LAB after high-pressuretreatment have been observed depending on the strain and the foodmatrix. When 2 inoculated LAB strains were treated at 400 MPa for 10 min, the cell counts were similar after 8 day of incubation, forhigh-pressure-treated and untreated samples. According to these reports(Patterson, 2005; Escriu and Mor-Mur, 2009) it becomes clear that gram positive bacteria (LAB) are more resistantthan gram negative bacteria (*Enterobacteria, Pseudomonas*); this is probably aresult of a more robust cell envelope in Gram positive bacteria that containsa high percentage of peptidoglycan and teichoic acids, assuggested by Escriu and Mor-Mur (2009).The contaminationof meat products with *Listeria monocytogenes* is an emerging publichealth problem of great concern. Thus, the effect of high-pressure treatment on this pathogen has been a main subject of numerous studies undergone in recent years (Simpsonand Gilmour, 1997; Garriga *et al.*, 2002; Hayman *et al.*, 2004; Aymerich *et al.*, 2005; Chen, 2007; Koseki *et al.*, 2007; Marcos *et al.*, 2008a, b; Porto-Fett *et al.*, 2010).Generally, a treatment at 400 MPa is necessary to significantly reducethe *Listeria* count(Simpson and Gilmour, 1997; Chen, 2007).A reduction of 1.6to e" 5 log CFU/g can be achieved in salami by high-pressuretreatment at 600 MPa and 18°C for 1 to 7 min or at 483 MPaand 18°C for 5 to 12 min, depending on the $a_w$of the productand on the treatment strength (Porto-Fett *et al.*, 2010).In chicken baters, containing a different type of fat, a reductionof 1.5 to 3log CFU/g of *Listeria innocua* was achieved after ahigh-pressure treatment at 400 MPa and 20°C for 2 min (Escriuand Mor-Mur, 2009). However the main risk associated with *Listeria*is its potential recovery during cold storage despite application of different methods that reduces its count. According to Marcos *et al.*(2008a), pressure treatment of cooked ham at 400 MPaand 17°C for 10 min and cold storage cannot prevent *Listeria*recovery. It is also suggested that pressures of 600 MPa and 10°C for 5 min followed by cold storagebelow 6°C is necessary to ensure cooked ham's safety for up to3 months (Jofr´e *et al.*, 2008).The same treatments that can remove *Listeria* risk are also generallyeffective at inactivating *Salmonella* spp., which show less potentialfor recovery during subsequent storage (Garriga *et al.*, 2002) as indicated by another study by Jofr´e *et al.*(2009) in which a treatment at600 MPa for 6 min at 31°C sufficiently reduced the bacterial countbelow the limit of detection in cooked ham, dry-cured ham, and marinated beef loin inoculated with mixture of *Salmonella* strains. A treatment at 500 MPa and 50°Cfor 10 min can lead to a reduction of more than 7log CFU/g ininoculated poultry sausages (Yuste *et al.*, 2000).After the application of a 600 MPa treatmentfor 6min at 31°C, the reduction in *S. aureus* counts inmeatproducts at 3.5log CFU/g was 2.7 log units in beef loinand 1.1 log units in cooked ham. As reported by Garriga *et al.* (2002) two different strains of *Staphylococcus aureus* were resistant to a treatment of 400 MPa at 17°C for 10 minutesin a meat sample, and in pork, *S. aureus* was affected little by a pressure treatment at 400 MPa and 20°C for 30 min according to Moerman (2005). Investigations have also been carried out with the Gram negative

Campylobacter and more frequently with *C. jejuni*. Martinez- Rodriguez and Mackey (2005) treated different *Campylobacter jejuni* strains inoculated into chicken meat at pressures 200 to 400 MPa and 25°C for 10 min. This microorganism has a relatively high sensitivity to pressure depending on the strain, as a 200-MPa treatment could sometimes reduce the microbial count by 2 log CFU/g. Lori *et al.* (2007) showed that treatment at 450 MPa and 15°C for 30 seconds was sufficient to inactivate more than 6 log CFU/g in chicken slurry thus confirming to low resistance to pressure of *C. jejuni*. Jofr´e *et al.* (2009) also reported that *Campylobacterjejuni* was the most inactivated microorganism among *Salmonella, Listeria, Staphylococcus, Escherichia coli,* and LAB that were spiked into different types of meat products. As of yet, no study has addressed *Campylobacter* recovery after a high-pressure inactivation. Simpson and Gilmour (1997) suggested that, regardless of bacterial species the inactivation of bacteria is highly dependent on the strain; sometimes a difference of 3-log CFU can be observed in the inactivation of two strains of the same bacterial species for the same treatment. This finding suggests the importance of using a multiple strains as target bacteria in further studies of meat and meat products (Garriga *et al.*, 2002; Jofr´e *et al.*, 2009). In the United States, the presence of *Vibrio vulnificus* in molluscan shellfish causes the highest fatality rate among food-borne pathogens (Cook, 2003). HPP is found successful in killing *Vibrio parahaemolyticus* and*V. vulnificus*in oysters without compromising their sensoryattributes (Cook, 2003;Kural, 2008;Lopez-Caballero, M.E., 2000).Pressures of 205–275 MPa at temperatures of 10–30°Cand treatment times of 1–3 min are typically used for rawoysters. Kural (2008) analyzed that the pressure treatmentof e"350 MPa for 2 min at temperatures between 1 and 35°C and e" 300 MPa for 2 min at 40°C is required for a 5-log reduction of pressure-resistant strainsof *V. parahaemolyticus* in live oysters. The HPP product maintained the sensory characteristics offresh oysters for an extended shelf life.Recently, the State of California recognized HPP as avalid process to reduce pathogenic *Vibrio* bacteria in fishproducts (Raghubeer,2007). Moreover, in a study of the Hepatitis Avirus and Calicivirus, Kingsley *et al.* (2002) reported that HPPhas the potential of making raw shellfish free of infectiousviruses. HPP also denatures the oyster's adductor muscleand induces the shell to open spontaneously. HPPshucking reduces the need for manual shucking andincreases the quantity of meat removed from the shell (Murchie, 2005). A heat shrink plastic band is placed around eachoyster shell prior to the HPP treatment to keep the shellclosed during distribution and storage.Pressures between 250 and 500 MPa can be successfully employed to treat other types of seafood, such as lobsters, thereby improving the microbiologicalquality and product yields.

Bacterial spores can only be destroyed at very high pressure (> 800 MPa) around ambient temperature. To overcome the use of very high pressure for inactivation of spores, temperature can be increased to 50-70°C or perhaps can be decreased to 0°C or even below.*Clostridium botulinum* is considered as most heat resistant bacteria and its spores are one of the most pressure resistant microorganisms known(Gould, 1995). Since vegetative cells are more sensitive to heat or pressure treatment and alternative practice can be followed whereby the spores are allowed to germinate by low pressure treatment (50-300 MPa) and then the vegetative cells obtained can be killed by suitable heat treatment or high pressure treatment (Smelt, 1998). However, pressure of about

600MPa in combination with a temperature range of 80-110°C have been used to inactivate spore forming *Bacillus cereus* (Van Opstal *et al.*, 2004).

Fungi

It includes yeasts(unicellular fungi) and moulds (hyphae producing). Yeasts consist of members of ascomycetes and imperfect fungi. As reported by Smelt (1998) pressures of 400 MPa for few minutes is sufficient for inactivation of most of the yeasts. At about 100 MPa the nuclear membrane of yeasts was destroyed and at about 400-600 MPa alterations in mitochondria and cytoplasm occurred. Also pressure of about 300- 600 MPa can inactivate most of the moulds. The ascospores are extremely resistant to high pressure (Chapman *et al.*, 2007).

Viruses

Viruses of food borne significance includes Calcivirus (group of enteric virus) and Hepatitis (hepatitis A and hepatitis E).However other viruses of concern are rota virus and human astrovirus. Due to high degree of structural diversity there is a wide range of pressure resistances (Smelt, 1998).The mode of inactivation of virus by HPP has not been fully explored, however it appears that viral envelope is one of the targets.Pressure treatment for about 275 MPa for 5 minutes can cause complete inactivation of suspensions of feline calcivirus, adenovirus and hepatitis A (Kingsley *et al.*, 2002). The destruction of envelopes in human immunodeficiency virus (HIV), Cytomegalo virus (CMV) at 300 MPa is reported by Nakagami *et al.*(1992). According to Gasper *et al.* (2002) high pressure can also induce minor changes in viral structures without disassembling the whole particle.

Prions

Prions are responsible for certain neurological disorders like bovine spongiform encephalopathy (BSE) in cattle and Creutzfeld-Jacob disease (CJD) in humans. Destruction of prions is more difficult than bacterial spores as it is reported to survive autoclaving at 134°C (Taylor,1999). However it was Brown *et al.* (2003) who reported that high pressure treatment of about 690-1200 MPa and 121-137°C reduces the infectivity of prions contained in hot dogs contaminated with prions. Pressure about 1000 to 1250 MPa at 135 to 142°C could also be applied to meat products that containother TSE agents (Cardone *et al.*, 2006). However, accordingto Heindl *et al.*(2008), TSE infectivity can already bedecreased by up to 6 to 8 log units with shorter treatments (5 min)at 800 MPa and 80°C. The authors stated that the decrease ininfectivity with this treatment was faster than prion protein degradation and that these treatments may be less aggressive than heattreatments that are known to denature prion proteins.

Effect of HPP on Enzymes

Due to ageing process post-rigor tenderization occurs within hours after slaughter and it depends upon the muscle fibre type, muscle type, species and the extent of proteolysis of key target myofibrillar proteins such as actin, myosin, tropomyosin, troponin-T, nebulin and titin, along with cytoskeletal desmin and costamere vinculin by proteolytic enzymes after death. Little or no changes are reported in connective tissue as a consequence of pressure treatment (Suzuki, 1993). Systems responsible for muscle proteindegradation are various exo- and endoproteases. The beststudied are the calpain/calpastatin system and the lysosomalcathepsins.

Calpains are the most extensively studied proteaseswith regard to ageing process, and their contribution to meattenderization is widely accepted (Koohmaraie and Geesink,2006; Sentandreu *et al.*, 2002) Calpainsconsists of cytoplasmic cysteine Ca^{2+} dependent proteases. Degradation of myofibrils by calpains has been correlatedwith post-mortem proteolysis and meat tenderization bysome authors. Little is known about the effect of HPP oncalpains. However, Cheret *et al.* (2006)reported that m and µ calpainsfrom sea bass subjected to HPP arereadily inactivated at 300 MPa, which may be due to structuralmodification and dissociation of calpain subunits. Evidencesuggests that inhibition of calpain activity by high pressureprevents the degradation of cytoskeletal proteins such asdesmin, reducing the water-holding capacity of musclethus increasing the drip loss. As rigor progresses, the space for water to be held in the myofibrils is reduced, and fluid can be forced into the extra-myofibrillar spaces where it is more easily lost as drip as a consequence of lateral shrinkage of the myofibrils during rigor; this can be transmitted to the entire cell if proteins that link myofibrils together and myofibrils to the cell membrane (such as desmin) are notdegraded (Kristensen and Purslow, 2001; Melody *et al.*, 2004).Water holding capacity is strictly related to post-mortem events such as pHdecline, proteolysis and protein oxidation (Huff-Lonergan *et al.*, 1996) and it isvery important in fish from a quality, nutritional andconsequently commercial points of view.Campus(2010a) reported areduction in degradation of desmin correlated with decreased water-holding capacity in HPP treated (300 and 400 MPa) sea bream muscle.Davis *et al.* (2004) also showed similar results in enhanced pork loins, in which reduced degradationof desmin was correlated with lower retention of fluids by muscle. Cathepsins, a group of enzymes including both exoandendo peptidases known to be expressed inmuscle tissue include six cysteine peptidases (cathepsins B,L, H, S, F and K) and one aspartic peptidase (cathepsin D) according to Sentandreu *et al.* (2002). They are mostly activeat acidic pH. Being located inside lysosomes, their role inmeat tenderization is still a subject of debate, although Calkins and Seidman (1988) and Johnson *et al.* (1990) correlated free cathepsinactivity (especially that of cathepsins B, H and L) with meat tenderness from the early post-mortemperiod to the end of the ageing period.Afterdeath, lysosomal enzymes become accessible to musclestructural proteins due to the progressive disruptionof lysosomes through membrane breakdown, which occursby the decrease in pH at high post-mortem temperature(Moeller *et al.*, 1977) or by the failure of ion pumps in lysosomal membranesduring rigor development following the depletionof ATP stores (Hopkins,2000). The application of high pressure to fresh meat enhances the endogenous cathepsin proteolytic activity involved in meat conditioning (Ohmori *et al.*, 1991).Kurth (1986) reportedthe maintenance of cathepsin B activity (in solution) under pressure (150 MPa) and even an increase in some pressure–heat combinations. As reported by Homma *et al.* (1994) anincreased activity of cathepsins B and L and inactivation of aminopeptidase B (also called RAP) and cathepsin H, an aminoendopeptidase in bovine muscle treated at pressureof about 100–500 MPa for5 min.Campus *et al.*(2008) observed reduced activity of cathepsins B and Lat pressure of 400 MPa, along with the loss of activity ofother enzymes (aminopeptidases and dipeptidylpeptidases).Proteolytic enzymes related to fish spoilage are generallymore susceptible to HPP than their mammaliancounterparts, since fish are adapted to cold habitats, andtheir enzymes tend to have a more flexible structure (Low and Somero, 1974). HPP of sheep'shead and bluefish cathepsin C at 300 MPa for 30 min resulted innear total inactivation whereas the treatment had little or no effect onbovine cathepsin C (Campus, 2010). In conclusion, the effect of high pressure on enzymeactivity will depend on the enzyme itself, on the nature ofthe medium (substrate availability, pH, ionic strength, etc.) and on the processing conditions (pressure, temperatureand time), and this will affect the texture and taste of theproduct (Campus, 2010).

Effect of HPP on Quality Attributes of Meat

Colour

It is an important sensory attribute that forms the basis of acceptance of meat products by the consumers.It is a known fact that high pressure denatures some protein resulting in change in physical functionality and rawmeat's color. An increase in lightness (*L*″), for pressures above

200 MPa is the most often reported modification of raw meat color (Korzeniowski *et al.*, 1999; Beltran *et al.*, 2004a;DelOlmo *et al.*, 2010; Marcos *et al.*, 2010). The increase in whitening effect due to*L*″ results has been observed in chickenmeat treated at 400 to 500 MPa and 5 to 10°C (Beltran *et al.*, 2004a; Del Olmo *et al.*, 2010), in pork meat treated at 200to 400 MPa and 20°C (Korzeniowski *et al.*, 1999), and inbeef meat treated at 200 to 600 MPa and 10°C as reported by Carlez *et al.*(1995) and Marcos *et al.* (2010). This whitening effect may be due to two reasons first being protein coagulation with a resulting loss ofsolubility of sarcoplasmic and/or myofibrillar proteins that affectstructure and surface properties according to Goutefongea *et al.* (1995); secondly it may be because of globin denaturation and heme group displacement or releaseas reported by Carlez *et al.* (1995). In beefmuscle, a significant effect of highpressure on the (redness)*a*″ value has also been observed. Jung *et al.*(2003) studied the effect of pressure treatments (50 to 600 MPa and 10°C for 5 min) on raw beef muscle and reported an increase in *a*″ values at pressureup to approximately 350 MPa and decrease in *a*″ values of beef meat at pressures up to 600 MPa. as also reported by Marcos *et al.* (2010) and Carlez *et al.* (1995).The increase in *a*″ values at low pressures(300 MPa)was attributed to the possible activation of the enzymatic systemresponsible for metmyoglobin reduction. The decrease in *a*″ valuesat pressures above 350 to 400 MPa was linked to the oxidation of ferrous myoglobin to ferric metmyoglobin (Carlez *et al.*, 1995; Jung *et al.*, 2003; Cava *et al.*, 2009) and possibly to further denatured myoglobin ferric species (Wackerbarth *et al.*, 2009).Meat discoloration is significantly, but only slightly, influencedby the treatment duration and can be observed after only 1 minof pressure exposure (Jung *et al.*, 2003; Del Olmo *et al.*, 2010). The color of cured products is less affected by pressure thanis raw meat's color (Karlowski *et al.*, 2002; Rubio *et al.*, 2007)however; Andr´es *et al.* (2004) reported a decrease in redness in dry-cured ham above 200 MPa. Goutefongea *et al.* (1995)

recommended cooking before high-pressure treatment to avoid color changes in cured meatproducts. Resistance of thenitrosylmyoglobin pigment to oxidation is probably the reason forthe color stability of cured meat products containing nitrates or nitrites(Goutefongea *et al.*, 1995; Farkas *et al.*, 2002; Pietrzak *et al.*,2007). HPP of white meat (or fish) does not cause any serious colour problems (Cheftel and Culioli, 1997).HPP of raw meat imparts cooked appearance and it cannot be sold as fresh meat. However, packaging of meat under vacuum with oxygen scavenger partly protects the meat colour.Recent advances in the use of HPP to improve thequality of cold-smoked salmon were reviewed byLakshmanan *et al.* (2003), although a lot of substantial amount ofwork needs to be done before we conclude that HPP is usefulin improving the quality of cold-smoked fish withoutaffecting its sensory profile.

Aroma and Taste

The effect of high pressure treatment on meat aroma has rarely been investigated. Schindler *et al.* (2010) studied the effects of high-pressure processing(400 to 600 MPa for 15 min at 5°C) on the aromaprofiles of raw meat (chicken and beef) by measuring the volatilecompounds that were released upon opening the package and bysensory evaluation of the odour. The authors found that pressure treatment of vacuum-packed beefand chicken below 600 MPa does not induce significant changesin the raw aroma profile of either meat type during 14 d of chilledstorage as compared to fresh untreated meat. Similar results were found by Rivas-Ca˜nedo *et al.* (2011) who evaluated the volatile profile of high pressuretreated(400 to 600 MPa, 12°C, 5 to 10 min) cooked pork meat.The presenceof amino acids and peptides (responsible for maillard reaction) is linked to the development of meat flavor. Campus *et al.* (2008) showed that high pressure of about 300 to 400 MPa and 20°C for 10 min reduced the content of volatile compoundsoriginating from the Maillard reaction in dry-cured loin. Thus, high pressure may influence the taste development ofmeat. Ohmori *et al.* (1991) reported an increase in overall autolytic activity of raw meat treated at high pressure between 100 and 300 MPa for 10 min at 25°C resulting in higher concentration of free amino acids.However, with higher pressure treatments at 400 MPa and 500 MPa, the concentration of free amino acids was identical tothat of the control sample during one week of chilled storage.Campus *et al.*(2008) showed that highpressure treatment (300 to 400 MPa for 10 min at 20°C), can stabilize the free amino acid content during storage of dry-cured loins due to reduction in theactivity of aminopeptidases. According to Suzuki *et al.* (1994), high pressure treatments (200 to 400 MPa at ambient temperature) do not influence subsequent changes in the componentsresponsible for the flavor of meat (the amount of amino acids, peptides,and nucleotides) during 7 d of chilled storage. It is customary in Japanthat the production of inosinic monophosphate (IMP) in meat is not delayed by the pressure treatment just after slaughter. IMP is one of the components responsible for "umami" taste. Mori *et al.* (2007) found that the IMP content in rabbit muscles after slaughter was instantly increased by a high-pressure treatment of 300 MPa for 5 min at 4°C and that adenosine triphosphate deaminase activitywas maintained at 70 per cent of its initial activity.Lower pressures had almost no effect on IMP production, and the authors concludedthat high pressure does not appear to reduce desirable tastecharacteristics in comparison with conditionedmeat. Saltiness,another taste characteristic is also modified by high-pressure treatment.As reported by Saccani *et al.*, 2004; Fulladosa *et al.*, 2009; Clariana *et al.*, 2011, the perception of saltiness detected by a trained panel wasincreased in high-pressure treated dry-cured ham compared to untreated products possibly due to modified interactionsbetween Na+ and proteins.

Crehan *et al.* (2000) mentioned that salt content can be decreased from 2.5 per cent to 1.5 per cent in frankfurter sausages, with an improvement in overall acceptability, by subjecting the meat batters to high pressure at 150 MPa before manufacture and cooking. In low-salt beef sausage with 1 per cent NaCl, all of the sensory attributes evaluated with a trained panel were improved by using batter that had been submitted to a high-pressure treatment at 200 MPa for 2 min at 10°C before cooking (Sikes *et al.*, 2009).

Sensory Acceptance

Sensory acceptance of high-pressure-treated meat products depends on color, texture, aroma, and taste modifications induced by the process. Problems of sensory acceptance occur with raw products, mainly because of visible color changes. Cooked and cured products are less modified by pressure, and good sensory acceptance is generally reported despite some changes in aroma and taste profiles (Simonin, 2012).

Hayman *et al.*(2004) compared consumer acceptance of untreated(7 d of chilled storage) and pressure-treated (600 MPa, 20°Cfor 80 s with 98 day of chilled storage) commercial ready-to-eatbeef products *i.e.* Strassburg beef and Cajun beef. The authors found that no difference in consumer acceptance between untreated productsat day 7 of chilled storage and pressure-treated products after98 day of chilled storage was reported. Rivas-Canedo *et al.* (2011) studied thesensory acceptance of the odor of high-pressure-treated (400 to 600 MPa and 12°C, 5 to 10 min) cooked pork meat and showedthat odor acceptability decreased in the same way in both control andpressure-treated samples during refrigerated storage.Uenaka *et al.*(2006) did sensory evaluation of raw visceral meat (bovine liver,chicken liver and gizzards) treated at 400 MPa at roomtemperature for 30 min without losing its tenderness and taste and he reported deterioration in these attributes for treatmentat 500 MPa.High pressure (e" 500 MPa and 50 to 65°C) treated cooked sausageswere generally preferredto conventionally heat-pasteurized products (80 to 85°Cfor 40 min) because of their better appearance, taste, and especially texture (Mor-Mur and Yuste, 2003; Chattong and Apichartsrangkoon, 2009).

Meat Tenderization and Texture

In early times of the invention of HPP it was mostly used for meat tenderization. Therefore, we can find number of documentations related to effect of HPP on meat tenderization. It is well known that tenderization of fresh meat occurs due to the following changes in the muscle during conditioning (mainly as a result of the activity of endogenous proteases): weakening of actin–myosin interactions, fragmentation of myofibrils into short segments as a result of Z line disintegration, degradation of elastic filaments consisting of connectin and the weakening of connective tissue (Koohmaraie, 1994). It was reported by Macfarlane (1973) and Bouton *et al.* (1977) that pre-rigor high pressure treatments of muscles at approximately 100 MPa and

30°C generally result in substantial shortening of the muscle (approximately 35 per cent) and an improvement in tenderness after cooking. According to Locker *et al.* (1960); Davey *et al.* (1967) such a degree of shortening (35 per cent) would be expected to result in considerable toughening. The improved tenderness has been suggested to be linked to the effect of pressure on the contraction state of the muscle (Macfarlane, 1973). In fact, the physical disruption of sarcoplasmic reticulum membranes under pressure leads to an increase in cytosolic Ca^{2+} (Okamoto *et al.*, 1995) which results in intense muscle contraction, an acceleration of postmortem glycolysis and a rapid pH decrease (Macfarlane, 1973) due to activation of Ca^{2+}dependent phosphorylases involved in the regulation of glycogen breakdown (Horgan and Kuypers, 1983). Macfarlane (1973) explained the basis of tenderizing effect, according to him the combining effects of muscle contraction and pressure during the treatment lead to breakage of myofibrillar structure, forcing myosin filaments of severely contracted muscles into Z discs, resulting in tenderization of meat. Tenderization has also been attributed to the activation of two enzymatic systems involved in the tenderization of meat during aging, namely cathepsins and calpains. Cathepsins are released from lysosomes when the muscle is high-pressure treated at 100 MPa just after animal death, and they can be absorbed rapidly by the myofibrils (Kubo *et al.*, 2002). Calpain activity in pressuretreated muscle is increased by pressure up to 200 MPa due to the activation of the calpain proteinase system by Ca^{2+} released from the sarcoplasmic reticulum and to the inactivation of the inhibitor calpastatin under pressure (Homma *et al.*, 1996). As is the case for pre-rigor meat, ithas been reported that pressure induces the release of cathepsinsfrom lysosomes in post-rigor muscles and provides a gradual increase in activity with increasing pressure up to 400 MPa (Homma *et al.*,1994). Jung *et al.*(2000) also observed an increasein cathepsin D activity related to the breakdown of lysosomalmembranes in post-rigor beef muscle treated at 520 MPa for260 s in comparison with untreated muscle. The activity of all enzymatic systems involved in meat tenderizationmay also be modulated by pH variation. ThesepH changes may also contribute to the influence of pressure on enzymatic activity in meat; for example, the drop in pH under pressure could contribute to the activation of cathepsins, which have optimal pH below 5.2 (Faustman, 1994). However, although some key enzymes in the tenderization process can be activated by high pressure at or below room temperature in post-rigor meat, this activation is not necessarily accompanied by a decrease in meat toughness as observed in pre-rigor meat. In the work of Jung *et al.*(2000), the activation of catheptic activity at 560 MPa had no conclusive effect on the tenderness of meat; instead, an increase in the shear force was observed in comparison with nontreated samples. Ma and Ledward (2004) also reported increase in meat hardness after high-pressure treatment at or above 200 MPa at 20°C in beef muscle and turkey breast fillets (Del Olmo *et al.*, 2010). They had reported in 2004 that there was no improvement in tenderness observed after cooking high pressure treated beef muscle at 70°C in comparison with an untreated cooked sample. It is not clear why these results differ from those previously reported, but the difference may result from the different methods used to evaluate meat tenderness. Suzuki *et al.*(1992) used a conical plunger to evaluate raw meat hardness; Schenkova *et al.*(2007) do not mention cooking before Warner-Bratzler shear force measurements. In all other mentioned studies addressing beef, Warner-Bratzler shear force measurements were performed on cooked meat. This difference in results could also be caused by variability in the aging of the high-pressure-treated meat.

However, it is finally concluded that long exposure to high pressures at moderate temperatures has been shown to be necessary to improve the meat tenderness of post-rigor muscles. Exposureof 150 to 200 MPa for 20 to 30 minutes at 60°C in beef muscleis necessary to improve tenderness (Macfarlane, 1973; Bouton *et al.*, 1977; Locker andWild, 1984; Ma and Ledward, 2004; Sikes *et al.*, 2010). This tendernessmay be attributed to increased protease activity at 50 to 60°C under pressure (in comparison with high-pressure treatments at lower temperatures (Ma and Ledward 2004; and Sikes *et al.*, 2010). However, high pressure has been shown to improve meat tenderness when applied to pre-rigor meat. In post-rigor meat, tenderization can be only achieved after long (20 to 30 min) exposure to 150 to 200 MPa at 60°C or more. It is believed to be due to the activation of some proteases involved in meat ageing. HPP performed at ambient temperatures do not improve the tenderness of post-rigor meat and it may be because of coagulation, aggregation, or gelation of sarcoplasmic proteins and myofibrils.

Meat Texture

Ashie and Simpson (1996) performed a puncture test and reported adecrease in "strength values" when blue fish was subjectedto pressure above 200 MPa for over 10 min and also in fishtreated at pressure above 300 MPa. They alsoreported a decrease in elasticity with increasing pressurejust after treatment. A decreasein elasticity of sea bream muscle with increasing pressurejust after treatment was reported by Campus *et al.* (2010b), but the elasticity was maintained duringstorage in samples treated at higher pressures. This is attributed to reduced degradation ofcytoskeletal proteins (assayed by western blotting) due toblockade of proteolytic activity by HPP. The myofibrillar proteins and their gel forming properties are used for development of processed muscle-based food by employing pressure-induced texture modifications. Highpressure can affect molecular interactions (hydrogenbonds, hydrophobic interactions and electrostatic bonds)and protein conformation, leading to protein denaturation,aggregation or gelation (Messens *et al.*, 1997). Depending on the source ofmuscle and other parameters such as protein concentration,pH and ionic strength, various changes occur in myofibrillarproteins depending on the pressure–temperature conditions(Acton and Dick, 1984; Hamm,1981;Sano *et al.*, 1994). Jime´nez Colmenero (2002) reviewed pressure induced gelation of muscle protein systems. According to Okazaki *et al.* (1997) pressure treatment of 100–500 MPa for 10 minute at 0°C hasbeen found to favour the formation of structures withgreater breaking strength in gels from fish meat mince,Alaska pollack surimi (above 200 MPa) and minced chumsalmon meat.Ashie and Lanier (1999) analyzed and implemented texturizing effects of HPP to increase the gel strength of uncooked surimi by two- to threefold by making protein substrates more accessible to transglutaminase, which increases intermolecular cross-link formation and gel strength. These improvements in texturalcharacteristics have created a high demand for HPP products from both the food service and retail sectors.Other potential uses of HPP technology that may beapplicable to the seafood industry include pressure-assistedfreezing (Kalichevsky *et al.*, 1995) and HPP thawing(Murakami *et al.*, 1992; Rouille´ *et al.*, 2002; Schubring *et al.*, 2003).

Effect of HPP on Nutrients

Unlike thermal processing, HPP has very little effect on flavour, vitamins and pigments asit modifies only non covalent bonds and does not affect low molecular weight compounds (Simonin, 2012). The effect of high pressure *i.e.* 600 MPa in combination with different elevated temperatures on broccolli, tomato and carrot was studied by Butz *et al.* (2002).His findings revealed that the broccoli treated with high pressure for 40 minutes at 75°C showed no loss of cholorophyll a orcholorophyll b as compared to untreated samples. Similarly tomatoes treated at ahigh pressure at 25°C for 60 min, did not show a change in either lycopene or carotenoid content and were similar to heat treated samples 95°C for 60 minutes. In addition the loss in antioxidant capacity of a water soluble portion of pressure treated carrot and tomato (500 - 800 MPa for 5 minutes) was very little compared to untreated samples. Quaglia *et al.* (1996) observed high retention of ascorbic acid (82 per cent) in green peas treated with 900MPa for 5 - 10 minutes at 20°C. Water soluble vitamins (Vitamin B complex and Vitamin C), contained in a model multivitamin system and natural food systems rich in ascorbic acid were subjected to high pressures of 200 – 600MPa for 30 minutes to determine the effect of HPP on vitamin retention. It was observed that ascorbic acid losses were about 12 per cent in model multivitamin system where as the loss was insignificant in food system. Sancho *et al.* (1999) also proposed that thiamine and pryridoxal in model system were unaffected by HPP. Thus, HPP retained vitamins better than conventional methods of sterilization.

Advantages of High Pressure Processing (HPP)

1. HPP is attractive for consumers, meets demand for freshness and minimal processing and also helps in avoidingirradiations, chemical additivesoruse of high temperatures.
2. HPP is suitable for low, medium and even high moisture products and canbe modified for both batch processing and semi-continuousprocessing.
3. HPP can be used to process rawproduct without significantly altering their flavor, texture orappearance.
4. HPPis preferred for packed food stuff as it withstands a change in volume up to 15 per cent followed by a return to its original shapewithout losing seal integrity or barrier properties.
5. HPPis a reliable means of preservation which inactivates spoilage causing organisms and enzymes, germinate or inactivate some bacterial spores, extend shelf life, reduce the potential for food borne illness, promoteripening of cheese and minimize oxidativebrowning.
6. HPPis applied evenly hence does not cause any change in geometry of food.
7. HPP reduces the processing time,retains freshness, flavor, texture, appearance and colorin food. Besides it also prevents vitamin C loss, reduces ice crystal damage andreduces functionally alteration that usually occurs in traditional thermal processing.

Limitation of High Pressure Processing

1. HPP is an expensive technique, requires high capital.
2. HPP system requires extremely high precision in its construction, useand maintenance.
3. HPP unit immediately becomes rate limiting steps in mostprocessing operation.
4. Inability of HPP to inactivate spores without combination of high temperature.

References

Abe, F. (2007) Exploration of the effects of high hydrostatic pressure on microbial growth, physiology and survival: perspectives from piezophysiology.*Biosci BiotechBiochem*, **71**(10): 2347–57.

Acton, J.C and Dick, R.L. (1984) Protein-protein interaction in processed meats. *RecipMeat Conf Proc*, **37**: 36–42.

Andr´es, A.I., Møller, J.K.S., Adamsen, C.E. and Skibsted, L.H. (2004) High pressure treatment ofdry-cured Iberian ham. Effect on radical formation, lipid oxidation and colour.*EuroFood Res Technol*, **219**(3): 205–10.

Ashie, I.N. and Simpson, B.K. (1996) Application of high hydrostatic pressure to control enzyme related seafood texture deterioration.*Food Res Int*, **29**: 5–6.

Ashie, I.N.and Lanier, T.C.(1999) High pressure effects on gelation of surimi and turkey breast muscle enhanced by microbial transglutaminase. *J Food Sci*, **64**: 704–708.

Aymerich, T., Jofre, A., Garriga, M.and Hugas, M.(2005) Inhibition of *Listeria monocytogenes* and *Salmonella* by natural antimicrobials and high hydrostaticpressure in sliced cooked ham. *J Food Prot*, **68**(1): 173–7.

Balasubramaniam, V.M., Farkas, D. and Turek, E. J. (2011) Review on preserving foods through high pressure processing. *J Food Tec*, **8**: 32 – 38.

Beltran, E., Pla, R., Capellas, M., Yuste, J.and Mor-Mur, M. (2004) Lipid oxidation and colour in pressure and heat-treated minced chicken thighs. *J Sci Food Agri*, **84**(11): 1285–9.

Bouton, P.E., Ford, A.L., Harris, P.V., Macfarlane, J.J.and O'Shea, J.M. (1977) Pressure-heat treatment of postrigor muscle: effect on tenderness *J Food Sci*, **42**(1): 132–5.

Brown, P., Meyer, R., Cardone, F.and Pocchiari, M.(2003) Ultra-high-pressure inactivation of prion infectivity in processed meat: a practical method to prevent human infection.*Procat NatAcad Sci USA***100**(10): 6093–7.

Butz. P., Edenharder, R., Garcia, A.F., Fister, H., Merkel, C.and Tauscher, B. (2002) Changes in functional properties of vegetables induced by high pressure treatment. *Food res Int*, **35**: 295-300.

Calkins, C.R., and Seidman, S.C. (1988) Relationship among calcium dependent protease, cathepsin B and H, meat tenderness and the response of muscle to aging. *J Ani Sci*, **66**: 1186–1193.

Campus, M., Flores, M., Martinez, A.and Toldra, F. (2008) Effect of high pressure treatment on colour, microbial and chemical characteristics of dry cured loin.*Meat Sci*, **80**(4): 1174–81.

Campus, M. (2010a) Review on High Pressure Processing of Meat, Meat Products and Seafood. *Food Eng Rev*, **2**: 256–273.

Campus, M., Addis, M.F., Cappuccinelli, R., Porcu, M.C., Pretti, L., Tedde, V., Secchi, N., Stara, G.and Roggio, T. (2010b) Stress relaxation behaviour and structural changes of muscle tissues from Gilthead Sea Bream (Sparus aurata L.) following high pressure treatment. *J Food Eng*, **96**: 192–198.

Cardone, F., Brown, P., Meyer, R., and Pocchiari, M. (2006) Inactivation of transmissible spongiform encephalopathy agents in food products by ultra-high pressure-temperature treatment.*BBA—Proteins Proteo*, **1764**(3): 558–62.

Carlez, A., Rosec, J.P., Richard, N., and Cheftel, J.C. (1994) Bacterial growth during chilled storage of pressure-treated minced meat.*LWT-Food Sci Technol***27**(1): 48–54.

Carlez, A., Veciana-Nogues, T., and Cheftel, J.C. (1995) Changes in colour andmyoglobin of minced beef meat due to high pressure processing.*LWT-FoodSci Technol*, **28**(5): 528–38.

Cava, R., Ladero, L., Gonz´alez, S., Carrasco, A., and Ram´ýrez, M. R. (2009) Effect of pressure and holding time on colour, protein and lipid oxidation of sliced dry-cured Iberian ham and loin during refrigerated storage. *Innov Food Sci Emerg Technol*, **10**(1): 76–81.

Chapman, B., Winley, E., Fong, A.S.W., Hocking, A.D., Stewart, C.M., and Buckle, K.A. (2007) Ascospore inactivation and germination by high-pressure processing is affected by ascospore age. *Inno Food Sci EmergTechnol*, **8**(4): 531–4.

Chattong, U.and Apichartsrangkoon, A. (2009) Dynamic viscoelastic characterisation of ostrich-meat yor (Thai sausage) following pressure, temperature and holding time regimes.*Meat Sci*, **81**(3): 426–32.

Cheftel, J.C. (1995) Review: High-pressure microbial inactivation and food preservation. *Food Sci Technol Int*, **1**: 75 – 90.

Cheftel, J.C.and Culioli, J. (1997) Review on Effects of high pressure on meat. *Meat Sci*, **46**(3): 211–36.

Chen, H.Q. (2007) Temperature-assisted pressure inactivation of *Listeria monocytogenes* in turkey breast meat. *Int JFood Micro*, **117**(1): 55–60.

Cheret, R., Ara´nzazu, H.A., Delbarre-Ladrat, C., Lambarrie, M.and Verrez-Bagnis, V. (2006) Proteins and proteolytic activity changes during refrigerated storage in sea bass (Dicentrarchus labrax L.) muscle after high-pressure treatment. *Euro Food ResTechnol*, **222**: 527–535.

Clariana, M., Guerrero, L., Sa´rraga, C., D´ýaz, I., Valero, A.and Garc´ýa-Regueiro, J.A. (2011) Influence of high pressure application on the nutritional, sensory and microbiological characteristics of sliced skin vacuum-packed dry-cured ham. *Inno Food Sci Emerg Technol*, **12**(4): 456–65.

Cook, D.W. (2003) Sensitivity of Vibrio species in phosphate buffered saline and in oysters to high hydrostatic pressure processing. *J Food Prot*, **66**: 2277–2282.

Crehan, C.M., Troy, D.J., and Buckley, D.J. (2000) Effects of salt level and high hydrostatic pressure processing on frankfurters formulated with 1.5 and 2.5 per cent salt. *Meat Sci*, **55**(1): 123–30.

Davis, K.J., Sebranek, J.G., Huff-Lonergan, E., and Lonergan, S.M. (2004) The effects of aging on moisture-enhanced pork loins. *Meat Sci*, **66**: 519–524.

Davey, C.L., Kuttel, H., and Gilbert, K.V. (1967) Shortening as a factor in meat ageing. *Int J Food SciTechnol*, **2**(1): 53–6.

Del, Olmo, A., Morales, P., A´ vila, M., Calzada, J., and Nun˜ez, M. (2010) Effect ofsingle-cycle and multiple-cycle high-pressure treatments on the colour andtexture of chicken breast fillets. *Inn Food SciEmergTechnol*, **11**(3): 441–4.

Diez, A. M., Santos, E.M., Jaime, I., and Rovira, J. (2008) Application of organic acid salts and high-pressure treatments to improve the preservation of blood sausage. *Food Micro*, **25**(1): 154–61.

Dunne, C.P. (2005) High pressure keeps food fresher. Available at http: //www. natick.army.mil/about/pao/05/05-22.htm. Accessed October 8, 2008.

Escriu, R.and Mor-Mur, M. (2009) Role of quantity and quality of fat in meat models inoculated with *Listeria innocua* or *Salmonella Typhimurium* treated by high pressure and refrigerated stored. *Food Micro*, **26**(8): 834–40.

Farkas, J., Hajos, G., Szerdahelyi, E., Andrassy, E., Krommer, J., and Meszaeos, L. (2002) Protein changes in high pressure pressurized raw sausage batter. *Proc Int Cong Meat SciTechnol*, **25**–30 August 2002, Rome, Italy: p 180–91.

Faustman, L. (1994) Postmortem changes in muscle food. In: Kinsman, D.M., Kotula, A.W., Breidenstein, D.C., editors. *Mus foods: meat, poultry, and seafood technol*, London: Chapman and Hall. p 63–78.

Fulladosa, E., Serra, X., Gou, P.and Arnau, J. (2009) Effects of potassium lactate and high pressure on transglutaminase restructured dry-cured hams with reduced salt content. *Meat Sci*, **82**(2): 213–8.

Garriga, M., Aymerich, M. T., Costa, S., Monfort, J. M., and Hugas, M. (2002) Bactericidal synergism through bacteriocins and high pressure in a meat model system during storage. *Food Micro*, **19**(5): 509–18.

Garriga, M., Gr'ebol, N., Aymerich, M. T, Monfort, J. M., and Hugas, M. (2004) Microbial inactivation after high-pressure processing at 600 MPa in commercial meat products over its shelf life.*Innov Food Sci Emerg Technol*, **5**(4): 451–7.

Gaspar, L.P., Silva, A.C.B., and Gomes, A.M.O (2002) Hydrostatic induces the fusion-active state of enveloped viruses. *J Bio Chem*. **277**: 8433 8439.

Gola, S., Mutti, P., Manganelli, E., Squarcina, N., and Rovere, P. (2000) Behaviour of *E. coli* 0157: H7 strains in model system and in raw meat by HPP: microbial and *technol*nological aspects. *High Press Res***19**(1): 91–7.

Gould, G.W. (1995) The microbe as a high pressure target. In High Pressure Processing of Foods (Ledward DA, Johnston DE, Earnshaw RG, Hasting APM, eds). Louughborough: *Nottingham Uni Pre*, pp. 27 – 36.

Goutefongea, R., Rampon, V., Nicolas, J., and Dumont, J.P. (1995) Meat color changes under high pressure treatment. *Proc 41st Int Cong Meat Sci Technol*, San Antonio, Texas, 20–25August 1995: p 384–5.

Hamm, R. (1981) Post-mortem changes in muscle affecting the quality of comminuted meat products. In: Lawrie R (ed) Development in meat science. *Else App Sci*, London, pp 93–124.

Hayman, M.M., Baxter, I., O'Riordan, P.J., and Stewart, C.M. (2004) Effects of high-pressure processing on the safety, quality, and shelf life of ready-to-eat meats. *J Food Prot*, **67**(8): 1709–18.

Heindl, P., Garcia, A.F., Butz, P., Trierweiler, B., Voigt, H., Pfaff, E., and Tauscher, B. (2008) High pressure/temperature treatments to inactivate highly infectious prion subpopulations. *Inn Food Sci Emerg Technol*, **9**(3): 290–7.

Hite, B. (1899)The effect of pressure in the preservation of milk. *West Virginia AgriExp Stat Bul*, **58**: 15 – 35.

Homma, N., Ikeuchi, Y., and Suzuki.(1994) Effects of high pressure treatment on the proteolytic enzymes in meat.*Meat Sci*, **38**: 219–228.

Homma, N., Ikeuchi, Y., Suzuki, A., Hayashi, R., and Balny, C. (1996) Effect of high pressure treatment on proteolytic system in meat. In: Hayashi R, Balny C, editors. *Progress in Biotechnol***13**: high pressure bioscience and biotechnology. *ProcInt Conf*on High Pressure Bioscience and Biotechnology, Kyoto, Japan: *Elsevier* p 327–30.

Hopkins, D.L. (2000) The relationship between actomyosin, proteolysis and tenderisation examined using protease inhibitors. PhD thesis, University of New England, Australia.

Horgan, D.J., and Kuypers, R. (1983) Effect of high-pressure on the regulation of phosphorylase-activity in pre-rigor rabbit muscle *Meat Sci*, **8**(1): 65–77.

Huff-Lonergan, E., Mitsuhashi, T., Beekman, D.D., Parrish, F.C., Olson, D.G., and Robson, M.(1996) Proteolysis of specific muscle structural proteins by mu-calpain at low pH and temperature issimilar to degradation in postmortem bovine muscle. *J AniSci*, **74**: 993–1008.

Jime´nez Colmenero, F. (2002) Muscle protein gelation by combined use of high pressure/temperature.*Trends Food Sci Technol*, **13**: 22–30.

Jofr´e, A., Garriga, M.and Aymerich, T. (2008) Inhibition of Salmonella spp., Listeria monocytogenes and Staphylococcus aureus in cooked ham by combining antimicrobials, high hydrostatic pressure and refrigeration. *Meat Sci*, **78**(1–2): 53–9.

Jofr´e, A., Aymerich, T., Grebol, N.and Garriga, M.(2009) Efficiency of high hydrostatic pressure at 600 MPa against food-borne microorganisms by challenge tests on convenience meat products. *LWT-Food Sci Technol*, **42**(5): 924–8.

Johnson, M. H., Calkins, C. R., Huffman, R. D., Johnson, D. D, and Hargrove, D. D. (1990) Differences in cathepsin B+L and calcium-dependent protease activities among breed types and their relationship to beef tenderness. *J Ani Sci*, **68**: 2371–2379.

Jung, S., Ghoul, M., and de Lamballerie-Anton, M. (2000) Changes in lysosomal enzyme activities and shear values of high-pressure-treated meat during ageing.*Meat Sci*, **56**(3): 239–46.

Jung, S., Ghoul, M., and de Lamballerie-Anton, M. (2003) Influence of high pressure on the color and microbial quality of beef meat.*LWT-Food Sci Technol*, **36**(6): 625–31.

Kadam, P. S., Jadhav, B. A., Salve, R. V., and Machewad, G. M. (2012) Review on the High Pressure *Technol*nology (HPT) for Food Preservation. *J Food ProcTechnol*, **3**: 135.

Kalichevsky, M. T., Knorr, D., and Lilliford, P. J. (1995) Potential food applications of high-pressure effects on ice-water transitions.*Trends Food Sci Technol*, **6**: 253–259.

Karlowski, K., Windyga, B., Fonberg-Broczek, M., Sciezynska, H., Grochowska, A., Gorecka, K., Mroczek, J., Grochalska, D., Barabasz, A., Arabas, J., Szczepek, J., and Porowski, S. (2002) Effects of high pressure treatment on the microbiological quality, texture and colour of vacuum-packed pork meat products. *High Press Res*, **22**(3): 725–32.

Kingsley, D. H., Hoover, D. J., Papafragkou, E., and Richards, G. P. (2002) Inactivation of hepatitis A virus and calicivirus by high hydrostatic pressure. *J Food Prot*, **65**: 1605–1609.

Knorr, D. (1993) Effects of high-hydrostatic-pressure processes on food safety and quality.*Food Technol*, **47**(6): 156–61.

Koohmarai, M. (1994) Muscle proteinases and meat ageing. *Meat Sci*, **36**(1–2): 93–104.

Koohmaraie, M.and Geesink, G.H. (2006) Contribution of postmortem muscle biochemistry to the delivery of consistent meat quality with particular focus on the calpain system.*Meat Sci*, **74**: 34–43.

Korzeniowski W, Jankowska B and Kwiatkowska A. (1999) The effect of high pressure on some technological properties of pork. *Electron J Pol Agric Univ*, **2**(2): 1–8.

Koseki, S., Mizuno, Y., and Yamamoto, K. (2007) Predictive modelling of the recovery of *Listeria monocytogenes* on sliced cooked ham after high-pressure processing. *Int J Food Micro*, **119**(3): 300–7.

Kristensen, L., and Purslow, P. P. (2001) The effect of ageing on the water-holding capacity of pork: role of cytoskeletal proteins. *Meat Sci*, **58**: 17–23.

Kubo, T., Gerelt, B., Han, G. D, Sugiyama, T., Nishiumi, T., and Suzuki, A. (2002) Changes in immunoelectron microscopic localization of cathepsin D in muscle induced by conditioning or high-pressure treatment. *Meat Sci*, **61**(4): 415–8.

Kural, A. G, Shearer, A. E. H., Kingsley, D. H.and Chen, H. (2008) Conditions for high pressure inactivation of Vibrio parahaemolyticus in oysters. *Int J Food Micro*, **127**: 1–5.

Kurth, L. B. (1986) Effect of pressure-heat treatment on cathepsin B1 activity. *J Food Sci*, **51**: 663–664.

Lakshmanan, R, Piggott, J. R., and Paterson, A. (2003) Potential applications of high pressure for improvement in salmon quality.*Trends Food Sci Technol*, **14**: 354–363.

Locker, R. H. (1960) Degree of muscular contraction as a factor in tenderness of beef.*J Food Sci*, **25**(2): 304–7.

Locker, R. H., and Wild, D. J. C. (1984) Tenderization of meat by pressure-heat involves weakening of the gap filaments in the myofibril. *Meat Sci*, **10**(3): 207–33.

L´opez-Caballero, M.E., Carballo, J., and Jim´enez-Colmenero, F. (1999) Microbiological changes in pressurized, prepackaged sliced cooked ham. *J Food Prot*, **62**(12): 1411–5.

Lopez-Caballero, M. E., Perez-Mateos, M., Montero, P.and Borderias, A. J. (2000) Oyster preservation by high-pressure treatment.*J Food Prot*, **63**: 196–201.

Lori, S., Buckow, R., Knorr, D., Heinz, V. and Lehmacheri, A. (2007) Predictive model for inactivation of *Campylobacter* spp. by heat and high hydrostatic pressure. J Food Prot, 70(9): 2023-2029.

L´opez-Caballero, M., Carballo, J.and Jim´enez-Colmenero, R. (2002) Microbial inactivation in meat products by pressure/temperature processing.*J Food Sci*, **67**(2): 797–801.

Low, P. S.and Somero, G. N. (1974) Temperature adaptation of enzymes.A proposed molecular basis for the different catalytic efficiencies of enzymes from ectotherms and endotherms.*CompBiochem Physio*, **49**: 307–312.

Ludwig, H.and Schreck, C. (1997) The inactivation of vegetative bacteria by pressure. In high Pressure Research in the Biosciences and Biotechnology (Heremans K, ed.). Leuven: Leuven University press.

Ma, H. J.and Ledward, D. A. (2004) High pressure/thermal treatment effects on the texture of beef muscle. *Meat Sci*, **68**(3): 347–55.

Mark de Boevere (2010) Personal communication. NC Hyperbaric, Spain.

Macfarlane, J. J. (1973) Pre-rigor pressurization of muscle: effect on pH, shear value and taste panel assessment. *J Food Sci*, **38**: 294–298.

Marcos, B., Aymerich, T., Monfort, J. M.and Garriga, M. (2008) High-pressure processing and antimicrobial biodegradable packaging to control *Listeriamonocytogenes* during storage of cooked ham. *Food Micro*, **25**(1): 177–82.

Marcos, B., Kerry, J. P.and Mullen, A. M. (2010) High-pressure-induced changes on sarcoplasmic protein fraction and quality indicators. *Meat Sci,* **85**(1): 115–20.

Martinez-Rodriguez, A.and Mackey, B. M. (2005) Factors affecting the pressure resistance of some *Campylobacter* species. *Lett Appl Micro,* **41**(4): 321–6.

Mc Clements, J.M.J., Patterson, M.F. and Linton, M. (2001) The effect of growth stage and growth temperature on high hydrostatic pressure inactivation of some pychrotrophic bacteria in milk. *J Food Prot,* **64**(4), 512 – 522.

Melody, J. L., Lonergan, S. M., Rowe, L. J., Huiatt, T. W., Mayes, M. S.and Huff-Lonergan, E. (2004) Early postmortem biochemical factors influence tenderness and water-holding capacity of three porcine muscles. *J Ani Sci,* **82**: 1195–1205.

Messens, W., Van Camp, J., and Huyghebaert, A. (1997) The use of high pressure to modify the functionality of food proteins. *Trends Food Sci Technol,* **8**: 107–112.

Moeller, P. W., Field, P.A., Dutson, T. R., Landmann, W. A., and Carpenter, Z. L. (1977) High temperature effects on lysosomal enzymes distribution and fragmentation of bovine muscle. *J Food Sci,* **42**: 510–512.

Moerman, F. (2005) High hydrostatic pressure inactivation of vegetative microorganisms, aerobic and anaerobic spores in pork Marengo, a low-acidic particulate food product.*Meat Sci,* **69**(2): 225–32.

Mor-Mur, M.and Yuste, J. (2003) High pressure processing applied to cooked sausage manufacture: physical properties and sensory analysis.*Meat Sci,* **65**(3): 1187–91.

Morales, P., Calzada, J., Avila, M.and Nunez, M. (2008) Inactivation of *Escherichia coli O157: H7* in ground beef by single-cycle and multiple-cycle high-pressuretreatments.*J Food Prot,* **71**(4): 811 -5.

Mori, S., Uchida, A., Yamamoto, S., Sultana, A., Tatsumi, R., Mizunoya, W., Suzuki, A.and Ikeuchi, Y. (2007) Effect of high pressure on the accumulation of IMP and on the stability of AMP deaminase in rabbit skeletal muscle. *J Food Biochem,* **31**(3): 328–42.

Moussa, M., Espinasse, V., Perrier-Cornet, J-M., and Gervais, P. (2009) Pressure treatment of *Saccharomyces cerevisiae* in low-moisture environments. *Appl Microbiol Biotechnol,* **85**(1): 165–174.

Murchie, L. W., Cruz-Romero, M., Kerry, J., Linton, J., Patterson, M., Smiddy, M.and Kelly, A. (2005) High pressure processing of shellfish: a review of microbiological and other quality aspects. *Innov Food Sci Emerg Technol,* **6**: 257–270.

Murakami, T., Kimura, I., Yamagishi, T., Yamashita, M., Sugimoto, M. and Satake, M. (1992) Thawing of frozen fish by hydrostatic pressure. In: Balny C, Hayashi R, Heremans K, Masson P (eds) High pressure and biotechnology. Colloque INSERM/J. Libby Eurotext Ltd, London, pp 329 331.

Nakagami, T., Ohno, H.and Shigehisa, T. (1992) Inactivation of herpes virus by high hydrostatic pressure. *J virolog methods,* **38**: 255 – 262.

Noma, S.and Hayakawa, I. (2003) Barotolerance of *Staphylococcus aureus* is increased by incubation at below 0oC prior to hydrostatic pressure treatment. *Int J Food Microbio,* **80**: 261 – 264.

Ohmori, T., Shigehisa, T., Taji, S.and Hayashi, R. (1991) Effect of high pressure on the protease activities in meat.*Agric Biol Chem*, **55**: 357–361.

Okamoto, A., Suzuki, A.and Ikeuchi, Y. M. S. (1995) Effects of high-pressure treatment on Ca^{2+}release and Ca2+ uptake of sarcoplasmic reticulum. *Biosci Biotechnol Biochem*, **59**(2): 266–70.

Okazaki, E., Ueda, T., Kusaba, R., Kamimura, S., Fukuda, Y.and Arai, K. (1997) Effect of heating on pressure-induced gel of chum salmon meat. In: Heremans K (ed) High pressure research in the bioscience and biotechnology. Leuven University Press, Leuven, Belgium, pp 371–374.

Olsson, S. (1995) Production equipment for commercial use. In high pressure processing of foods (Ledward D.A, Johnston D.E, Earnshaw R. G. Hastings A.P.M, eds). Nottingham: *Nottingham Uni Press*, Ch 12, pp. 167 – 180.

Otereo, L.and Sanz, P.D. (2000) High – pressure shift freezing. Part 1: amount of ice instantaneously formed in the process. *Biotechnol Prog*, **16**(6), 1030 – 1036.

Patterson, M. F., Quinn, M., Simpson, R., Gilmour, A. (1995) Sensitivity of vegetative pathogens to high hydrostatic pressure treatment in phosphate buffered saline and foods.*J Food Prot*, **58**(5), 524-529.

Patterson, M. F. (2005) Microbiology of pressure-treated foods. *J Appl Microbiol*, **98**(6): 1400–9.

Pietrzak, D., Fonberg-Broczek, M., Mucka, A. and Windyga, B. (2007) Effects of high-pressure treatment on the quality of cooked pork ham prepared with different levels of curing ingredients. *High Press Res*, **27**(1): 27–31.

Porto-Fett, A. C. S., Call, J. E., Shoyer, B. E., Hill, D. E., Pshebniski, C., Cocoma, G. J.and Luchansky, J. B. (2010) Evaluation of fermentation, drying, and/or high pressure processing on viability of *Listeria monocytogenes*, *Escherichia coli* O157: H7, *Salmonella* spp., and *Trichinella spiralis* in raw pork and Genoa salami. *Int J Food Microbiol*, **140**(1): 61–75.

Quaglia, G.B., Gravina, R., Paperi, R. and Paoletti, F. (1996) Effect of high pressure treatments on peroxidase activity ascorbic acid content and texture in green peas. *Lebensmittel Wissenschaft Technol*.**29**: 552 – 555.

Raghubeer, E. V. (2007) High hydrostatic pressure processing of seafood.Technical note "Seafood white paper".http: //www.avure.com/archive/documents/ Food- products/seafood-white-paper.pdf.

Rivas-Ca˜nedo, A., Juez-Ojeda, C., Nu˜nez, M.and Fern´andez-Garc´ýa, E. (2011) Effects of high-pressure processing on the volatile compounds of sliced cooked pork shoulder during refrigerated storage. *Food Chem***124(**3): 749–58.

Rouille´, J., Lebail, A., Ramaswamy, H. S.and Leclerc, L. (2002) High pressure thawing of fish andshellfish.*J Food Eng***53**: 83–88.

Rubio, B., Mart´ýnez, B., Garc´ýa-Cach´an, M. D., Rovira, J.and Jaime, I. (2007) Effect of high-pressure preservation on the quality of dry cured beef "Cecina de Leon".*Innov Food Sci Emerg Technol*, **8**(1): 102–10.

Saccani, G., Parolari, G., Tanzi, E.and Rabbuti, S. (2004) Sensory andmicrobiological properties of dried hams treated with high hydrostatic pressure: *proc 50th int cong of meat sci technol*: Helsinki, Finland: p 726–9.

Sáiz, A.H., Mingo, S.T., Balda F.P. and Samson C.T. (2008) Advances in design for successful commercial high pressure food processing.*Food Australia,* **60**(4): 154-156.

Sancho, F., Lambert, Y., Demanzeau, G., Largeteau, A., Bouvier, J. M., and Narbonne, J.F. (1999) Effect of ultra high hydrostatic on hydrosoluble vitamins.*J Food Eng,* **39**: 247 - 253.

Sano, T., Ohno, T., Otsuka-Fuchino, H., Matsumoto, J. J.and Tsuchiya, T. (1994) Carp natural actomyosin: thermal denaturation mechanism.*J Food Sci,* **59:** 1002–1008.

Schenkova, N., Sikulova, M., Jelenikova, J., Pipek, P., Houska, M.and Marek, M. (2007) Influence of high isostatic pressure and papain treatment on the quality of beef meat.*High Press Res,* **27**(1): 163–8.

Schindler, S., Krings, U., Berger, R. G.and Orlien, V. (2010) Aroma development in high-pressure-treated beef and chicken meat compared to raw and heat treated. *Meat Sci,* **86**(2): 317–23.

Schubring, R., Meyer, C., Schlu¨ter, O., Boguslawski, S.and Knorr, D. (2003) Impact of high pressure assisted thawing on the quality of fillets from various fish species. *Innov Food Sci Emerg Technol,* **4**: 257–267.

Sentandreu, M. A., Coulis, G.and Ouali, A. (2002) Role of muscle endopeptidases and their inhibitors in meat tenderness. *Trends Food SciTechnol,* **13**: 398–419.

Sikes, A. L, Tobin, A. B.and Tume, R. K. (2009) Use of high pressure to reduce cook loss and improve texture of low-salt beef sausage batters.*Innov Food Sci Emerg Technol,* **10**(4): 405–12.

Sikes, A., Tornberg, E., and Tume, R. (2010) A proposed mechanism of tenderizing post-rigor beef using high pressure-heat treatment.*Meat Sci,* **84**(3): 390-399.

Simonin, H., Duranton, F. and Lamballerie, M. de (2012).Comprehensive Review on New insights into the high-pressure processing of meat and meat products.*Food Sci Food Safety*.**11**: 285 – 306.

Simpson, R. K.and Gilmour, A. (1997) The resistance of *Listeria monocytogenes* to high hydrostatic pressure in foods. *Food Microbiol***14**(6): 567–73.

Smelt, J.(1998) Recent advances in the microbiology of high pressure processing.*Trends in Food SciTechnol,* **9**(4): 152-158.

Suzuki, A., Watanabe, M., Ikeuchi, Y.and Saito, M. (1992) Pressure effects onvthe texture, ultrastructure and myofibrillar protein of beef skeletal muscle. *Proc 38th Int CongMeat Science Technol,* 23–28 Ao^ut 1992. France: Clermont-Ferrand. p 423–6.

Suzuki, A., Watanabe, M., Ikeuchi, Y., Saito, M., and Takahashi, K. (1993) Effects of high pressure treatment on the ultrastructure and thermal behaviour of beef intramuscular collagen.*Meat Sci,* **35**: 17–25.

Suzuki, A., Homma, N., Fukuda, A., Hirao, K., Uryu, T.and Ikeuchi, Y. (1994) Effects of high pressure treatment on the flavour-related components in meat.*Meat Sci,* **37**(3): 369–79.

Taylor, D.M. 1999. Inactivation of prions by physical and chemical means.*J HospInf,* **43**(suppl.), S69 – S76.

Ting, E.Y. and Marshall, R.G. (2002) Production issues related to UHP food. In Engineering and food for 21st century (Welti-Chanes J, Barbosa- Canovas GV, Aquilera JM, eds). *Food Preser Technol,* Series, Boca Raton: CRC Press, Ch 44, pp 727- 738.

Uenaka, T., Mori, T., Yabuno, .Y.and Sumi, K. (2006) Sterilization of livestock visceral meat by high pressure treatment. *J Jpn Soc Food Sci Technol,* **53**(12): 651–4.

Van Opstal, I., Bagamboula, C.F., Vanmuysen, S.C.M., Wuytack, E. Y.and Michiels, C. W. (2004) Inactivation of *Bacillus cereus* spore in milk by mild pressure and heat treatments. *Int J Food Microbio,* **92**(2): 227-234.

Wackerbarth, H., Kuhlmann, U., Tintchev, F., Heinz, V.and Hildebrandt, P. (2009) Structural changes of myoglobin in pressure-treated pork meat probed by resonance Raman spectroscopy.*Food Chem,* **115**(4): 1194–8.

Yuste, J., Pla, R.and Mor-Mur, M. (2000)*Salmonella enteritidis* and aerobic mesophiles in inoculated poultry sausages manufactured with high-pressure processing. *Lett Appl Microbiol,* **31**(5): 374–7.

Yuste, J., Pla, R., Capellas, M., Sendra, E., Beltran, E.and Mor-Mur, M. (2001) Oscillatory high pressure processing applied to mechanically recovered poultry meat for bacterial inactivation. *J Food Sci,* **66**(3): 482–4.

Yuste, J., Capellas, M., Pla, R., Llorens, S., Fung, D. Y. C. and Mor-Mur, M. (2003). Use of conventional media and thin agar layer method for recovery of food borne pathogen from pressure treated poultry products. *J Food Sci,* **69**: 2321-2324.

2016, Dairy and Food Product Technology
Editors: Birendra Kumar Mishra and Subrota Hati
Published by: BIOTECH BOOKS, NEW DELHI
Pages 171–180

Chapter 11

Regulations in Trade and Export of Meat and Meat Products

Meena Goswami[1] *and Heena Sharma*[2]

[1]*Department of Livestock Products Technology, College of Veterinary Sciences and Animal Husbandry, DUVASU, Mathura – 281 001, U.P.*

[2]*Division of Livestock Products Technology, Indian Veterinary Research Institute, Izatnagar, Bareilly – 243 122, U.P.*

Introduction

India is agriculture based country and Livestock sector is one of the important components of agriculture in India. Tremendous progress made in agriculture, food processing and changing food habits in population coupled with long pending demand from stakeholders for and integrated food laws. An impressive progress has been made by meat industry under agriculture, evolving from backyard venture to a fully fledged commercial agro industrial business in India. Meat and meat products being one of the important components of balanced diet, this sector warrants action on many fronts to increase their availability at affordable price. Growing population and rising income along with fast India has a large livestock population ranking globally first in cattle (210 million-15 per cent), buffalo (111million -57 per cent) and goat (154 million-17 per cent), second in sheep (74 million-7 per cent). Over the last census 2003-07, cattle, buffaloes, sheep, goat and poultry have recorded a positive growth rate of 1.83, 1.84, 3.87, 3.10 and 7.33 per cent respectively. Overall livestock had recorded a growth rate of 2.23 per cent during the same period. The estimated poultry population in 2007 was 648.8 million comprising 617.7 million fowls, 27.6

million ducks and 3.5 million diversified poultry birds such as Japanese quails, turkeys, guinea fowl, emu etc. (BAHS, 2010). India ranks first in milk and fish production, second in egg production and third in broiler meat production. Though, meat output grew tremendously since eighties, still productivity of most meat producing species is low. The meat industry in India has not taken its due share. There are many reasons for the slow growth of the meat industry, attitude of public towards meat including religious and socio-political considerations. India has established a few most modern, mechanized environment friendly abattoirs-cum-meat processing plants in various states based on slaughtering buffaloes and sheep. These plants utilize all the slaughter house byproducts like meat cum bone meal, tallow, bone chips etc. and also produce value added products. This would not only require the enhanced resources allocation for the development of livestock sector but also on human resource development to take up the challenge of delivering the livestock services at grass root level. Therefore, a multipronged approach is required to supplement the efforts of country.

There are two distinct systems under which meat is being produced, distributed and consumed namely domestic market and export market:

- ✰ Domestic meat market- there are about 4000 slaughter houses registered with local bodies and more than 25000 unregistered premises where animals are slaughtered. Traders/individual butchers but their animals from weekly livestock markets and supply/bring them to the slaughter houses which cater to the wet markets. For domestic market meat is produced in municipal slaughter houses which are old fashioned and meat inspection is carried out by local vetenarians. The meat is sold hot during the day time after morning slaughter.
- ✰ Export meat market- there are about 20 integrated abattoirs cum meat processing plants where animals are received from the suppliers who procure the animals from the weekly market. Strict ante mortem and post mortem inspection along with HACCP are followed at these slaughter houses. The quality and safety of meat for international market is governed by the sanitary and phyto sanitary requirements of the importing countries/ OIE guidelines to produce quality safe meat. Mostly the meat is deboned and deglanded, frozen both in plate freezers or blast freezers, packed in cartons and exported in refrigerated containers.

The global trade is expanding rapidly and significantly due to increase in consumer demands linked to growing education and awareness of consumers, internationalization of tastes and habits, developments in science and technology, and improvement in communications and transportation. Wholesome and safe meat products have become very important in international trade coupled with the breaking down of tariff barriers and quantitative restrictions. Governments have also realized their role in protecting health and safety of their populations by imposing stringent regulations based on health, safety and environmental considerations so that standards and regulations do not create unnecessary barriers to trade. The need for a strong import control mechanism is quite obvious. The need for an equally strong export control mechanism is a natural corollary of such import control systems of importing countries, which have to have a provision for recognition of export certification systems of their trading partners through equivalence agreements. In recognizing that quality and safety can be assured through application of proper or well-designed food control systems (exports and imports), the Codex Alimentarius Commission established the Codex Committee on Food Import and Export Inspection and Certification Systems (CCFICS) to develop principles and guidelines in this area.

Although food control should cover both export and import as is evident from the terms of reference of this Committee and most of the documents it developed, most governments have emphasized on development and strengthening of import control systems with a view to protecting their populations and to prevent dumping of inferior quality products into their country. However, the situation in India and some other exporting countries has been somewhat different, with export inspection and certification being compulsory in certain food items.

Authentication and Quality Certification of Meat and Meat Products

With an expanding world economy, liberalization of food trade, growing consumer demand, developments in food science and technology, and improvements in transport and communication, international trade in fresh and processed food will continue to increase. Access of countries to food export markets will continue to depend on their capacity to meet the regulatory requirements of importing countries. Creating and sustaining demand for their food products in world markets relies on building the trust and confidence of importers and consumers in the integrity of their food systems. The SPS Agreement permits member countries to impose measures to protect human, animal and plant life or health. Furthermore, the Agreement, through its provision for adherence/adoption of Codex standards, which in turn provide for legislative framework for imports and the role for official/government inspection/ certification agencies and recognition of such agencies at the exporting country's end through equivalence agreements, permits members to establish formal systems of import control to ensure the appropriate degree of protection for their populace.

Some of the benefits of such export control systems are highlighted below:

(i) Minimize impediments to trade by reducing the time for inspection and testing at the importing end.

(ii) Minimize and even eliminate rejection or non-compliance at the point of import.

(iii) Avoid duplication of inspection, sampling and tests at the exporting and importing ends and lead to usage of collective resources more efficiently and effectively.

(iv) Are financially more effective as cost of recall, cost of testing at importing end and cost of destruction of consignments is minimized.

(v) Take care of variation in quality due to production by small farmers, fishermen or enterprises.

(vi) Help in building up the image of the country, as ensures that inferior quality products are not exported by unscrupulous one-time or fly-by-night operators. Such problems can be minimized with mandatory export certification. For example, in the Indian dairy sector, export certification has become mandatory and it is obligatory for exports to take place only from material processed in an approved unit implementing food safety management systems.

(vii) Enable official inspection/health certificates to be given as the same are often required by the buyers.

(viii) Help in 'Capacity Building' in a country with respect to product as well as systems. With a mandatory export certification system, the country identifies the weaknesses and focuses on correcting these.

(ix) Decisions on a country's products that are exported are taken by the country itself rather than by the importing country. For example, if the product does not meet an importing country's requirement, the exporter can, in consultation with the official certifying body send it to a third country, which permits the same, rather than the importing country deciding that it is not fit for consumption as its requirements are not met and therefore needs to be destroyed.

(x) Facilitate negotiating Agreements/MoUs for recognition of food control systems and certification by the importing country.

(xi) Provide protection to the consumer of the importing country as the broad objective of the exporting country is to ensure that requirements of the importing country are met.

(xii) Facilitate implementation of various forms of voluntary certification which address the entire chain from farm to table. This is simplified as a major part of the total chain, namely processing is already covered and only additional areas such as those at farm level need to be certified.

Indian Scenario of Export Control and Certification Systems

There is a rapid change in meat eating trend s and consumption pattern throughout the world and hence changes are taking place in meat production practices accordingly. India has been operating export control systems since 1963 which have been well-defined and established under the Export (Quality Control and Inspection) Act, 1963. The Act empowers the Central Government to notify commodities for pre-shipment inspection and certification, specify the minimum standards (generally recognizing international, importing countries' standards and contractual specifications), and prescribe the manner of export inspection and certification, whether compulsory or voluntary. The export control system is operated by the Export Inspection Council of India (EIC), India's official export certification body, through its field organizations, the Export Inspection Agencies, having head offices in Chennai, Delhi, Kochi, Kolkata and Mumbai with 41 sub-offices including laboratories around the country.

The abattoirs and meat plants engaged in export production are required to be registered with APEDA, which is compulsory vide DGFT notification No. 12/(2004-05) dated 21 December, 2004. The plants are periodically inspected by a committee consisting of officials of various Government departments.

a) Standards have been laid down for export of meat and meat products under export (quality control and inspection) act, 1963. The export of raw meat (chilled and frozen) shall be allowed subject to the provision specified to the Gazette notification 1993 on raw meat (chilled or frozen) under export act, 1963. The notification lays down the standards for abattoirs, meat processing plants and the products. Offal's too are subjected to same conditions of quality control and inspection.

b) All consignments of raw meat (chilled and frozen) and export of canned meat products need to accompany with a pre-shipment inspection certificate for which the government has designated three agencies i) All states Directorates of Animal Husbandry ii)Export Inspection Agency iii) Directorate of Marketing and Inspection/FSSAI, Government of India, to carry out the inspection in accordance with either the standards prevalent in the exporting country or standards prescribed under Meat Food Products Order, 1973 under Export (Quality control and Inspection) act, 1963 or orders made there under.

c) Export of meat and meat products will be allowed subject to the exporter furnishing a certificate to the customs at the time of export that above items have been obtained/sourced from an abattoir/meat processing plant registered with APEDA.

Export Policy and Strategy

Export of meat and meat products from India is, inter-alia, governed by the provisions of following Acts, rules and orders (Source: Indian Meat Industry; Red Meat Manual, APEDA,) issued by Government of India:

a) The Export (Quality Control and Inspection) Act, 1963

b) The export of raw meat (chilled/frozen) (Quality control and Inspection) rules, 1992

c) Provision contained in Government of India order issued vide S. O. 203, dated 15-01-1993

d) ITC (HS) classification of import and export items

e) Prevention of Cruelty act, 1960 and Rules 2001

f) Various mandatory Indian standards such as IS:1982-1971; Code of Practice for ante mortem and post mortem inspection of meat animals: IS:2537:1995: meat and Meat Products-Beef and Buffalo meat fresh, chilled and frozen technical requirements; and Codex Alimentarius: Judgment code for slaughter animals and meat

Export of Animal Products from India

Quantities in MT **Value in Rs. Crore**

Products	Year					
	2008-09		2009-10		2010-11	
	Quantity	Value	Quantity	Value	Quantity	Value
Buffalo meat	462749	4839.70	495019	5480.60	709437	8412.68
Sheep and Goat meat	37790	493.37	52868	747.20	11908	253.18
Poultry products	1057016	422.05	1016783	372.11	619150	301.32
Animal casings	1823	8.84	2020	31.52	1809	34.14
Processed meat	857	10.14	716	9.58	1366	21.05
Swine meat	817	9.17	1117	10.34	1115	10.51
Dairy products	70146	980.96	34380	402.68	36867	533.89
Natural honey	15587	148.96	13311	146.65	31675	249.58
Total	1646790	6913	1616216	7201	1413330	9817

Source: APEDA, 2012.

The effective manual Quarantine and Certification Services are necessary to ensure that international trade in animal products takes place without incurring unacceptable risk to human and animal health. The animal disease situation in the exporting country, in the transit country and importing country must be in compliance with SPS requirements need to be critically examined for permissions.

The import of pig fat, poultry fat, fat of bovines, ovines, caprines etc is totally prohibited whereas meat of bovines, their edible offals is restricted for import including guts, bladders, stomachs of bovine, caprine and bovine. The import if permitted against a license from DGFT after recommendation from Department of Animal Husbandry, Dairying and Fisheries (DADF). the emat and products, edible offalsof swine, caprine, ovine, poultry, rabbit, liver, tongue etc, ig bristles, hairs, yak tail comes under the free items and import is permitted against a Sanitary Import permit from DAD, Ministry of Agriculture, GOI.

Export Regulation under EXIM Policy

The exports of beef of cows, oxen and calf and offals is prohibited and not permitted to be exported. Meat of buffalos (both male and female) and offals of buffaloes except gonads and reproductive organs is free to export subject to the condition that exporter has to produce a certificate from the designated veterinary authority of the state, from which the emat or offals emanate, to the effect that the meat or offals are from buffaloes not used for breeding and milch purposes. The quality control and inspection under (a) and (b) stipulated above rules and regulations for export are required to be fulfilled. The meat and offals of Indian sheep and goat is free for export subject to fulfilling the conditions (a) and (b) and (c) stipulated above. Tallow, fat and/or oils of any animal origin excluding fish oil are prohibited for export.

Export food control activity is a multi-disciplinary activity covering a number of aspects such as food science, microbiology, analytical chemistry, plant pathology, veterinary science, etc. A number of agencies would normally be involved in any country including various departments of government, control organizations, promotional bodies, research institutions, agricultural institutions, farming community, trade associations, non-governmental organizations (NGOs), consumers etc. There needs to be a suitable documented export control food strategy with clear objectives, including the countries of focus, well designed plan of action with role clarity provided for different players and clear networking of the organizations within the country.

Legislative Framework for Export

Food legislation includes acts, regulations, and requirements or procedures prescribed by the government relating to export of foodstuffs to meet requirements of the importing country while ensuring conditions of fair trade. Food control needs to be simple, complete, covering various aspects of the food chain as needed and address requirements of importing country - both issues of safety and quality. It should provide authority to carry out controls at all stages of the food chain. Furthermore, it should be flexible to allow taking into account new technologies, developments and changing trade needs. It also needs to be WTO compatible and as far as possible based on Codex standards, guidelines and recommendations, but depending on importing country's requirements. Legislation may also include provisions for registration of establishments or listing of certified processing plants, establishment approval, licensing or registration of traders or agents, equipment design approval, penalties, coding requirements and charging of fees. Necessary provisions need to be included for ensuring integrity, impartiality and independence of the official and officially recognized inspection and certification systems.

Government Interventions for Export Control

Inspection services should design control programmes based on precise objective and appropriate risk analysis. HACCP or a similar quality and safety assurance and management system based approach should be encouraged with responsibility for meeting the food quality and safety regulatory requirements of importing country resting with the food industry with all segments of the food chain having responsibility for establishing food safety and quality controls. The responsibility of food control regulators is to ensure, through a surveillance system of industry and other components of the food chain that they meet the requirements specified by the importing country. Elements of a control programme should include the following:

- Inspection;
- Sampling and analysis;
- Checks on hygiene, including personal cleanliness and clothing;
- Examination of routine and other records;
- Examination of the results of any verification systems operated by the establishment;

- Audit of establishments by the national competent authority responsible for export control;
- National audit and verification of the control programme.

An administrative procedure should ensure that controls by the inspection systems are carried out regularly proportionate to the degree of risk, where non-compliance is suspected and in a coordinated manner between different authorities (if several exist). Control should also cover, as appropriate, the establishment, installations, means of transport, equipment and material; raw materials and ingredients for preparation and production of food stuffs; semi-finished and finished products; cleaning and maintenance products; processes for manufacture or processing of foodstuffs; preservation methods; labeling integrity and claims etc. Formal documentation of the export control programmes is also necessary.

Decision Criteria and Action

Control programmes should target the appropriate stages of operation depending on specific objectives. The frequency and intensity of controls should be designed to take into account the risk as well as the reliability of controls already carried out by those handling the products at various other stages *i.e.* production, manufacturing, etc. In case of rejected products, information should be sought by the export food control authority from the importing country as per CAC/GL 25-1997 - Codex Guidelines for the Exchange of information between countries on rejections of imported foods. Such information on rejections should be provided at the earliest opportunity, by the export control authority, to the exporter, the manufacturer, producer and any related department depending on the situation.

Other Facilities for Export Inspection

Adequate facilities including equipment, transportation and communication facilities should be available to ensure delivery of export inspection and certification services.Laboratories are the backbone of the inspection and certification activity. In order to test to requirements prescribed by the importing countries, the laboratories should have state-of-the-art equipment and manpower that is qualified and trained to operate such equipment. The laboratories used by the export inspection and certification services need to be accredited as per international standard ISO 17025 under officially recognized programmes to ensure that adequate quality controls are in place to provide for reliability of test results. Internationally accepted quality assurance techniques should be implemented to ensure reliability of analytical results.

Official inspection and certification services should have access to a sufficient number of qualified personnel in food science, technology, chemistry, biochemistry, microbiology etc. The personnel should be trained in areas of inspection and certification systems, audit techniques, risk analysis techniques, testing, technological aspects etc. and have a status that ensures impartiality and no direct commercial interest in the products or establishments being inspected or certified.

Certification and Official Accreditation

Certification should provide assurance of conformity of a product to importing country requirements by checks on each product or a batch of products or by approval of the system being implemented by the processor with regular checks by the inspection and certification service on various aspects of the system being implemented. The competent authorities should take all necessary steps to ensure the integrity, impartiality and independence of official or officially recognized certification systems.

Export inspection or certification bodies may be officially accredited to provide services on behalf of official agencies. These bodies shall comply with criteria laid down in international standards such as ISO/IEC 17020, ISO Guides 62 and 65, as well as Codex Guidelines for the Design, Operation, Assessment and Accreditation of Food Import and Export Inspection and Certification Systems with specific reference to competence, independence and impartiality of personnel. The performance of such inspection and certification bodies should be regularly assessed by the competent authority.

Assessment and Verification of Inspection and Certification Systems

The export inspection and certification system should be subject to audit separate from routine inspection, may be self-evaluation or by third parties. Internationally recognized assessment and verification procedures should be used. Guidelines for conducting assessment and verification of an exporting country by an importing country have been given in the Annex to CAC/GL 26-1997 and an importing country may undertake a review of the exporting country's systems, if so agreed.

For decisions relating to food export control systems, both developmental as well as implemental, there is a need to have information as well as data that is scientifically collected, shared with decision makers and implementers as well as processors. This would include information on regulatory requirements of importing countries; data on residues and other parameters which would help in framing regulatory requirements as well as being used for decision making on implementation actions to prevent food-borne hazards, plan food control activities etc. While ensuring transparency, any constraint of professional and commercial confidentiality should be respected.

Some other issues to be addressed

- ☆ Various meat and animal feed testing laboratories should be accredited with the laboratories of developed countries o facilitate the export as well as supply of safe and wholesome products to the domestic consumers.
- ☆ Information about linked animal diseases, veterinary drug residues and feed toxins as well as their maximum limits set by CAC and consequences of their presence beyond permissible limits to human health, trade and industry needs to be disseminated to farmers, feed manufacturers, field veterinarians, animal products processors and the consumers by state.

- ☆ The climate change and effect of biotic and abiotic stresses on livestock specially buffalo meat production performance and needs special attention for the research.
- ☆ National Meat and Poultry Processing Board may lead to act as facilitator between meat industry and research organizations, assist in technology transfer and adoption and organize customized training programmes for the meat workers to observe proper sanitation and hygiene while handling meat in slaughterhouses, processing plants and consumer/retail outlets.
- ☆ Processing of meat into value added products, establishment of cold chains, modern storage facilities and packaging of products in durable and attractive packaging materials are important areas which require great attention by the industry near future to meet growing demand of ready to eat products.
- ☆ To minimize the multiplicity of food laws, there is need for harmonization of meat and food standards with CODEX standards.

References

APEDA, (2012). Statistics for export of agricultural products- Ministry of Commerce and Industry, GOI.

Bredahl, M. E., J. R. Northen, A. Boecker, and M. A. Normile. (2001). "Consumer demand sparks the growth of quality assurance schemes in the European food sector." In A. Regmi (ed.), Changing Structures of Global Food Consumption and Trade (pp. 90S102). Agricultural and Trade Report No. WRS-01-1, USDA/ Economic Research Service, Washington, DC.

Buzby, J. C., and P. D. Frenzen. (1999). "Food safety and product liability." *Food Policy*. 24: 637-651.

Caswell, J. A., and N. H. Hooker. (1996). "HACCP as an international trade standard." *American Journal of Agricultural Economics*.78: 775-779.

FAO (2008). FAO of United Nations. Animal Production Year Book.

Henson, S., and G. Holt. (2000). "Exploring incentives for the adoption of food safety controls: HACCP implementation in the UK dairy sector." Review of Agricultural Economics 22, 407S420.

Meat food products order (1973). Extraordinary Gazette of India, Part II-Section 3-Sub-section (ii) dated 28.3.1973 available at www.fssai.gov.in/./0/./MFPO per cent 201973-Amended per cent 20_English_.pdf.

Shashi Sareen. *Food export control and certification*. Second FAO/WHO Global Forum of Food Safety Regulators. Available at http: //www.fao.org/docrep/meeting/ 008/y5871e/y5871e0m.html.

Tanner, B. (2000). "Independent assessment by third-party certification bodies." *Food Control*. 11: 415-417.

Taylor, E. (2001). "HACCP in small companies: Benefit or burden?" *Food Control*. 12: 217-222.

2016, Dairy and Food Product Technology *Pages 181–191*
Editors: **Birendra Kumar Mishra and Subrota Hati**
Published by: **BIOTECH BOOKS, NEW DELHI**

Chapter 12

Ozone Utility in Food Processing Sector

Kumaresh Halder

Assistant Professor, NIFTEM, Plot No. 97, Sector 56, HSIIDC Industrial Estate, Kundli, Sonipat – 131 028, Haryana

Introduction

Food products should be manufactured with proper care and within a hygienic zone as it is consumed by all age group in our society. It is mandatory to clean, sanitize and disinfect the equipments as well as whole processing area after every production to meet all legal food quality and safety requirements. Disinfection is an important step of cleaning operation which is traditionally done by using chlorine, quaternary ammonium compound (QAC), H_2O_2 or other halogen compound. Because of their several disadvantages researches are going on to find out an alternative effective environment friendly disinfectant. Ozone is proven as an effective solution as it is an environment friendly, strong oxidant as well as potent disinfecting agent and in United States Food and Drug Administration granted generally recognized as safe (GRAS) status for use of ozone in bottled water in 1982. In India use of ozone in waste water treatment and purification of drinking water has been recognized but no example of using ozone in dairy or food industry is found till now. If we ignore some minor disadvantages of ozone it is the most acceptable disinfectant available around us.

Existing Method of Disinfection

The methods currently available for disinfection and sterilization are through the use of oxidative and non-oxidative biocides. Chlorine is the most widely used chemical as disinfectant generally in the form of hypochlorite. However, quaternary ammonium compound, H_2O_2 or other halogens are also used. Disinfection is done after completion of acid and alkali cleaning to make it free from any organism or residual matter present on the cleaning surface. Sometime thermal method is also applied for disinfection but it is very costly.

Problems with Current Methods and Alternative Methods

The non oxidative biocides (QAC) are not reliable in disinfection because the process of disinfection often may not be completed (in the concentrations used) or some bacteria may be resistant and the preferred procedure is the use of oxidative biocides. Using non oxidative biocides in food preparation is also limited in that they may impart some undesirable taste and cause problems in preparation such as frothing etc. Chlorine has traditionally been the sanitizer of choice in the food processing industry, but experts share a growing concern about the dangerous by-products such as trihalomethanes or dioxins produced when chlorine reacts with organic matter in the water. These substances are known carcinogens and are regulated in drinking water by the U.S. Environmental Protection Agency. Firstly some organisms may survive even after the process as using higher concentrations which may be impractical may only kill them. Further Chlorine itself imparts some flavor not accepted by consumers. The alternatives are required to minimize the usage of those above mentioned old practices. Hydrogen peroxide is one, which has also been viewed as an alternative, in that it is a reasonable oxidant and the byproduct is water, but the overall costs make it rather less attractive alternative. Ozone is known since from 20 years back, but now it is proven to be an effective alternative disinfectant in food processing and cold storages.

Ozone and its Properties

When we think about ozone then first thing come into our mind about the protective layer that exists at the outer surface of earth protects us from harmful a ultra-violet ray. Ozone was first discovered by the European researcher C. F. Schonbein in 1839. Ozone is formed at stratosphere layer of atmosphere by the action of UV rays or by electrical arc on photochemical smog. Ozone found at a very low concentration in atmosphere and readily decomposes at room temperature. It is colourless at room temperature but at -112°C can be condensed into dark bluish liquid. It is soluble in cold water but at higher temperature decomposes very fast. Ozone has a pungent, characteristic odor described as similar to "fresh air after a thunderstorm". It is readily detectable at 0.01–0.05 ppm level. Ozone has a longer half-life in the gaseous state than in aqueous solution. Liquid ozone is easily exploded by electrical sparks or by sudden changes in temperature or pressure. Ozone composed of three oxygen atoms in its natural form. Ozone is a tri-atomic molecule. To produce ozone one free oxygen atom shall react with diatomic oxygen molecule. Resonance structure of ozone molecules has been shown in Figure 12.1.

In order to generate free oxygen atom it is required to break the O-O bonding by Ultraviolet radiation or by corona discharge methods. In order to generate commercial levels of ozone, the corona discharge method is usually used.

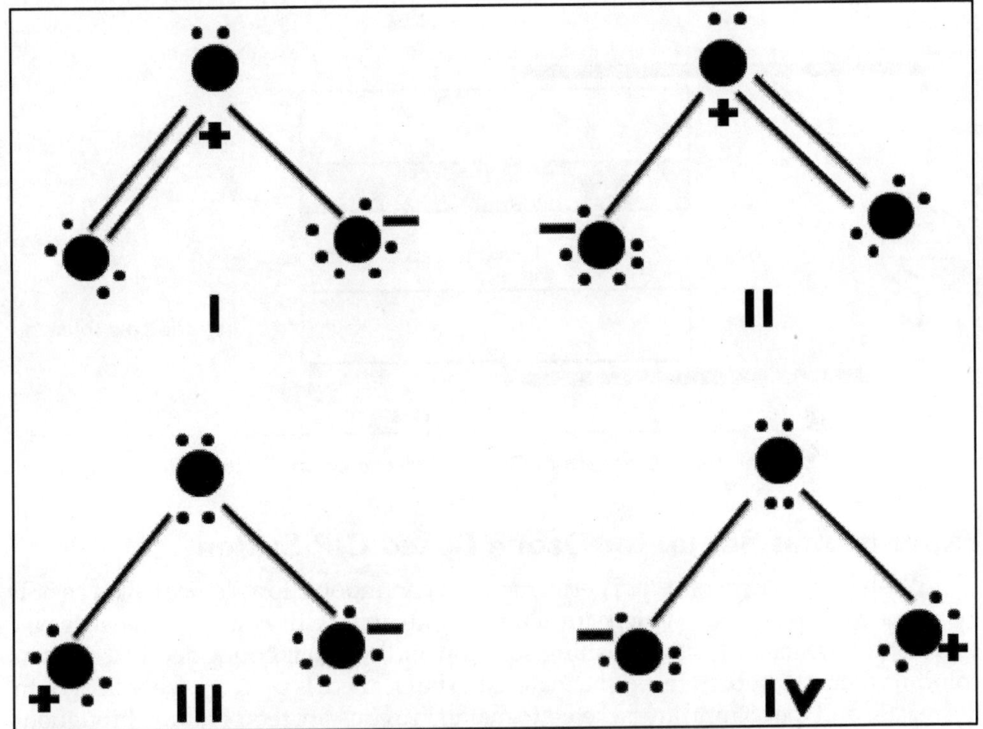

Figure 12.1: Resonance Structures of Ozone Molecule.

There are two electrodes (Figure 12.2) in corona discharge, one of which is the high-tension electrode and the other is the low-tension electrode (ground electrode). Those are separated by a ceramic dielectric medium and narrow discharge gap is provided. When the electrons have sufficient kinetic energy (around 6–7 eV) to dissociate the oxygen molecule, a certain fraction of these collisions occur and a molecule of ozone can be formed from each oxygen atom. If air is passed through the generator as a feed gas, 1–3 per cent ozone can be produced. Ozone is an effective oxidant because of its electron drawing capability from a source. Its reactive nature is even more than oxygen because it can react readily with substrate on its own. Several researches have proved that the bactericidal effects of ozone on wide range of microorganisms including Gram positive and Gram negative bacteria as well as spores and vegetative cells. Gram positive bacteria as *Listeria monocytogenes, Staphylococcus aureus, Bacillus cereus, Enterococcus faecalis,* and such Gram negative bacteria as *Pseudomonas aeruginosa, and Yersinia enterocolitica*, yeasts *Candida albicans* and *Zygosaccharomyces* bacilli and spores of *Aspergillus niger*. Effect of ozone on Bacillus spp. spores has been found more detrimental than hydrogen peroxide. The potent oxidation capacity that makes ozone very effective in destroying microorganisms. Ozone has been used experimentally to control mold on Cheddar cheese surfaces and in cheese rooms. At high ozone concentrations ozone appeared to destroy the molds present. It is also effective in reduction of the population of coliforms, *E. coli*, and *Salmonella spp*. from food samples.

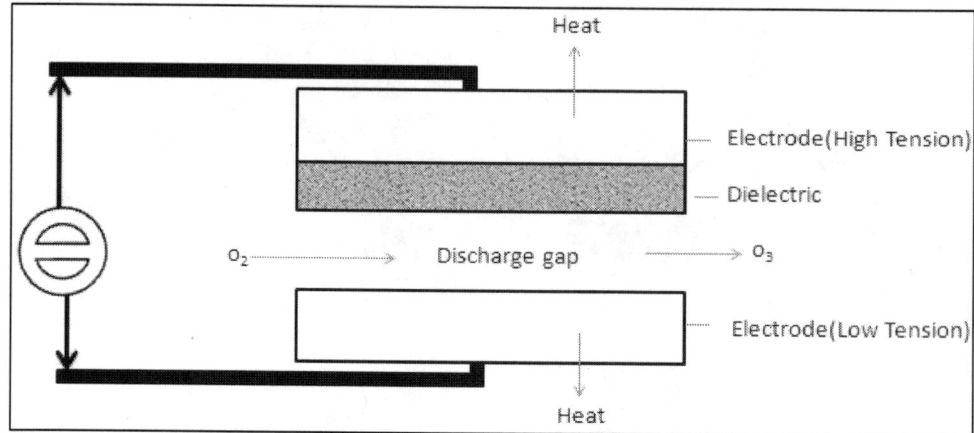

Figure 12.2: Schematic Diagram of Ozone Generation by Corona Discharge.

Experimental Set-up for Ozone Based CIP System

Washing experiments were performed in a continuous flow device called as BSF by a group of researcher, which simulates an industrial CIP system. It consists of a stirred tank which holds the washing solution and which is connected to a packed column where the substrate, *i.e.* the material to be cleaned, is placed once soiled. Both the reactor and the column are jacketed to maintain a constant temperature throughout the experiment. To carry out the washing experiments in the presence of ozone the BSF device was slightly modied by placing a sintered-glass diffuser at the bottom of the stirred tank in order to bubble the ozone-carrying gas into the washing bath. The modied system is presented below in Figure 12.3.

Suggested Mechanisms of Antimicrobial Action of Ozone

Studies and researches indicate that the ozone destroys microorganisms by the progressive oxidation of vital cellular components. The bacterial cell surface has been suggested as the primary target of ozonation. Two major mechanisms have been identified in ozone destruction of the target organisms. First mechanism is that ozone oxidizes sulfhydryl groups and amino acids of enzymes, peptides and proteins to shorter peptides. The second mechanism is that ozone oxidizes polyunsaturated fatty acids to acid peroxides. Ozone degradation of the cell envelope unsaturated lipids results in cell disruption and subsequent leakage of cellular contents. Double bonds of unsaturated lipids are particularly vulnerable to ozone attack. In Gram-negative bacteria, the lipoprotein and lipopolysaccharide layers are the first sites of destruction resulting in increases in cell permeability and eventually cell lysis. Chlorine selectively destroys certain intracellular enzyme systems; ozone will cause widespread oxidation of internal cellular proteins causing rapid cell death. Cellular death can also occur due to the potent destruction and damage of nucleic acids. Thymine is more sensitive to ozone than cytosine or uracil. Ozone also destroys viral RNA and alters polypeptide chains in viral protein coats.

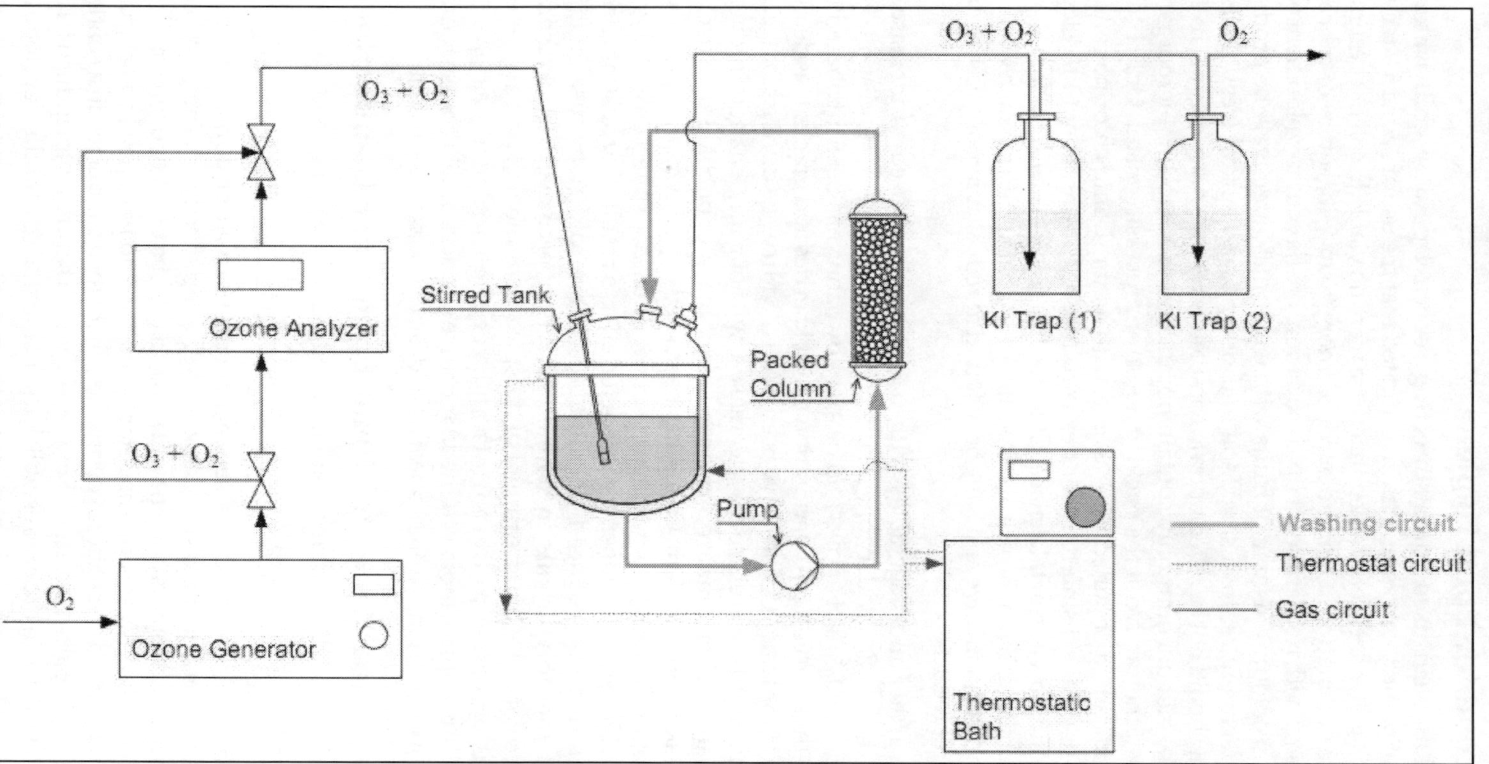

Figure 12.3: Schematic Diagram of the Modied BSF Device.

Use of Ozone for Food Hygiene

Ozone has been used for disinfecting recycled poultry chill water and disinfection of poultry carcasses. It is investigated that the use of ozone for recycling of poultry chill water and determined that there were no viable E. coli or presumptive Coliform after ozonation. Ozone destroyed more than 2 log-units of all carcass microorganisms with no significant lipid oxidation, off-flavour development or loss in carcass skin colour. Washing of fruits with water may effectively be produce in water can effectively remove sand, soil and other derbies from fresh fruits and vegetables but should not be relied upon to completely remove microorganisms. Fruits and vegetables not uncommonly contain population of 10^4 to 10^6 microorganisms per gram when they arrive at the packing house or processing plant. Only 1 to 2 log unit reduction can be obtained by washing with water. Using chlorine cannot be totally relied upon as a disinfectant for fresh produce as has only about 1 to 2 log unit reduction of microorganism take place. Chlorine has never been useful for removing virus contamination. Even the use of trisodium phosphate (TSP) in wash water have not been found very effective and have been found to reduce microorganisms only to about 2 log unit reduce.

Treating fruits and vegetables with ozone has been used to increase shelf life.Treatment of apples with ozone resulted in lower weight loss and spoilage. An increase in the shelf-life of apples and oranges by ozone has been attributed to the oxidation of ethylene. Fungal deterioration of blackberries and grapes was decreased by ozonation of the fruits. Onions have been treated with ozone during storage. Mold and bacterial counts were greatly decreased without any change in chemical composition and sensory quality Shredded lettuce in water bubbled with ozone gas had decreased bacterial content. Ozone has been used experimentally as a substitute for ethylene oxide for the decontamination of whole black peppercorns and ground black pepper. No new components were detected as a result of ozonation. Because ozonation successfully reduced microbial loads and did not cause significant oxidation of the volatile oils in whole black peppercorns, this method was recommended for industrial treatment of the spice. A number of patents have been issued for using ozone to treat fruits and vegetables. Ozone has been used experimentally to control mold on Cheddar cheese surfaces and in cheese rooms. At high ozone concentrations ozone appeared to destroy the molds present.

Use of Ozonated Water for Plant Equipment Disinfection

Several chemical disinfectants as mentioned earlier are being used traditionally by dairies and food processing plants for cleaning operations. Chlorine is the most widely used disinfectant. It is proved that dairy soil can effectively be removed by ozone than chlorine. The capacity assessment of ozone has been done to facilitate the removal of fatty soils from hard surfaces, and to develop effective cleaning procedures involving ozone, surfactants and enzymes. To do this the effect of ozone on a model fatty material (pork lard), several surfactants (anionic and non-ionic) and a lipase for detergents was studied. Washing experiments where then conducted in a continuous flow device (called Bath- Substrate- Flow, BSF), which simulates an industrial clean-in-place (CIP) system, in order to establish a suitable strategy for the development of cleaning processes involving ozone. Due to its strong oxidation potential it is very much effective for killing of microorganisms. Some food industries use thermal sanitation methods and/or irradiation. Thermal sanitation is very effective in destroying contaminating microorganisms; however, steam and hot water are quite expensive to generate and excessive heat can be damaging to food processing equipment. Radiation methods are not practical for use in food processing plants because of the inherent hazards associated with radioactive materials. Chemical sanitation methods are the most commonly utilized sanitizers in the food industry. Chlorinated agents are used worldwide for disinfecting water, wastewater and for sanitizing food processing plant equipment. Even though chlorine sanitizers have several disadvantages including being harmful and irritating in high concentrations, being prone to forming carcinogenic compounds and being toxic to the environment, these compounds are economical bactericides that inactivate all types of vegetative cells. Chlorine has been a preferred disinfectant in the food and water industries for many years. Although chlorination efficiently decreases the spread of food borne infectious disease, chlorine combines with many organic compounds to form toxic by-products. These by-products are released in drinking water and adversely affect public health as well as the environment. Among these by-products, trihalomethanes (THM) and haloacetic acids (HAA) are mutagenic and carcinogenic. Releases of chlorinated compounds into the environment can result in formation of carcinogenic THM in rivers, streams and lakes. Potential drinking water can be affected along with native aquatic and terrestrial species. Consequently, efforts are being made to minimize releases of chlorine into the environment. Although the hazards of carcinogen/mutagen formation from chlorine are extensive, lack of suitable sanitizer/ disinfectant would result in much greater illness and loss of life due to microbial contamination of foods and water. Therefore, chlorine use continues, even though it is known to be hazardous. Food researchers are searching for alternative cleaning and sanitizing agents effective against food spoilage and pathogenic bacteria, yet harmless to humans and the environment. Additionally, these agents must also be non-corrosive to expensive food processing equipment. Ozone is a potential alternative to chlorine for use in the food industry. Stainless steel plates were used to simulate dairy plant equipment. It has been found that ozone was as effective as chlorination against dairy surface attached bacteria, both treatments reduced bacterial populations by 99 per cent. Ozone, chlorine and heat were compared for killing effectiveness against food spoilage bacteria in synthetic broth. If the cleaning process is ineffective, microorganisms may form biofilms on equipment surfaces. Microorganisms accumulate and proliferate in the porous biofilm material. Once embedded in this matrix, the microorganisms often become resistant to the action of sanitizers. Ozone had been proved effective in this particular case also. Soiled stainless steel coupons were treated with ozonated water as a pre-rinse. Studies on the effect of ozone and chlorine on the several gasket materials, which are used in fluid food processing, has been done. Both treatments did not significantly affect the tensile strength. The elasticity of ozone treated PTFE gaskets was significantly affected by ozone treatment. Both treatments had a bleaching effect on black gaskets such as Buna N, EPDM, and Viton. The tensile strength of EPDM and Viton was decreased by ozone treatment but not significantly when compared to chlorine treatment. From the results of this experiment, it is hypothesized that ozone use in the pre-rinse stage may allow for decreased detergent use in the cleaning solution recirculation step. Ozone improves the bacteriological quality of the recalculating chilled water for dairy plant operations as given in United States Grade "A" Pasteurized Milk Ordinance. Ozone was suggested for using in chilled water to reduce bacterial content. Studies on the effect of ozone on metals such as Aluminum, copper, carbon steel, 304 stainless steel and 316 stainless steel in dairy chilling water systems was done. Those were examined to record durability of metals against ozone. Metals have been analyzed according to weight loss of metals and by using scanning electron microscopy. Carbon steel has the significantly different weight loss than the control. There was no significant difference among other metals within control and ozone treatments. There was severe pitting on ozone treated copper samples.

Large quantities of water are required for cleaning and disinfection in the food industry. The wastewater quantity is largely dependent on production and cleaning patterns. Wine, beer and dairy processing plants installations use considerable amounts of water with the amount depending on the type and size of equipment to be cleaned and the materials processed; 60-80 per cent of the total water consumption is used for cleaning activities. Cleaning, disinfection and several other unit operations produce wastewater. These wastes are likely treated by biological methods. The primary concern with these food-processing wastes is that organic matter provides a food source for microbial growth. Microorganisms can therefore proliferate rapidly and subsequently cause a decreasing amount of dissolved oxygen in the water. BOD and COD values of wastes are very important. Worldwide, hazardous cleaner and sanitizer compounds are also routinely dumped into sewage systems. Eventually, these chemicals enter the environment with deleterious effects on delicate ecosystems. These compounds complicate waste degradation and also maybe toxic to the environment. The dairy sector uses a vast amount of water, and generates a huge amount of wastewater in maintaining the required level of hygiene and cleanliness (between 25 and 40 per cent of the total water consumed). The largest proportion of wastewater is cleaning water. This is used for equipment cleaning, *e.g.* line purging at product change-over, start-up, shut-down and change-over of HTST pasteurisation units as well as some product washing. Breweries use significant amounts of water and energy and produce wastewater and solid residues. Suspended solids in the wastewater originate from the discharge of by- products, diatomaceous earth and possible label pulp from the bottle cleaner. Nitrogen originates mainly from detergents used for tank cleaning, from the malt and from additives. Phosphorus may come from the cleaning agents used. Heavy metals are normally present in very low concentrations. Wear of the machines, especially conveyors in packaging lines, may be the source of nickel and chromium. In the wine industry, wastewater is generated in nearly all process steps, *e.g.* cleaning of containers, reactors and filters. The highest concentrated wastewater is produced during fermentation, fining and ageing/ racking due to the washing out of the sediments, marcs and lees. The semisolid fractions can be separated for further dewatering, drying, processing or disposal rather than being washed with water, due to their high organic load. Wine bottles are cleaned before filling, and consequently washing water enters the wastewater treatment plant or is recycled. Even after the recovery process, the wastewater shows an acidic character (pH 4.6) except when caustic solutions are used in the elimination of tartrate or during the conditioning of bottles. The most polluting wastewater during wine production is generated during the fermentation and racking (especially first racking) operations. This typically contains soluble organic material, SS, nitrate, nitrite, ammonia and phosphate from product remnants and removed deposited soil. It also contains residues of cleaning agents, *e.g.* acid or alkali solutions. In principle, the cleaning and disinfection agents that are used are discharged via the wastewater, either in their original state or as reaction products. Wastewater may have a high or low pH due to the use of acid and alkaline cleaning solutions. The use of phosphoric and nitric acids will increase the phosphate and nitrate content of the wastewater. Badly designed systems and inadequate product removal prior to the start of cleaning may lead to large quantities of product entering the cleaning water.

Adopting ozone in cleaning and disinfection processes can bring various advantages over commonly employed disinfectants. Ozone breaks down quickly into oxygen without leaving undesirable residues. This is an advantage both from the point of view of food safety and to improve the quality of wastewaters by avoiding the presence of harmful chlorine compounds. Replacing chemical products with ozone also lowers the concentration of salts and, therefore, the electrical conductivity of discharges. The use of ozone can save water in comparison to other biocides, as it is faster-acting. Additionally, since it does not leave residues it does not require a final rinse to remove any residual disinfectant that might remain in the treated medium. Another advantage, provided adequate microbiological controls are implemented, is that the ozonated water that has been used for disinfection can potentially be re-used for the initial cleaning stages, either directly or after reozonation to attain the required quality. Wastewaters are oxygenated by ozone conversion, so ozone use will improve the performance of aeration tanks and biological wastewater treatment processes. This is also an advantage from the point of view of reducing odour generation. Artificial dairy waste samples were treated with ozone; ozonation treatment lowered the biochemical oxygen demand of dairy waste by 15 per cent.Ozone peroxidises organic material so that it is easier for biodegradation. With lowered treatment needs, required sewage treatment capacities and levied surcharges could possibly be reduced. The bactericidal effects of ozone have been documented on a wide variety of organisms, including Gram-positive and Gram-negative bacteria as well as spores and vegetative cells. As it does not leave any residue due to quick decomposition of its structure, restrictions should be applied to human exposure to ozone.

Worldwide Acceptability of Ozone

Wide spread acceptability of Ozone has made it very popular in food processing also. The properties of Ozone has allowed its unrestricted and any food processing application.Ozone has been demonstrated as the most effective Biocide with significantly increased lethality.Ozone requires very low contact time compared to Chlorine and requires much lower concentrations for effective kill,compared to conventional biocides. In India CIP system using chlorine or thermal method of disinfection is generally recommended for the dairy and food plants. Utilization of ozone has not been recognized anywhere in plant cleaning operation but several examples of ozone disinfection have found in drinking water purification and waste water treatment in India. Expert or consultants in India never suggest to use ozone in plant cleaning. It may be due to lack of confidence on ozone or lack of reference users of ozone in plant cleaning operation in Indian subcontinent. However sone organizations in India who provides ozone solutions for drinking water purification, waste water treatment, food processing plant disinfection, cooling tower water purification, swimming pools water purification and several other processes. They have introduced Plug and Play ozone system for cleaning operation in food processing industry.

Limitation of Ozone in Plant Cleaning

At lower concentration ozone has no detrimental effect on human body but at but at higher concentration (*i.e.* at more than 0.2 ppm) it may appear as toxic to human respiratory and pulmonary organs. Experimentally it is proved that effects of ozone on gasket materials used in plant equipments are not impressive. However, on equipment surfaces it acts gently.

Conclusion

Cleaning, sanitation and disinfection operations are of key importance for the food industry, to maintain the required levels of food quality and safety. The industry is seeking for more effective solutions in terms of disinfection as well as environmental issues to ensure the safer operations and products. On the basis of sound advantages of ozone it may successfully be applied in food industry as discussed above such as to maintain food surface hygiene, sanitation of food plant equipment, reuse of wastewater, treatment and lowering BOD and COD of food plant waste.

Abbreviations

QAC: Quaternary Ammonium Compound
GRAS: Generally Recognized As Safe
H_2O_2: Hydrogen Peroxide
U.S.: United States
UV: Ultra Violet
eV: Electron Volt
BSF: Bath- Substrate- Flow
CIP: Cleaning-in-Place
RNA: Ribonucleic Acid
TSP: Trisodium Phosphate
THM: Trihalomethanes
HAA: Haloacetic acids
PTFE: Polytetrafluoroethylene
EPDM: Ethylene Propylene Diene Monomer

BOD: Biological Oxygen Demand

COD: Chemical Oxygen Demand

HTST: High Temperature Short Time

References

Arnold LJ 2009. The Development, Implementation and Validation of a Cleaning System within Dairy Processing Plant, Thesis, University of Wales Institute, Cardiff.

Arranz MAP, Schories G 2007. The use of aqueous ozone for cleaning operations in breweries, Ozone clean in place in food industries - ozone as an alternative sanitizing agent for cleaning operations in food industry, IOA Conference and Exhibition Valencia, Spain (October 29 – 31, 2007).

Baratharaj V. Ozone in waste water treatment. Ozone Technologies and Systems India Pvt.Limited, Chennai, India.

Baratharaj V. Use of ozone in food processing and cold storage. Ozone Technologies and Systems India Pvt. Limited, Chennai, India.

Canut A, Pascual A 2007. OzoneCip: Ozone Cleaning in Place in Food Industries. IOA Conference and Exhibition Valencia, Spain (October 29 – 31, 2007).

Chand R, Bremner DH, Namkung KC, Collier PJ, Gogate PR 2007. Water disinfection using the novel approach of ozone and a liquid whistle reactor. Biochemical Engineering Journal 35, pp.357-364.

Dosti B (1998). Effectiveness of ozone, heat and chlorine for destroying common food spoilage bacteria in synthetic media and biofilms, Thesis, Clemson Universtiy, Clemson, SC, pp. 69.

Food Safety Technology: A Potential Role for Ozone? 1998. Agricultural Outlook. Economic Research Service/USDA. Pp.13-15

Guzel-Seydim ZB, Greene AK, Seydim AC 2004. Use of ozone in the food industry. Lebensm.-Wiss. u.-Technol. 37.pp.453–460

Guzel-Seydim ZB, Wyffels JT, Greene AK, Bodine AB 2000. Removal of dairy soil from heated stainless steel surfaces: Use of ozonated water as a prerinse. Journal of Dairy Science, 83, pp.1887–1891.

Jurado-Alameda E, Garcia-Roman M, Altmajer-Vaz D, Jimenez-Perez JL 2012. Assessment of the use of ozone for cleaning fatty soils in the food industry. Journal of Food Engineering.Vol. 110. pp.44–52.

Pascual A, Llorca I, Canut A 2007. Use of ozone in food industries for reducing the environmental impact of cleaning and disinfection activities. Trends in Food Science and Technology 18, S29–S35.

2016, Dairy and Food Product Technology
Pages 193–202
Editors: Birendra Kumar Mishra and Subrota Hati
Published by: BIOTECH BOOKS, NEW DELHI

Chapter 13

Recent Trends in Nanocomposite Packaging Materials for Dairy and Food Applications

P. Narender Raju

Scientist, Dairy Technology Division,
National Dairy Research Institute, Karnal

Introduction

The rapid industrialization, population explosion and changing life-styles are leading to increased demand for processed and packaged foods. Although there is a wide variation in the level of processing of different food materials the size of semi-processed and ready-to-eat packaged food industry is over Rs. 4000 crores and is growing at over 20 per cent. Food packaging like any other packaging is an external means of preservation of food during storage, transportation and distribution. Hence, it forms an integral part of the product manufacturing process. Food packaging performs four major disparate functions *viz.* containment, protection, convenience and communication. In pursuit of achieving these goals, many materials have been discovered by man for use as food packaging materials *viz.* wood and paper, glass, metals, plastics, composites, etc. The inherent properties of these packaging materials make them either highly suitable or unsuitable for a particular food product or level of package. The food industry has been largely depending on the petroleum-based plastics for packaging materials. The major reasons for the popularity of plastic packaging materials are their excellent functionality, ease of processing, light weight and low cost. However, in spite of their versatility, a limiting property of most plastics in food packaging is their poor barrier to gases and vapors, including oxygen, carbon dioxide and organic vapors. Further, they are considered menace to environment once they reach municipal solid waste (MSW) and/or landfills or as they not biodegradable.

Food packaging is the largest user of plastics (~40 per cent). In India, as per the Central Pollution Control Board, approximately 15,300 tonnes of plastic waste is generated per day (The Hindu, 2011). The volume of plastics discarded annually creates a substantial waste which is causing a great threat to environment. Consequently, the approach of making materials from biodegradable materials that can be disposed of through composting or recycling got momentum. As a result a number of biodegradable materials such as naturally occurring polymeric materials, polymers made by polymerization of organic molecules and biodegradable polymers from petrochemicals have been investigated for use as alternative to plastics. Biopolymers from agricultural food stocks, food processing waste and other resources have the ability upon blending and/or processing to result in biopolymeric packaging material called as biodegradable polymers or bioplastics (Davis and Song, 2006). Unfortunately, so far the use of biodegradable films for food packaging has been strongly limited because of the poor barrier properties and weak mechanical properties shown by natural polymers. Also, biopolymers cannot meet the requirements of a cost-effective film with mechanical and barrier properties matching those of plastics (Kumar *et al.*, 2011). The most frequently used strategies to enhance barrier properties are blending of polymers, coating with high barrier materials and the use of multilayered films containing a high barrier film. Recently, a new class of materials represented by bio-nanocomposites has proven to be promising option in improving the mechanical, barrier and thermal properties of these biopolymer-based packaging materials. Polymers can also be added with suitable fillers to form composites for enhanced barrier properties.

Polymer Nanocomposites

Polymer nanocomposites are created by dispersing an inert, nanoscale filler throughout a polymeric matrix in which the filler has at least one dimension smaller than 100 nm. Filler materials could be either flakes, fibers, whiskers or nanoparticles. The mechanical, thermal and barrier of nano-composites are often remarkably different from those of non-reinforced biopolymer-based materials. Addition of relatively low levels of nanoparticles (less than 5 per cent) have been shown to substantially improve the properties of finished plastic, increasing the deformability and strength, and reducing the electrical conductivity and gas permeability. Further, polymers when incorporated with certain nanoparticles have the ability to interact with the food/ environment and package and confer antimicrobial properties and while some continuously monitor and detect changes in the package environment. Such nanocomposites have applications in active and smart packaging systems. Filler materials which have been used widely used by researchers include clay and silicate nanoplatelets, silica (SiO_2) nanoparticles, carbon nanotubes, grapheme, starch nanocrystals, cellulose-based nanofibers or nanowhiskers, chitin or chitosan nanoparticles, silver nanoparticles ($AgNO_3$), titanium nanoparticles (TiO_2), magnesium nanoparticles (MgO), copper nanoparticles (CuO), zinc (ZnO), etc.

Advantages of Nanocomposites in Food Packaging

Enhancement of Barrier Properties

Critical issues in food packaging are that of permeability and migration. None of the packaging materials used for food applications is completely impermeable to atmospheric gases and water vapour. In general, permeability of a polymer to oxygen or moisture is dependent on a large number of interrelated factors including polarity and structural features of polymeric side chains, hydrogen bonding characteristics, molecular weight and polydispersity, degree of branching or cross-linking, processing methodology, method of synthesis and degree of crystallinity (Duncan, 2011). Various inorganic nanoparticles have been recognized as possible additives to enhance the polymer performance. Among all, as of now the layered inorganic solids like clay have attracted attention by packaging industry. This is not only due to their availability and low cost but also due to their significant enhancements and relatively simple processability. The characteristic feature of clay minerals is the stacked arrangement of negatively charged silicate layers with a thickness of about 1 nm and lateral extensions of about 100 μm (Figure 13.1). Montmorillonite (MMT), hectorite, saponite and kaolinite are the commonly used layered silicates (Ray and Okamoto, 2003; Duncan, 2011). MMT comprises of highly anisotropic platelets separated by thin layers of water. Each platelet contains a layer of aluminum or magnesium hydroxide octahedral sandwiched between two layers of silicon oxide tetrahedral. The faces of each platelet have a net negative charge, which causes the interstitial water layer (known as gallery) to attract cations (Ca^{2+}, Mg^{2+}, Na^{+}, etc.) (Figure 13.1). The structural characteristics contribute to MMT's excellent utility as a filler material for polymer nanocomposites, typically giving rise to impressive increase in polymer strength and barrier properties. Three types of composites are created when clay is dispersed in polymer (Figure 13.2). Clay nanoplatelets tend to agglomerate when dispersed in polymer leading to tactoid structure (microcomposites) with reduced aspect ratios and reduced barrier efficiencies. Hence, the key requirement for the generation of nanoclay composites is separation of the ultrafine layers, a process known as exfoliation. As the clay nanoparticles are essentially impermeable crystals, gas molecules must diffuse around them (tortuous path) rather taking a straight line path that lies perpendicular to the film surface (Figure 13.3). In general, the higher the degree of exfoliation of the nanoclay the higher the improvement in barrier and mechanical properties of the film.

When biopolymers are combined with nanoparticles, the resulting bio-nanocomposites exhibit significant improvements in the mechanical properties, dimensional stability and solvent or gas resistance with respect to the pristine polymer due to high aspect ratio and high surface area of nanoparticles. The application of nanocomposites promises to expand the use of edible and biodegradable films produced from agro-processing products and byproducts. Starch-clay is the most often cited biodegradable nanocomposites investigated for various applications including food packaging (Park *et al.*, 2002; Avella *et al.*, 2005; Chen and Evans, 2005; Yoon and Deng, 2006; Cyras, *et al.*, 2008; Tang *et al.*, 2008). Significant improvements in mechanical properties were reported with the addition of montmorillonite (MMT)

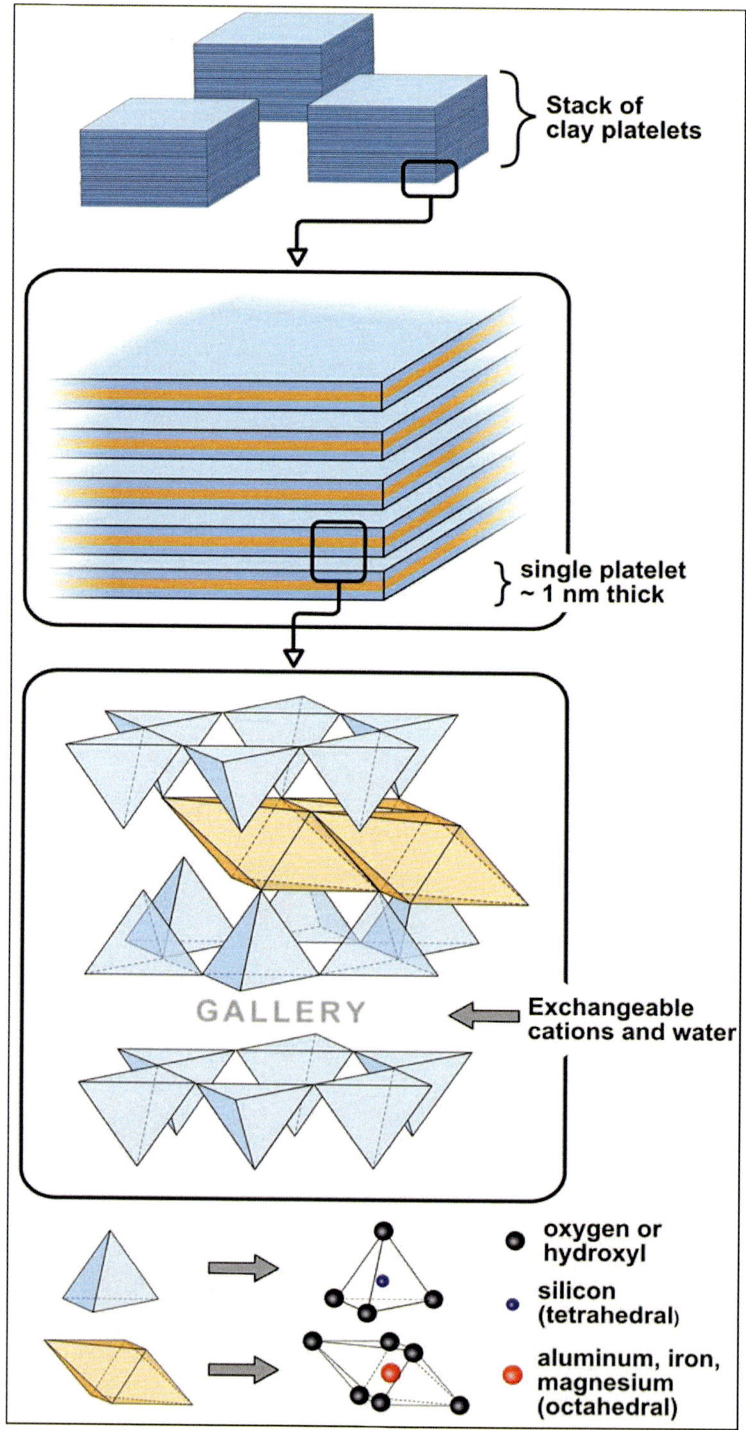

Figure 13.1: Structure of Montmorillonite.
(*Source*: Duncan, 2011)

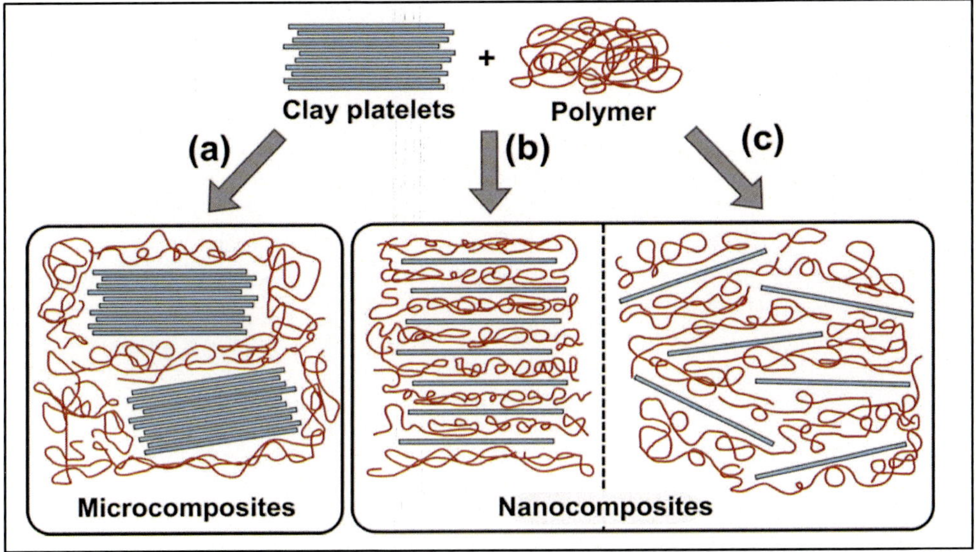

Figure 13.2: Polymer-clay Composite Morphologies. (a) Tactoid; (b) Intercalated; (c) Exfoliated (*Source*: Duncan, 2011).

clay. Azeredo *et al.* (2009) incorporated cellulose nanofibers in mango puree films to improve functional properties *viz.* tensile properties, water vapour permeability and glass transition temperature. Das *et al.* (2011) studied the physico-chemical properties of the jute micro/nanofibril reinforced starch/polyvinyl alcohol biocomposite films.

Nanocomposites of amorphous polylactic acid (PLA) and chemically modified kaolinite were studied by Cabedo *et al.* (2006). The combination of PLA and montmorillonite layered silicate may result in a nanocomposite with barrier properties suitable for food packaging applications (Sinclair, 1996; Thellen *et al.*, 2005). Yu *et al.*

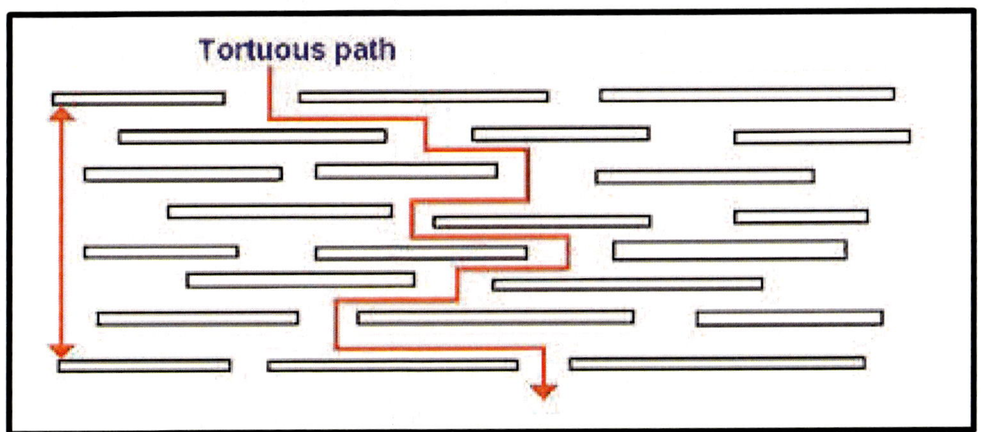

Figure 13.3: Illustration of Tortuous Pathway Created by Incorporation of Exfoliated Clay Nanoplatelets into a Polymer Matrix Film.

(2006) reported that through the use of high-powered ultrasonics, soy protein/clay nanocomposites could be produced, which exhibited improvement in modulus. Kumar *et al.* (2010) studied the effects of the pH of film forming solution, MMT content, and extrusion processing parameters on the structure and properties of soy protein isolate-MMT bio-nanocomposite films. Tunc *et al.* (2007) investigated that wheat gluten/MMT nanocomposite films could be prepared by casting. Voon *et al.* (2012) studied the effect of addition of halloysite nanoclay and SiO_2 nanoparticles on barrier and mechanical properties of bovine gelatin films. Biodegradable titanium dioxide (TiO_2)/whey protein isolate (WPI) blend films were made by casting denatured WPI film solutions incorporated with TiO_2 nanoparticles (Zhou *et al.*, 2009). Similar studies regarding interactions on ZnO-whey protein nanocomposites have been reported (Shi *et al.*, 2008). The reported improvements in barrier properties of biopolymer-based nanocomposites are given in Table 13.1.

Table 13.1: Barrier Properties of some Bionanocomposite Films

Film Type	*Water Vapour Permeability@ ($g\ mm\ hr^{-1}\ m^{-2}$)*	*Oxygen Permeability* ($m^3\ mm^{-2}\ s\ Pa$)*	*Reference*
Corn Starch + 0 per cent MMT	1.61		Tang *et al.* (2008)
Corn Starch + 3 per cent MMT	1.42		
Corn Starch + 6 per cent MMT	1.06		
Corn Starch + 9 per cent MMT	0.77		
Potato Starch + 0 per cent MMT	1.81		Tang *et al.* (2008)
Potato Starch + 3 per cent MMT	1.22		
Potato Starch + 6 per cent MMT	0.98		
Potato Starch + 9 per cent MMT	0.84		
Soy protein isolate (SPI)	3.1[a]		Rhim *et al.* (2005)
SPI + MMT	2.9[a]		
SPI+ Bentonite	1.5[a]		
SPI + Talc	2.3[a]		
SPI + Zeolite	2.0[a]		
Poly lactic acid (PLA)			Lagaron *et al.* (2005)
PLA		11×10^{-19}	
PLA + 4 per cent MMT		10×10^{-19}	
PLA + 4 per cent Kaolinite		6×10^{-19}	

* 21°C, 40 per cent RH; @ 25°C, 75 per cent; a: $ng.m.m^{-2}.s.Pa$

Enhancement of Antimicrobial Properties

In addition to proving as a passive barrier, packaging can contribute to the control of microbial growth in food products, which cause spoilage or in case of pathogens, diseases and illness. Antimicrobial (nisin, silver oxide, zinc oxide, magnesium oxide) nanoparticle and antioxidant coatings in the matrix of the packaging material can reduce the development of bacteria on or near the food product, inhibiting the microbial growth on non-sterilized foods, maintain sterility of pasteurized foods by preventing post-contamination and prevent oxidative changes in food thereby improving the quality. Most antimicrobial activities of nanocomposites have centered on nanoparticles of silver and zinc oxide. Silver has a long history of being used as an antimicrobial agent in food and beverage storage applications. Silver nanoparticles have been found to be potent agents against numerous species of bacteria including *Escherichia coli, Enterococcus faecalis, Staphylococcus aureus, Vibrio cholera, Pseudomonas aeruginosa, Shigella flexneri, Bacillus subtilis, Salmonella enterica, Micrococcus luteus, Listeria monocytogenes* and *Klebsiella pneumonia* (Duncan, 2011). Cellulose is a good carrier of silver nanoparticles. Fernandez *et al.* (2010) developed cellulose-silver nanoparticle hybrid materials (cellulose-based silver loaded absorbent pads) and demonstrated that when fresh cut "Piel de Sapo" melon was placed over the pads, the pads released silver ions after the melon juice impregnated the pad and controlled the spoilage-related bacteria.

Titanium dioxide (TiO_2) coated packaging film has shown to considerably reduce *E. coli* contamination on food surfaces. TiO_2 is a photocatalytic disinfecting material for surface coatings (Chawengkijwanich and Hayata, 2008). Page *et al.* (2007) reported that combining titanium dioxide with silver has been shown to improve the disinfection process through improving photocatalysis. Zinc oxide also exhibits antibacterial activity and its antimicrobial property increases with decreasing particle size (Yammamoto, 2001). Recently, a starch/ZnO-carboxymethylcellulose sodium nanocomposite was prepared using ZnO nanoparticles stabilized by carboxymethylcellulose sodium (CMC) as the filler in glycerol plasticized-pea starch (Yu *et al.*, 2009). Busolo *et al.* (2010) developed a novel silver-based nanoclay as an antimicrobial in PLA food packaging coatings and reported that such compounds when incorporated at either 1 per cent, 5 per cent or 10 per cent exhibited strong biocidal activity (99.9 per cent CFU reduction) against *Salmonella* spp. when the sample taken was 1.5 g.

Smart Packaging

A package that continuously monitors the internal environment and responds or communicates the changes to external environment and/or consumer, beyond performing the basic functions, is called as a smart package and the technique as smart packaging. The unique chemical and electro-optical properties of nanoscale particles enable them to be part of a smart package. Such nanoparticles can be used to detect the presence of gases, aromas, chemical contaminants and pathogens. Mills (2005) reported development of a promising photoactivated indicator ink for in-package oxygen detection based on nanosized TiO_2 or SnO_2 particles and a redox-active dye (methylene blue). This detector gradually changes colour in response to even minute changes in oxygen. Luechinger *et al.* (2007) reported development of porous metal films for optical humidity sensing from copper nanoparticles protected by a 2-3 nm carbon coating. The film is reported to have exceptional sensitivity with optical shifts in the visible light range of up to 50 nm for a change of 1 per cent in relative humidity. A non-invasive method of measuring CO_2 content in modified atmosphere packages has been devised by von-Bultzingslowen *et al.* (2002). It was reported to be based upon lifetime analysis of luminescent dyes standardized by fluorophore-encapsulated polymer nanobeads. Many such applications of nanoparticles in smart packaging are being reported.

Future Scope

With the worldwide growing scientific evidence of the potential benefits of nanotechnology in food packaging, there exists huge scope for us to get benefitted in India as such applications are at nascent stage. Biopolymers from agricultural food stocks, food processing waste and other resources could be exploited for developing nanocomposite films for packaging applications. Also, development of sensor and antimicrobial based smart packaging materials could be developed.

References

Avella, M., De-Vlieger, J.J., Errico, M.E., Fischer, S., Vacca, P. and Volpe M.G. (2005). Biodegradable starch/clay nanocomposite films for food packaging applications. *Food Chemistry*, Vol.93: 467-474.

Azeredo, H.M.C., Mattoso, L.H.C., Wood, D. Williams, T.G., Avena-Bustillos, R.J. and McHugh, T.H. (2009). Nanocomposite edible films from mango puree reinforced with cellulose nanofibers. *Journal of Food Science*, Vol. 74(5): N31-N35.

Busolo, M.A., Fernandez, P., Ocio, M.J. and Lagaron, J.M. (2010). Novel silver-based nanoclay as an antimicrobial in polylactic acid food packaging coatings. *Food Additives and Contaminants*. Vol. 27(11): 1617-1626.

Cabedo, L., Feijoo, J.L., Villanueva, M.P., Lagaron, J.M. and Gimenez, E. (2006). Optimization of biodegradable nanocomposites based on a PLA/PCL blends for food packaging applications. *Macromol. Symp.* Vol. 233(1): 191-197.

Chawengkilwanich C. and Hayata, Y. (2008). Development of TiO_2 powder-coated food packaging film and its ability to inactivate *Escherichia coli* in vitro and in actual tests. *International Journal of Food Microbiology*, Vol.123(3): 288-292.

Chen, B.Q. and Evans, J.R.G. (2005). Thermoplastic starch-clay nanocomposites and their characteristics. *Carbohydrate Polymers*, Vol. 61: 455-463.

Cyras, V.P., Manfredi, L.B., Ton-That, M.T. and Vazquez, A. (2008). Physical and mechanical properties of thermoplastic starch/montmorillonite nanocomposite films. *Carbohydrate Polymers*, Vol.73(1): 55-63.

Das, K., Ray, D., Bandyopadhyay, N.R., Sahoo, S., Mohanty, A.K. and Misra, M. (2011). Physico-mechanical properties of the jute micro/nanofibril reinforced starch/polyvinyl alcohol biocomposite films. *Composites Part B: Engineering*, Vol. 42(3): 376-381.

Davis, G. and Song, J.H. (2006). Biodegradable packaging based on raw materials from crops and their impact on waste management. *Industrial Crops and Products*, Vol.23: 147-161.

Duncan, T.V. (2011). Applications of nanotechnology in food packaging and food safety: barrier materials, antimicrobials and sensors. *Journal of Colloid and Interface Science*, Vol. 363: 1-24.

Fernandez, A., Picouet, P. and Lloret, E. (2010). Cellulose-silver nanoparticle hybrid materials to control spoilage related microflora in absorbent pads located in trays of fresh-cut melons. *International Journal of Microbiology*, Vol.142: 222-228.

Kumar, P., Sandeep, K.P., Alavi, S. and Trong, V.D. (2011). A review of experimental and modeling techniques t odetermine properties of biopolymer-based nanocomposites. *Journal of Food Science*, Vol.76(1): E2-E14.

Kumar, P., Sandeep, K.P., Alavi, S., Truong, V.D. and Gorga, R.E. (2010). Preparation and characterization of bio-nanocomposite films based on soy protein isolate and montmorillonite using melt extrusion. *Journal of Food Engineering*, Vol.100: 480-489.

Luechinger, N.A., Loher, S., Athanassiou, E.K., Grass, R.N. and Stark, W.J. (2007). High sensitive optical detection of humidity on polymer/metal nanoparticle hybrid films. *Langmuir*, Vol.23(6): 3473-3477.

Mills, A. (2005). Oxygen indicators and intelligent inks for packaging food. *Chemical Society Reviews*, Vol. 34(12): 1003-1011.

Page, K., Palgrave, R.G., Parkin, I.P., Wilson, M., Savin, S.L.P and Chadwick, A.V. (2007). Titania and silver-titania composite films on glass-potent antimicrobial coatings. *Journal of Materials Chemistry*. Vol.17(1): 95-104.

Ray, S.S. and Okamoto, M. (2003). Polymer/layered silicate nanocomposites: a review from preparation to processing. *Progress in Polymer Science*, Vol. 28: 1539-1641.

Shi, L., Zhou, J. and Gunasekaran, S. (2008). Low temperature fabrication of ZnO-whey protein isolate nanocomposite. *Materials Letters*, Vol. 62(28): 4383-4385.

Sinclair, R.G. (1996). The case for polylactic acid as a commodity packaging plastic. *Journal of Mass Spectroscopy and Pure Applied Chemistry*. Vol. 33(5): 585-597.

Tang, X., Alavi, S. and Herald, T.J. (2008). Barrier and mechanical properties of starch-clay nanocomposites films. *Cereal Chemistry*, Vol. 85(3): 433-439.

The Hindu (2011). Cutting plastic waste. *The Hindu*, editorial section published on 28th February.

Thellen, C., Orroth, C., Froio, D., Lucciarini, J., Farrell, R., D'Souza, N.A. and Ratto, J.A. (2005). Influence of montmorillonite layered silicate on plasticized poly(l-lactide) blown films. *Polymer*. Vol.46(25): 11716-11727.

Tunc, S., Angellier, H., Cahyana, Y., Chalier, P., Gontard, N. and Gastaldi, E. (2007). Functional properties of wheat gluten/montmorillonite nanocomposite films processed by casting. *Journal of Membrane Science*, Vol. 289(1-2): 159-168.

Voon, H.C., Bhat, R., Easa, A.M., Liong, M.T. and Karim, A.A. (2012). Effect of addition of halloysite nanoclay and SiO_2 nanoparticles on barrier and mechanical properties of bovine gelatine films. *Food and Bioprocess Technology*. Vol.5(5): 1766-1774.

Yammamoto, O. (2001). Influence of particle size on the antibacterial activity of zinc oxide. *International Journal of Inorganic Materials*, Vol.3: 643-646.

Yoon, S. and Deng, Y. (2006). Clay-starch composites and their application in packaging. *Journal of Applied Polymer Science*, Vol. 100(2): 1032-1038.

Yu, L., Dean, K. and Li, L. (2006). Polymer blends and composites from renewable resources. *Progress in Polymer Science*, Vol. 31(6): 576-602.

Zhou, J.J., Wang, S.Y. and Gunasekaran, S. (2009). Preparation and characterization of whey protein film incorporated with TiO_2 nanoparticles. *Journal of Food Science*, Vol. 74(7): N50-N56.

2016, Dairy and Food Product Technology
Editors: Birendra Kumar Mishra and Subrota Hati
Published by: BIOTECH BOOKS, NEW DELHI

Pages 203–220

Chapter 14

"Probiocap": A Novel Scientific Approach

S. Makhal[1], S. Hati[2] and S. Bera[3]

[1]ITC Ltd., R&D Center, Bangalore
[2]Anand Agricultural University, Anand – 388 110, Gujarat
[3]Faculty of Dairy Technology, WBUAFS, Mohanpur – 741 252, West Bengal

Introduction

Probiotics, live organisms that have multiple functions and effects on the human body, play an important role in maintaining the precise balance of desirable and undesirable bacteria in the human digestive system The potential health-promoting effects of dairy products, which incorporate probiotic organisms, such as *Lactobacillus* spp. and *Bifidobacterium* spp. has stimulated a foremost research exertion in recent years. All of us carry in our intestinal tracts a complex ecosystem of microbes. These bacteria are highly important to our health, providing us with protection against intestinal infections, supplying us with additional nutritional value from the food we eat, and contributing to the development of our immune system. Probiotic foods are the most important discipline of functional foods, which are defined as "foods containing live microorganisms, which actively enhance the health of consumers by improving the balance of microflora in the gut when digested live in sufficient numbers" (Fuller, 1992b). There have been long-term interests in the use of cultured milks products with various strains of lactic acid and other probiotic bacteria to improve health of humans (Salminen *et al.*, 1998a). Studies relate the possible health benefits of consuming cultured and culture containing milks. Understanding the relationship between microbial populations in the colon and health continues to increase. Clearly, it is important to comprehend the effects of colonic bacteria on host health in order to fully exploit potential applications of prebiotics and probiotics. The colon can be both an organ of health and of disease, especially with regard to the microbiota. Thus the interest in probiotics and prebiotics arises, in part, from the desire to manipulate or enhance the 'beneficial' gut microbiota in a manner that decreases the peril of developing bacterial associated colonic diseases.

For probiotic foods to be beneficial for human health, the probiotic strains should maintain their viability in cheese until the time of consumption. We are still at the early state in the development of probiotic cheese for human application in terms of its viability and survivability during processing, preservation and ripening as well as GI transit. The development of spray dried skim milk powders harbouring probiotics has been proved useful as direct-vat inoculation, thereby providing more convenient way of incorporation of beneficial biocultures into cheese. One of the novel challenges in developing probiotic foods is to get survivability during storage period without hampering the normal cultures responsible for the development of proper body and texture and flavour as well as in time of preservation treatments given used to enhance the shelf life of the products. In addition, it is obvious that their ability to reproduce in the GI tract is an important necessity for their overall efficacy. A radical application of microencapsulation, "Probiocap technology" for the improved stabilization of probiotics is now being evolved, which has been seemed to dramatically enhance the viability of probiotic bacteria. That is why technolologlies to develop "Probiocap" and other such may be used as promising itinerary for transporting high concentration of biomass ensuring maximum protection of the biological integrity of the probiotic products.

Probiotic Foods

In the present changing scenario, consumer awareness has increased towards the health driven foods, hence recent advances in functional foods demonstrate much promise in new product development using probiotics to derive health benefits. Probiotic foods come under functional foods which, in addition to their basic nutritive value and natural being, will contain the proper balance of ingredients, which will help as functional, therapeutic and nutraceutical agents enabling the human population to prevent the illness and diseases. Functional foods serve to endorse health or help to prevent diseases and in general, the term is used to indicate a food that contains some health promoting components beyond traditional nutrients (Berner and O'Dannell, 1998). Probiotic foods are the most important discipline of functional foods, which are defined as "foods containing live microorganisms, which actively enhance the health of consumers by improving the balance of microflora in the gut when ingested live in sufficient numbers" (Fuller, 1992b). Because the human gut microbiota can play a major role in host health, there is currently a dynamic interest in the manipulation of the gut flora toward a potentially remedial community with the application of prebiotics. A prebiotics is a non-digestible food ingredient that beneficially affects the host by stimulating the growth and/or activity of one or a limited number of bacteria in the colon and thus improves host health (Gibson and Roberfroid, 1995). The prebiotics are mainly non-digestible carbohydrates (oligo and poplysaccharides), some peptides and proteins, and certain lipids and because of their chemical structure, these compounds are not absorbed in the upper part of the GIT or hydrolizesd by human digestive enzymes. There have been long-term interests in the use of cultured milks products with various strains of probiotic bacteria to improve health of humans (Salminen *et al.*, 1998a,b).

Understanding the relationship between microbial populations in the colon and health continues to increase. Clearly, it is important to comprehend the effects of colonic bacteria on host health in order to fully exploit potential applications of prebiotics and probiotics. The colon can be both an organ of health and of disease, especially with regard to the microbiota. Thus the interest in probiotics and prebiotics arises, in part, from the craving to manipulate or enhance the 'beneficial' gut microbiota in an approach that decreases the peril of developing bacterial associated colonic diseases. This may occur by provision of substrates that subvert the toxigenic potential, *e.g.* switching a proteolytic species towards a more saccharolytic type of fermentation.

Therapeutic Dose

Since now days consumers are more diverted towards the health driven foods, there is promising prospective to develop foods with some therapeutic values using probiotics. Currently, there are no legal recommendations for consumption of probiotics in foods. However, it is generally accepted that health benefits from consumption of probiotic strains should be demonstrated through controlled clinical trials and that the manufacturer should provide advice on the minimum dose and duration of use of each individual strain or product (FAO/WHO, 2001). Adequate numbers of viable cells, namely the "therapeutic minimum" needs to be consumed regularly for transfer of the probiotic effect to consumers. The international Dairy Federation (1997) proposed that in probiotic foods, "the specific microorganisms shall be viable, active and abundant at the level of at least 10^7 cfu/g in the product to the date of minimum durability" (Ouwchand and Salminen, 1998).

It is evident that there are difficulties involved in defining a general minimum effectual dose for all probiotic cultures, given that variations occur depending on the particular strain or delivery system used. It has been suggested that approximately 10^9 cfu/day of probiotic microorganisms is necessary to elicit health effects. Based on daily consumption of 100 g of a probiotic food, it has been suggested that a product should contain at least 10^7 cells/g, a level paralleling current Japanese recommendations (Ishibashi and Shimamura 1993), but considerably higher numbers have been proposed by others (Lee and Salminen, 1995).

Survival of these bacteria during shelf life and until consumption is an imperative consideration. Kurmann and Rasic (1991) suggested to achieve optimal potential therapeutic effects, the number of probiotic organisms in a probiotic food should meet a suggested minimum of > 10^6 cfu/ml as satisfactory level. This criterion is referred to as the "therapeutic minimum" (Davis *et al.*, 1971 and Rybka and Kailasapathy, 1995). One should aim to consume 10^8 live probiotic cells per day. Regular consumption of 400-500 g/week of bio-yoghurt, containing 10^6 viable cells per ml would provide these numbers (Tamime *et al.*, 1995). Fermented Milks and Lactic Acid Bacteria Beverage Association of Japan has developed a standard, which requires a minimum of 10^7 viable bifidobacteria cells/ml to be present in fresh dairy products (Ishibashi and Shimamura, 1993). The National Yoghurt Association (NYA) of the United States specifies 10^8 cfu/g of lactic acid bacteria at the time of manufacture as a prerequisite to use the NYA "Live and Active culture" logo on the containers of products (Kailasapathy and Rybka, 1997).

Probiotics: A Diet-Health Alliance

There is a universe of life within the digestive tract. Microorganisms populate the intestines and colon in numbers 10 times greater than the total number of cells in the body itself - over 10 billion per gram of stool. One half the dry weight of stools is a microorganism. This population of organisms is being increasingly found to have profound effects on health. There is a delicate balance, however, between those organisms which contribute to health in assisting digestion, synthesizing nutrients, and inhibiting cancer-causing biochemicals, for example - and those which can cause disease. Beneficial (probiotic) organisms in the diet can help rebalance the digestive tract. It is only recently, that the interrelationship between intestinal microorganisms and the health benefits deriving from it is beginning to be understood. At present, it is generally recognized that an "optimum" balance in microbial population in our digestive tract is associated with good nutrition and health (Rybka and Kailasapathy, 1995). The microorganisms primarily associated with this balance are lactobacilli and bifidobactera. Increasing evidence indicates that consumption of probiotic microorganisms can help maintain such a favourable microbial profile and results in several therapeutic benefits (Lourens-Hattingh and Vijoen, 2001). In recent years, probiotic bacteria have increasingly been incorporated into foods as dietary adjuncts.

The human intestinal tract constitutes a complex bionetwork of microorganisms. More than 400 bacterial species have been identified in the faeces of a single subject (Finegold *et al.*, 1977). The bacterial populations in the large and small intestine are very high and reach maximum counts of 10^{12} cfu/g and $10^4 - 10^8$ cfu/g, respectively (Hoier, 1992). All these intestinal microflora exist in dynamic balance with one another. Some are useful, some detrimental and some neutral to the physiological functions of the body. Thus, intestinal microflora can influence health in a number of ways, both positive and negative (Sandine, 1979). These include impacts on nutrition and physiological functions. For instance, one of the short chain fatty acids produced by colonic bacteria (butyrate) is important in determining the rate of colonic cell growth and differentiation, drug efficacy, carcinogenesis, immunological responses, resistance to infection and resistance to endotoxins and other stresses (O'Sullivan, 1996 and Buttriss, 1997). The beneficial bacteria tend to predominate during periods of good health. If the ecological balance of the gut is disturbed due to prolonged disease, dispossession from foods and water, travel (especially by air), antibiotics, radiation etc. (Hanevaar and Huis in't Veld, 1992), certain microorganisms with negative roles in the human system may dominate. In such cases, there is an increase in the products of putrefaction, toxins and carcinogens. Pathogenic bacteria, which are normally present at low levels, also boom, if the resistance of the body is lowered for any reason, manifesting their pathogenicity and causing diseases. Living probiotic cultures help uphold the critical balance and can stabilize a disturbed intestinal flora.

There is mounting evidence of health benefits of probiotic foods and the reverberation in market is to combine these health benefits with product appeal and versatility without any procrastination. Current clinical applications of probiotic bacteria in the well document areas, such as treatment of acute rota virus diarrhoea, lactose maldigestion, constipation, colonic disorders and side-effects of pelvic radiotherapy and more recently food allergy including milk hypersensitivity and changes associated with colon cancer development (Salminen *et al.*, 1998). There are myriad evidences to support the view that oral administration of some *Lactobacillus* and *Bifidobacterium* species are able to restore the normal balance of probiotic populations in the intestine. In addition to their established role in GI therapy, the probiotic foods are claimed to serve several nutritional and therapeutic benefits, such as antimicrobial properties (Shah, 2000), antimutagenic properties (Lankaputra and Shah, 1998), anticarcinogenic properties (Mitsuoka, 1989), improvement in lactose metabolism (Vesa *et al.*, 1996), reduction in serum cholesterol (Fukushima and Nakano, 1996) and immune system simulation (Schiffrin *et al.*, 1994). Therefore, in the near future probiotic foods will be seen in many different markets beyond what is seen today. LFRA's 1996 report valued the global market at $ 6.6 billion in 1994,with Japan accounting for just under half of that. Some forecasts suggested that the market would reach $17 billion by the year 2000,with the fastest growth rates in the US (Young, 1996).

The claimed beneficial effects from the consumption of fermented milks were once a very contentious issue. Research conducted since the turn of the past century has, however, enhanced the understanding of the resulting therapeutic effects and it is currently recognized as wholesome. The consumption of probiotic products is helpful in maintaining good health, restoring body vigour, and in skirmishing intestinal and other diseases (Mital and Garg, 1992). Fuller (1989) listed out the claimed beneficial effects and therapeutic application of probiotic bacteria in humans, which includes:(I) Beneficial effects, such as maintenance of normal intestinal realm, augmentation of immune system, reduction of lactose intolerance, reduction in serum cholesterol levels, anticarcinogenic activity and improved nutritional value of foods, and (ii) Therapeutic applications, such as prevention of urogenital infections, mitigation of constipation, protection against travellers diarrhoea, prevention of infantile diarrhoea, reduction of antibiotic of induced diarrhoea, prevention of hypercholesterolaemia, protection against colon/bladder cancer and prevention of osteoporosis. Probiotic bacteria, thus, offer new dietary alternatives for the management of such conditions through stabilization of intestinal microflora, promotion of colonization resistance, regulation of the immune response and preservation of intestinal integrity (Salminen *et al.*, 1998a).

Probiotics: Challenges

The hurdles of getting a product from the field to the bottle to store shelves to the end user are numerous and immense. If the purchaser is lucky enough to buy a fresh nutraceutical whose contents are still intact, this doesn't guarantee that they will receive a beneficial dose. Even the freshest, most biologically active substances can lose much of their potencies in the long passage from mouth to stomach. If the final destination is the intestines (as is the case with probiotics), that makes it even more of a long shot. Probiotic foods formulators, now are confronted with a myriad of obstacles in getting an effective product to market, of which most consumers are blissfully unaware. Stability, heat and cold sensitivity, rancidity, unpalatable flavors, unappealing colors, odors, and adulteration are just the beginning.

While manufacturers have long been fortifying products with probotics, they have faced significant processing challenges regarding the stability and survivability of probiotics during processing and preservation treatments as well as during their passage through the stomach. In fact, many active culture die even before the consumer receives any of the health benefits. If these beneficial microorganisms stay alive during food processing conditions, they must also survive during their passage from the mouth to intestine. Specially, probiotics are extremely susceptible to environmental conditions *viz.* water, oxygen, processing and preservation treatments, acidity and salt concentration, which collectively affect the overall viability of probiotics. Probiotics can lose upto 95 per cent of their viability when unprotected. This is why the concentration of viable cells in the duodenum rarely exceeds 10^5 cfu/g of product, while as high as 10^8 and 10^{11} cfu/g have been found in the ileum and the colon, respectively (Siuta-Cruce and Goulet, 2001). These unprotected microflora are also typically destroyed during such processing treatments as sterilization, pasteurization, microwave treatments, thermization, retorting disinfections, and irradiation, washing and peeling.

Technologies for Improving Survivability of Probiotics

Probiotic foods or supplements do not necessarily meet a particular level or even state how much of the culture is present in viable form in the product. This unregulated "state of affairs" combined with increased scrutiny of supplement and food claims by regulatory agencies and scientific community has led to ensure that probiotic foods can provide a high enough concentration of live probiotics to practically afford a host of health benefits to an individual. Probiotic suppliers have developed a variety of proprietary techniques to preserve and protect the integrity of these tiny living organisms.

Freeze-drying

It has long been the main process used to preserve probiotic organisms. Generally, freeze-drying is the most commonly used method to protect probiotics from stomach acid and to deliver a high number of live organisms to the small and large intestines Probiotics are generally freeze-dried in a state of suspended animation.

Probio-Tec

In "Probio-Tec" proprietary process, bacterial cultures are coated with polysaccharides to ensure stability throughout the manufacturing process and to protect the organisms from stomach acid. To keep more bacteria alive longer, stabilizing agents (such as antioxidants) are also being used before and after freeze-drying. The main purpose is to protect the organisms from heat, light, air and moisture. These four factors affect stability. The more it can protect the product from these elements, the more stable it will be.

LiveBac

A new-patented process, based on Microencapsulation, is "LiveBac" that protects organisms from air and moisture, thus greatly extending product shelf life even without refrigeration. LiveBac probiotics are said to end up with the same number of organisms that they started with. Before developing this process, one would lose one logarithm a year. From an initial number of one million, the final resultant will be of 100,000. Some encapsulated LiveBac probiotics had zero left (non viable cells) at end of year. So far, by this technique, the investigators have not lost a single count in 10 months and they strongly suspect this will be the same at the end of year.

Controlled Delivery Technology (CDT)

Patented Controlled Delivery Technology (CDT) process has been developed that can be used on numerous supplements and formulations including probiotic products. Using this technology, tablets or capsules can be either "programmed" to release active ingredients into the GI tract at a constant rate or pulsed at precisely time intervals. They can also be used in products that contain several active ingredients, which can be distributed to specific locations in the digestive tract, thus optimizing uptake. An offshoot of this technology is Bio-Tract, a patented name for CDT applied to probiotics. "For instance, *Lactobacillus acidophilus* can be targeted to release in the upper intestine, while bifidobacteria is released in the lower intestine.

Packaging

Packaging is also an important consideration. Manufacturers should choose a material that is impervious to moisture, like glass. In Europe, aluminum tube containers are popular, but it is not found in the United States. Scientists also recommended that manufacturers provide desiccants, either in canister or in small sachets containing silicon dioxide or silica gel to reduce moisture. They said that while some of the new encapsulation techniques promise to extend shelf life, the long-term data remain to be seen. In addition, "Nutraceutix" is presently offering live probiotics in foil pouches or sachets.

Cold or Room Stability

Several manufacturers contended that nothing will preserve probiotics indefinitely at room temperature and that it's still best to refrigerate. Some scientists suggested the best means of preserving probiotics is freezing them, followed by refrigeration and room temperature is the least preferable option. One can preserve probiotics for a long time if they are kept frozen without any damage to the product, as frozen condition slows down the process of die-off considerably.

Enteric Coating

Institut Rosell of Montreal, Canada uses its STAR (Stomach Acid Resistance) enteric-coating technology to protect probiotics during the passage through the gastric barrier. The STAR enteric coating prevents the solubilization of capsules in the stomach and protects probiotics against acid shock. The process can be applied to any type of capsule and is approved as Generally Regarded As Safe (GRAS). Institut Rosell now uses a water-based coating instead of the solvent-based coatings used by pharmaceutical companies. The objective is to protect and coat the capsules so that the consumer can take it at any point in time, not just following a meal. For instance, 95 percent of capsules with STAR coating will survive the acidity of the stomach compared to only two percent without coating.

In the present state of art, such probiotic cultures are available that are stable at ambient temperatures by packaging them in such a manner so as to protect them from the adverse conditions of oxygen, moisture, light and heat, *e.g.* in gel capsules, similar to vitamin capsules. The important aspect of this technology is its ability to deliver high dose of viable probiotics in the jejunum and the ileum, which can escape the harsh processing and storage treatments, and harsh acidity of the stomach and bile in the intestine. Capsules have until now been considered the preferred delivery form for probiotics since they don't involve heating. Nutraceutix offers a new technology "Cryotableting" that is able to preserve the organisms in tablet form by quickly cooling down the tablets after they've been produced. The advantage of tablets over capsules is that they don't require refrigeration.

However, the application of such large capsules is unfeasible for incorporation in food systems and is luxurious too. The most apposite alternative is to develop "Probiocap" by applying microencapsulation technology. Therefore, microencapsulation is the most suitable alternative technology to offer the best protection to the probiotic cells resulting from the freeze-drying and milling. Ultimately, such microencapsulated probiotics can be used in numerous food systems.

Microencapsulation

Microencapsulation is a process where droplets of liquids, solids, or gases (core) are coated by thin films (coatings), which protect the core until it is needed (Sheu and Rosenberg, 1995). Different types of core materials: solids, liquid and gases or mixture of these, such as dispersions of solids in liquids, solutions and complex emulsions have microencapsulated. Some times the core materials may contain, in addition to core material, other substances, such as antioxidants, emulsifiers, preservatives, stabilizers, which are added to enhance the stability and functionality of the active materials. In the foods alone, a large number of substances have been microencapsulated, such as acidulants, amino acids, antimicrobials, bases, colourants, edible oils, flavour, enzymes, microorganisms, flavour enhancers, leavening agents, minerals, sugars, salts, and vitamins (Kanawjia *et al.*, 1992). The core can be released at different times as and when occasion demands by any desired mechanisms, such as disruption, dissociation, dissolution or diffusion and with any desired rates, such as instantaneous, delayed, controlled or sustained release (Kanawjia *et al.*, 1992) depending on the properties of the coatings that are applied. The coating on a core is semi-permeable and protects the core from severe conditions and controls substances flowing into the core and the release of metabolites from the core (Jackson and Lee, 1991). In this process, cells are retained within the encapsulating membrane to reduce the cell injury or cell loss (Shah, 1999).

During microencapsulation, specially designed equipment coats probiotic bacteria in a matrix of vegetable fatty acids. This will increase formulation possibilities, broadening the range of ingredients with which probiotics can be blended. Similarly, mildly elevated food processing temperatures, in which heat is required to melt or partially cook food components or to preserve the product, can destroy bacteria integrity. Institut Rosell/Lallemand's microencapsulation provides protection at 50°C (122°F) for several hours, giving a protection that is over three times more resistant than an unprotected control. Additionally, the passage through the gastric barrier is very stressful for probiotics, especially on an empty stomach where the pH can be as low as 1.5. The hydrophobic coating surrounding microencapsulated bacteria protects the fragile microbial cells, allowing them to pass into the intestine.

In order to make this technology successful to entrap probiotics, the protective wall materials should be such that it would afford protection to the probiotics against the processing and storage treatments, GI transient but release them in post-stomach in the human body. It has been investigated that some lipid coatings are not only an efficient protective barrier against chemical entities, such as moisture, oxygen, and acids but also a good protector against short exposure to high temperature and pressure (Siuta-Cruce and Goulet, 2001).

Materials for "Probiocap"

The structure formed by microencapsulating agent around the core material is called the wall material. The wall protects the core against deterioration, limits evaporation of volatile core materials (Kadian *et al.*, 1999) The encapsulating agents should have certain ideal characteristics, depending on the objectives and requirements, process of encapsulation, chemical characteristics of the core material, the intended use of the core material, the conditions under which the product will be stored, and the processing conditions to which it will be exposed (Kanawjia *et al.*, 1992). Some general characteristics of the encapsulating agent are that it is insoluble in and non-reactive with the core material, have solubility in the end-product food system, and be able to withstand high temperature of the spray-drying process. Some typical encapsulation agents are dextrans, gums, starches or proteins.

Table 14.1: Coating Materials Used to Produce Microcapsules

Class of Coating Material	*Specific Types of Coatings*
Gums	Gum arabic, agar, sodium alginate, carageenan
Carbohydrates	Starch, dextran, sucrose, corn syrup
Celluloses	CMC, methycellulose, ethylcellulose, nitrocellulose, acetylcellulose, cellulose acetate-phthalate, cellulose acetate-butylate-phthalate
Lipids	Wax, paraffin, tristearin, stearic acid, monoglycerides, diglycerides, beeswax, oils, fats, hardened oils
Inorganic materials	Calcium sulfate, silicates, clays
Proteins	Gluten, casein, gelatin, albumin

Source: Jackson and Lee, 1991.

Entrapment of Microorganisms: Some Focuses

Several studies have reported on the microencapsulation to develop "Probiocap" by using gelatin or vegetable gum to provide protection to acid sensitive bifidobacteria

(Rao *et al.*, 1989 and Ravula and Shah, 1999). Entrapment of living microbial cells in calcium alginate is simple and low cost. Furthermore, alginate is non-toxic so that it may be safely used in foods. Alginate gels can be solubilized by sequestering calcium ions thus releasing entrapped cells (Rao *et al.*, 1989). Ravula and shah (1999) have developed a microencapsulation technique with sodium alginate. Encapsulated organisms were incorporated in fermented frozen dairy desserts and investigated the viability of probiotic bacteria. The counts of *L. acidophilus* and *bifidobacteria* decreased to <10^3 cfu/g in the control batch, whereas the counts were >10^5 cfu/g in the products made with encapsulated organisms.

Rao *et al.* (1989) developed a technology for microencapsulation of *Bifidobacterium longum* with cellulose acetate phthalate (CAP) using phase searation coacervation method. Microbiological analysis indicated that microencapsulated *B. pseudolongum* survived the simulated gastric environment in large numbers than non-encapsulated *B. pseudolongum*. In another report, Kim *et al.* (1988) described a method for the preparation of stable microencapsulated lactic acid bacteria using polyvinyl acetate phthalate. Due to the presence of ionizable phthalate group, this polymer is insoluble in acid media at pH of 5 or below but is soluble when the pH is increased to 6 or higher. In addition, this compound is physiologically inert when administered *in vivo* and is, therefore widely used as an enteric coating material for release of drugs and other pharmaceutical substances in the intestine, This is successfully used to prepare microcapsules of active vial antigens and other proteins for oral consumption (Moharaj *et al.*, 1984). Microencapsulation of bifidobacteria in K-carrageenan appeared to increase the viability of bifidobacteria in yoghurt (Adhikari *et al.*, 2000).

In another investigation, *L.delbruecki* spp. *bulgarius* cells were entrapped in beads of calcium alginate and evaluated for their ability to survive freezing process (Sheu *et al.*, 1993). More entrapped cells survived than did cells that were not entrapped. Significantly, lower b-galactosidase activities were observed with Ca-alginate encapsulated cells (Sheu and Marshall, 1992). Forty per cent more lactobacilli survived freezing ice milk when they were entrapped in calcium alginate than when they are not entrapped (Sheu and Marshall, 1993). Hyndman *et al.* (1993) also microencapsulated *Lactococcus lactis* spp. *cremoris* with in gelatin membranes cross-linked with toluenc-2, 4-di-isocyanate at an oil/water interface. Microencapsulation resulted in acidification of milk to pH 5.5 within 2.8 hr, similar to that achieved in free cell fermentations. In the recent past, a team of Chinese scientists studied microencapsulation of *L. delbruecki* spp. *bulgarius* and *Streptococcus thermophilus* by spray drying and observed improved survival rates of microorganisms significantly, with survival rates of up to 50 per cent (Hua *et al.*, 1998).

Demos *et al.* (1998) attempted to microencapsulate different lactic acid bacteria by spray drying and observed that survival rate after spray drying was correlated with the outlet air temperature and storage temperature influenced the storage stability. Investigations have been conducted in the recent years to make encapsules of lactic acid bacteria with hydrogenated maize oil by injecting coat-core emulsions into a chilled dispersion fluid (Lee *et al.*, 1998). They also studied their stability using different emulsifiers and observed that the microcapsules were not disrupted after heat treatment of 37°C for 1 hr. Microencapsulation of three different strains of bifidobacteria in alginate or K-carrageenan beads has been also proved effective in improving the survival throughout the storage for 10 weeks at-20°C from 43-44 per cent to about 50-60 per cent with better survivability in alginate beads than in K-carrageenan beads (Kebary *et al.*, 1998). Addition of glycerol and mannitol during preparing of alginate beads increased the survival of bifidobacteria from 58.8 to 88.5 per cent.

With the aims to derive better retention and protection of the cells, recovery and reuse of the starter cultures, *Lactococcus lactis* spp. *lactis* was also microencapsulated within cross-linked chitosan membranes formed by emulsification/interfacial polymerization (Groboillot *et al.*, 1993). Champrgne *et al.* (1992) also investigated the cell release during fermentation by *Lactococcus lactis* entrapped in calcium alginate beads with poly-L-lysine (PLL) and observed that it did not significantly reduce the release of cells during consecutive fermentations. In another work, Khalil and Mansour (1998) evaluated the survival of bifidobacteria microencapsulated in alginate and their effect on the quality of mayonnaise. They observed that the viability of the free cells disappeared after two weeks, however encapsulated *B. bifidum* survived well for 12 and *B. infantis* for 8 weeks. Mayonnaise containing encapsulated bifodobacteria had lower total bacterial counts compared to other treatments. Even, sensory properties of mayonnaise were improved by the addition of encapsulated bifidobacteria.

The entrapment of *Lactococcus lactis* spp. *cremoris* CRA-1 in alginate/poly-L-lysine (Alg/PLL) nylon or cross-linked polyethyleneimine (PEI) membranes has been investigated (Larisch *et al.*, 1994). They reported that Alg/PLL encapsulation resulted in viable and active cell preparations, which acidified milk at a rate proportional to cell concentration, but at rates, less than that of free cell preparations. Hong (1997) studied the enhancing survival of lactic acid bacteria in ice cream by natural encapsulation and observed that *Streptococcus thermophilus* strains survived better than their non-encapsulated mutants did in reduced fat ice cream during freezing and frozen storage at –29 °C for 16 days. He reported that various factors that improved survival of *S. thermophilus* (ST 1068) in ice cream harvest of cells in late log phase of 37 °C, keeping overrun at 50 per cent and storage at –17 °C.

"Probiocap": A Novel Approach in Improving Probiotic Survivability

The greatest, most significant, precedent-breaking health-providing nutrient in the world would be totally useless if there was no way that it could be absorbed by the human body. In the graveyard of new nutritional substances lie dozens that failed because, there was no way for the active ingredient, such as probiotics to arrive intact from the bottle to the user's stomach or intestines. Raw materials are inherently sensitive and often react when combined. Supplement manufacturers have tried to overcome ingredient limitations in many ways. One of the most common practices has been to adjust the formulation with additional secondary ingredients included solely for reducing the impact of the side reactions caused by uncontrolled primary-ingredient reactions. Sometimes the simplest option is to put a less-than-impressive product on the market with compromised features and quality, or not to put a product on the market at all because of undesirable interactions. Often, expensive overdosing is used to overcome inherent limitations. In the future research work, we should look at the new technology by which probiotic foods manufacturers and suppliers may solve these and other problems to produce effective, potent nutraceuticals in consumer-friendly forms.

The probiotic market shows great potential for manufactures and has continued to gain unprecedented momentum, despite the complex processing challenges of developing probiotic foods with those beneficial microorganisms. The bacteria often die during manufacturing process, storage treatments to be given or during passage to the intestine. Shelf life is also unpredictable, and the industry has had difficulty baking up label claims. Manufacturers who want to inflate the use of probiotic must have to think over probiotic stability and survival rates during production, preservation and gastric transient to ensure actual benefits to the consumer. Manufacturers who want to inflate the use of probiotic must have to think over probiotic stability and survival rates during production, preservation and gastric transient to ensure actual benefits to the consumer.

Traditional freeze-dried probiotic bacteria are sensitive to high moisture, extreme temperatures and other physical and chemical stresses. This sensitivity limits their use in many applications. Their viability also decreases during digestion due to extreme gastric acidity, which has long been a concern. A revolutionary patent-pending microencapsulation technology, "Probiocap"®, for the improved stabilization of probiotics has now been evolved in Canada. This new technology dramatically enhances the viability of probiotic bacteria. "Probiocap"®: increases heat resistance improves resistance to compression enhances acid resistance extends shelf life specially designed equipment is used to coat and entrap probiotic bacteria in a matrix of food grade vegetable fatty acids. This allows for the use of probiotics in new applications: powders, cream type formulas, such as chocolate bars, and processed foods as it withstands high temperatures. "Probiocap" uses a food-grade fatty acid coating and therefore protects probiotic strains against the action of excessive stresses such as temperature, oxygen, stomach acidity, moisture, and pressure. In addition, "Probiocap" also expands probiotics into many new applications in the functional-food industry, including creamy formulas, powders, and nutritional bars," allowing them to be mixed with other food ingredients that could have inhibited their viability. The technology is triggered to separate when the release of this functional ingredient is desired for maximum benefit *viz.* in the intestines, directly after exposure to the low pH of the stomach. The coatings to be selected should allow the probiotic to pass through the gastrointestinal tract without being destroyed by gastric juices.

Recently, Albany, NY-based Balchem Encapsulates and Institut Rosell jointly announced a new microencapsulation technology, called "Probiocap" that applies a coating to the probiotic to protect it from moisture high humidity and acidity. The coating allows the probiotic to pass through the gastrointestinal tract without being destroyed by gastric juices and is triggered for release in the intestine based on pH conditions. To date, Balchem's technology has been applied successfully to several strains of probiotics including *L acidophilus, L rhamnosus* and *Bifidobacterium longum.* This technology has been developed to enlarge the usage of probiotics in different fields. The companies are currently in the initial phase of this project.

Probiotics are especially vulnerable to food processing techniques such as sterilization, pasteurization, disinfections, irradiation, washing and peeling which effectively destroys them. Up until now, there hasn't been the technology to stabilize probiotics in food. They are trying for providing the stabilized process to suppliers depending on the direction they wish to go – the food system they want to place them in. To date, clinical testing has revealed that tablets with the encapsulated probiotics had an unprecedented 100 percent recovery rate. In standard industry testing of tablet compression, 50-75 percent of probiotics die. Initial research suggests that Balchem's encapsulation technology may triple shelf life for probiotics in certain applications and help manufacturers deliver a reliable quantity of probiotics to meet label claims. This process would allow probiotic foods formulators to guarantee stability and to recover all the active bugs. It also eliminates the need to overdose. It has been the practice of manufacturers to recommend a dose of one to 10 billion live cells to ensure the delivery of at least 100 million live cells in the GI tract. Using "Probiocap" technology they wouldn't need to overdose to this extent. The technology is said to ensure a ten-fold increase in bacteria survival. Manufacturers using this technology can include lower concentrations, thereby reducing production costs. The objective of such project is to deliver the active ingredient *i.e.* probiotics to where you want it to be. The technology developed to protect the extremely sensitive probiotic bacteria so it can go through the stomach and reach the intestines intact.

Status of "Probiocap"

The mortality rate, as high as 90-99 per cent, is usually investigated for unprotected probiotics exposed to low pH conditions of the stomach and to the bite acids in the duodenum (Siuta-Cruce and Goulet, 2001) and other processing treatments. Hence, it is common practice to recommend the intake of 1 to 10 billion of live cells as to ensure a delivery of at least 10^8 live cells per dose in the GI tract. Microencapsulation of probiotics can surmount these major hurdles and provide the probiotics to escape mild processing conditions, such as milk heat treatment, microwave treatments, thermization etc. This technology uses to coat these microorganisms to extends shelf life, increase heat resistance, improve compression and enhance acid tolerance (Siuta-Cruce and Goulet, 2001). They reported that the technology has been utilized for several strains of probiotics for use in food applications, including *L. acidophilus, L. rhamnosus* and *Bifidobacterium longum*. The technology has also been applied to two animal probiotics geared toward animal feed, specifically *Pediococcus acidilactici* and *Enterococcus faecium*.

On laboratory scale, encapsulated probiotics have been successfully incorporated into a number of foods with promising results. These include yoghurt-covered raisins, nutrient bars, chocolate bars and tablets. Testing in chocolate bars and nutrient bars more exhibited a good recovery rate. With advances in microencapsulation technology, Institut Rosell/Lallemand, Montreal, Canada, now provides probiotics that can be a key ingredient in functional foods. Moreover, their addition to food systems is seen as an effective means of restoring some of the initial beneficial food-associated microflora that may be destroyed during microbial reduction treatments—such as pasteurization. The company offers several strains of lactic acid bacteria and is constantly innovating and researching new strains that are potentially applicable for various health conditions. Typical applications include confectionery, cereals, and snacks, such as nutrition bars, powdered foods, such as infant formulas, meal replacements and beverages.

"Probiocap" and Tomorrow

The probiotic market shows great potential for manufactures and has continued to gain unprecedented momentum, despite the complex processing challenges of developing probiotic foods with those beneficial microorganisms. The bacteria often die during manufacturing process, storage treatments to be given or during passage to the intestine. Shelf life is also unpredictable, and the industry has had difficulty baking up label claims. Manufacturers who want to inflate the use of probiotic must have to think over probiotic stability and survival rates during production, preservation and gastric transient to ensure actual benefits to the consumer. Probiotic products have been marketed in recent years, primarily in the fermented food category or as live microbial dietary supplements whose biological activity needed to be preserved. This tends to limit their consumption as supplements taken after meals or in fresh dairy foods. One technology that makes it possible to "have high dose of your probiotics and eat it alive, too!" is microencapsulation.

Microencapsulation has been investigated to be the best accessible technology to preserve the potency of probiotics to be ultimately delivered into the GIT. The novel application of microencapsulated probiotics would allow the beneficial microorganisms to be incorporated readily in high dosage and allow the probiotic food designers to provide esurience on viability and quantity of probiotics upto the GIT, even after the processing and some preservation treatments. Keep in mind that by encapsulating probiotics we can prevent unwanted ingredient interaction and increase shelf life. This process is opening up avenues previously not possible to manufacturers. It may also allow the medical industry in the field of bacteriotherapy, wherein very high doses of viable probiotic bacteria are required for the treatment of patients with chronic paucities.

Conclusion

In terms of probiotic applications, the list of benefits continues to grow. Probiotics hold a very promising future because researchers are realizing the importance of intestinal health, the importance of the internal ecosystem, and the fact that probiotic benefits reach beyond their effect in the GI tract. The newest research is focusing on the effect of probiotics on allergies. Allergies are mediated by immune responses, so if probiotics have an effect on immune modulation, they might have an effect on treating or preventing allergies. That's some of the most recent research going on right now and it seems to be the case.

Maintaining the viability, stability, and functionality of probiotics during processing, formulation and storage is essential to delivering the health benefits of these ingredients to consumers. In the future project, the effects of processing on probiotics should be explored and used to develop optimal process and formulation technologies to maintain the stability and functionality of probiotics. New processing techniques should be applied to the development of functionally enhanced probiotics with improved viability. Certainly, advances in stability, viability and possible applications are giving manufacturers exciting options for producing new and viable products. However, we need real-time stability testing; we can't do accelerated stability tests on probiotics. We should be careful of the advertising hype. Data are yet not published in scientific journals that show stability at day one, 30 and 365. Until to date, the powder form of probiotics remains the most stable and that the liquid form required for use in food is very unstable.

References

Adhikari, K., Mustapha, A., Grun, I.V. and Fernando, L. (2000). Viability of microencapsulated bifidobacteria in set yoghurt during refrigerated storage. *J. Dairy Sci.*, **83**: 1946-1951.

Berner, L.A. and O' Dannell, J.A. (1998). Functional foods and health claims legislation: Application to dairy foods. *Int. Dairy J.*, **8**: 355-362.

Buttriss, J. (1997). Nutritional properties of fermented milk products. *Int. J. Dairy Technol.*, **50** (1): 21-27.

Champagne, C.P., Gaudy, C., Poncelet, D. and Neufeld, R.J. (1992). *Lactococcus* release from calcium alginate beads. *Appl. Environ. Microbiol.*, **58**(5): 1429-1434.

Davis, J.G., Ashton, T.R. and McCaskill, M. (1971). Enumeration and viability of *Lactobacillus bulgaricus and Streptococcus thermophilus* in yoghurt. *Dairy Industries*, **36**: 569-573.

Demos, S., Zgoulli, S., Evrard, P., Roblain, D. and Destain, J. and Thonart, P. (1998). Spray drying of different lactic acid species. 12th Forum for Applied Biotechnology, Provincial court Brugge, Belgium, 24-25 Sept., 1998, Part I. Mededelingen – Facultcit – Landbouukundige – en- Toegcpuste – Biologische – Wetenschappen, Universitcit – Gent. 1998, **63**: N0. 4a: 1253-1261.

FAO/WHO (2001). Evaluation of health and nutritional properties of powder milk with live lactic acid bacteria. *Report from* FA/WHO *Expert consultation* 1-4 Oct, 2001, Cordoba, Argentina.

Finegold, S.M., Sutter, V.L., Sugihara, P.T., Elder, H.A., Lehmann, S.M. and Phillips, R.L.(1977). Fecal microbial flora in seventh day adventist populations and control subjects. *Am. J. Clin. Nutr.*, **30**: 1781-1792.

Fukushima, M. and Nakamo, M. (1996). Effect of a mixture of organisms, *L. acidophilus* or *S. faecalis* on cholesterol metabolism in rats fed on a fat and cholesterol enriched diet. *Br. J. Nutr.*, **76**: 857-867.

Fuller, R. (1989). Probiotics in man and animals. *J. Appl. Bacteriol.*, **66**: 365-378.

Fuller, R. (1992). *Probiotics-The Scientific Basis*. Chapman and Hall, London.

Gibson, G.R. and Roberfroid, M.B. (1995). Dietary modulation of the human colonic microbiota: Introducing the conceptof prebiotics. *J. Nutr.*, **125**: 1401-1412.

Groboillot, A.F., Champagne, C.P., Darling, G.D., Poncelet, D. and Neufeld, R.J. (1993). Membrane formation by interfacial cross-linking of chitosan for microencapsulation of *Lactococcus lactis. Biotech. and Bioengg.*, **42**: 1157-1163.

Hanevaar, R. and Huis in't Veld, J.H.J. (1992). Probiotics: A general view. In: *The Lactic Acid Bacteria in Health and Disease* (Ed. B.J.B. Wood), Elsevier Applied Science, London, NY. Pp: 151-170.

Hoier, E. (1992). Use of probiotic starter cultures in dairy products. *Food Aus.*, **44**(9): 418-420.

Hong, S.H. (1997). Enhancing survival of lactic acid bacteria in ice cream by natural encapsulation. *Dissertation Abstr. Int.*, **13**(9): 5407

Hua, W., Yan Qun, L., Jintteng, F., Yi, Z., HiuLing, W., MuLin, C., KaiYu, W., Xianlning, L., Wenxiu, D., Wci, H., YQ, L., Fu, J.H., Zeng, Y., Wu, J.L., Cai, M.L., Wu, K.Y., Li, X.M. and Ding, W.X. (1998). Preliminary study of the microencapsulation of lactis. *China Dairy Indus.*, No. **6**: 13-16.

Hyndman, C.L., Grobillot, A.F., Poncelet, D., Champagne, C.P. and Neufeld, R.J. (1993). Microencapsulation of *Lactococcus lactis* within cross-linked gelatin membranes. *J. Chem. Technol. Biotechnol.*, **56**(3): 259-263.

IDF. (1997). Standards for fermented milks. International Dairy Federation. D-Doc 316.

Ishibashi, N. and Shimamura, S. (1993). Bifidobacteria: Research and development in Japan. *Food Technol.*, **47**(6): 126-134.

Jackson, S.J., and Lee, K. 1991. Microencapsulation and the food industry. *Lebensmittel-Wissenschaft and Technologie*. 24(4): 289-297.

Kadian, A., Dabur, R.S. and Kapoor, C.M. (1999). Application of microencapsulation technology in dairy industry. *Indian Food Industry,* **18**(1): 57-63.

Kailasapathy, K. and Rybka, S. (1997). *L. acidophilus* and *Bifidobacterium* spp.–Their therapeutic potential and survival in yoghurt. *Aus J. Dairy Technol.*, **52**: 28-35.

Kanawjia, S.K., Pathania, V. And Singh, S. (1992). Microencapsulation of enzymes, microorganisms and flavour and other applications in foods. *Indian Dairyman,* **44**(6): 280-290.

Kebary, K.M.K., Hussein, S.A. and Badausi, R.M. (1998). Improving viability of bifidobacteria and their effect on frozen ice milk. *Egypt. J. Dairy Sci.*, **26**(2): 319-337.

Khalil, A.H. and Mansour, E.H. (1998). Alginate encapsulated bifidobacteria survival in mayonnaise. *J. Food Sci.*, **63**(4): 702-705.

Kim, H., Kamara, B.J., Good, I.C. and Enders, G. L., Jr. (1988). Method for preparation of stable microencapsulated lactic acid bacteria. *J.Ind. Microbiol.*, **3**: 253.

Kurmann, J.A. and Rasic, J.L. (1991). The health potential of products containing bifidobacteria. In: *Therapeutic Properties of Fermented Milks* (Ed. R.K. Robinson), Elsevier Applied Sciences, London. Pp: 117-158.

Lankaputra, W.E.V. and Shah, N.P. (1998). Antimutogenic properties of probiotic bacteria and organic acids. *Mutation Res.*, **397**: 169-182

Larisch, B.C., Poncelet, D., Champagne, C.P. and Neufeld, R.J. (1994). Microencapsulation of *Lactococus lactis* subsp. *cremoris. J Microencapsulation*, **11**(2): 189-195.

Lee, K.W., Baick, S.C., Kim, S.K. and Kim, H.U. (1998). Properties and yield of the microcapsules made with hydrogenated corn oil. *Korean J. Dairy, Sci.*, **20**: 169-176.

Lee, Y.K. and Salminen, S. (1995). The coming of age of probiotics. *Trends Food Sci. Technol.*, **6**: 241-245.

Lourens-Hattingh, A. and Vijoen, B.C. (2001). Yoghurt as probiotic carrier food. Int. Dairy J., **11** : 1-17.

Maharaj, I., Nairn, G.J. and Campbell, J.B. (1984). Simple rapid method for the preparation of enteric-coated microspheres. *J. Pharm. Sci.*, **73**: 39.

Mital, B.K. and Garg, S.K. (1992). Acidophilus milk products: Manufacture and therapeutics. *Food Rev. Int.*, **8**(3): 347-389.

Mitsuoka, T. (1989). *Microbes in Intestine.* Yakult Honsha Co. Ltd., Tokyo, Japan.

O'Sullivan, M.G. (1996). Metabolism of bifidogenic factors by gut flora -An overview. *Bull. Int. Dairy Fed.*, ***313:*** 23-30.

Ouwchand, A.C. and Salminen, S. (1998). The Health effects of cultured milk products with viable and nonviable bacteria. *Int. Dairy J.*, **8**: 749-758.

Rao, A.V., Shivnarain, N. and Maharaj, I. (1989). Survival of microencapsulated *Bifidobacterium pseudolongum* in simulated gastric and intestinal juices. *Can. Inst. Food Sci. Technol.*, **22**: 345-349.

Ravula, R.R. and Shah, N.P. (1999). Survival of microencapsulated *L. acidophilus and Bifidobacterium* spp. in fermented frozen dairy desserts. *J. Dairy Sci.*, **82**: 4(Abstr.).

Rosenthal, I., Rosen, B. and Bernstein, S. (1996). Surface pasteurization of Cottage cheese. *Milchwissenschaft*, **51** (4): 198-201.

Rybka, S. and Kailasapathy, K. (1995). The survival of culture bacteria in fresh and freeze dried. AB Yoghurt. *Aus. J. Dairy Technol.*, **50** (2): 51-57.

Salminen, S., Deighton, M.A., Benno, Y. and Gorbach, S.L. (1998a). Lactic acid bacteria in health and disease. In: *Lactic Acid Bacteria* (Eds. S. Salminen and A Von Wright), 2nd Ed., Mared Dekker, Inc., NY, Hong Kong. Pp: 24-253.

Salminen, S., Ouwchand, A.C. and Isolauri, E. (1998b). Clinical applications of probiotic bacteria. *Int. Dairy J.*, **8**: 563-572.

Schiffrin, E.J., Rochat, F., Link-Amster, J. and Aeschlimann, J.M. (1994). *REMS Immunol. Med. Microbiol.*, **10**, 55-64.

Shah, N.P. (1999). Probiotic bacteria: Selective enumeration and survival in dairy foods. *J. Dairy Sci.*, **83**: 894-907.

Shah, N.P. (2000). Probiotic bacteria: Selective enumeration and survival in dairy foods. *J. Dairy Sci.,* **83**: 894-907.

Sheu, T.Y. and Marshall, R.T. (1992). Encapsulation and conditions of conditions of freezing affect viability and beta-galactosidase activity of *Lactobacillus bulgaricus* in frozen dessert. *J. Dairy Sci.,* **75**(Suppl. 1): 117.

Sheu, T.Y. and Marshall, R.T. (1993). Microencapsulation of lactobacilli in calcium alginate gels. *J. Food Sci.,* **58**(3): 557-561.

Sheu, T.Y., and Rosenberg, M. (1995). Microencapsulation by spray drying ethyl caprylate in whey protein and carbohydrate wall systems. *J. Food Sci.,* **60**(l): 98-103.

Sheu, T.Y., Marshall, R.T. and Heymann, H. (1993). Improving survival of culture bacteria in frozen desserts by microentapment. *J. Dairy Sci.,* **76**: 1902-1907.

Siuta-Cruce, P. and Goulet, J. (2001). Improving probiotic survival rates. *Food Technol.,* **55**(10): 36-42.

Tamime, A.Y., Marshall, V.M.E. and Robinson, R.K. (1995). Microbiological and technological aspects of milks fermented by Bifidobacteria. *J. Dairy Res.,* **62**: 151-187.

Vesa, T.H., Marteau, P., Zidi, S., Potchart, P. and Rambaud, J.C. (1996). Digestion and tolerance of lactose from yoghurt and different semi-solid fermented dairy products containing *L. acidophilus* and *Bifidobacteria* in Lactose maldigesters: Is bacterial lactose important. *Eur. J. Clin. Nutr.,* **50**: 730-733.

Young, J.N.(1996). Functional foods: Strategies for successful product development.FT Management Report.

2016, Dairy and Food Product Technology Pages 221–243
Editors: Birendra Kumar Mishra and Subrota Hati
Published by: BIOTECH BOOKS, NEW DELHI

Chapter 15

Milk and Milk Products as Source and Carriers of Bio-Active Ingredients

Shaik Abdul Hussain[1], Anuj Kumar[2], Richa Badola[2], Sawale Pravin Digambar[2] and Manju Gare[3]

[1]Scientist, Division of Dairy Technology, NDRI, Karnal
[2]Ph.D. Scholar, Division of Dairy Technology, NDRI, Karnal
[3]Ph.D. Scholar, Dairy Microbiology Division, NDRI, Karnal

Introduction

The fast moving lifestyle of today's environment has burdened us with many ailments and disorders *viz.*, coronary heart disease, diabetes, ageing etc. and in the process consumers have also become progressively more aware about direct contribution of foods toward health. There is a continual demand for enhanced functionality foods which are useful against lifestyle maladies either by preventive effect or by recuperative effect.

The term "Functional Foods" was first introduced in Japan in the mid-1980s and referred to processed foods containing ingredients that aid specific body functions, in addition to being nutritious. Currently, there is particularly no accepted term/definition for functional food and a variety of other synonymous products *i.e.* nutraceuticals, dietary supplements, fortified foods, medical foods, designer foods etc. have also evolved carrying the same uncertainty. The European Commission's Concerted Action on Functional Food Science in Europe defined 'functional foods' as food which improves health and well-being of humans and/or causes disease risk reduction by beneficially affecting the bodily functions. These foods, instead of a pill, capsule or dietary supplement, are consumed as part of normal diet. The whole action of functional foods is particularly dependent upon 'physiologically active components' which are the active compounds imparting the health benefits.

The development of functional foods is one of the most intensive areas of food product development worldwide. Functional foods have been developed in virtually all food categories. More recently, food industries have taken further steps to develop food products that offer multiple health benefits in a single food product (Sloan, 2004). Global functional foods market is dominated by dairy products, with sales valued at $ 9.23bn in 2010 (http://www.leatherheadfood.com/, Accessed on 5th April, 2013).

Milk and milk products have been praised for their nutritional quality and are well accepted by all the age groups. With advent of many sophisticated analytical, biochemical, and cell-biological research tools, the presence of many physiologically active components in milk have been well demonstrated. This has offered opportunities to isolate, concentrate, or modify milk components for their further application. Vitamins, minerals and other plant, animal, microbial derived ingredients have also been incorporated in milk and milk products to impart some health benefit. This chapter deals with the various bioactive ingredients native to milk and application of various functional ingredients in the manufacture of functional dairy foods.

Functional Ingredients Derived from Milk

Milk has been known as nature's most complete food. Aside from nutritional aspects, milk proteins *i.e.* casein and whey proteins have been found to be increasingly important for biochemical functions necessary for human metabolism and health. Milk proteins are currently the main source many biologically active peptides, even though other animal and plant proteins contain potential bioactive sequences (Wu and Ding, 2002). Milk also contains various other functional components.

Milk Protein Based Bioactive Constituents

Dairy products contribute significantly to daily protein intake due to higher nutritional value of milk proteins. The multiple functional properties of major milk proteins are now largely characterized (Mulvihill and Ennis, 2003). The major bioactive protein components of bovine colostrum and milk are presented in Table 15.1.

Casein Derived Bioactives

Caseins are a group of phosphate-containing proteins named α_{s1}-, α_{s2}-, β-, and κ-casein which comprise about 80 per cent of the milk proteins. Caseins are generally regarded as the ion carriers (Ca, PO_4, Fe, Zn, Cu) and are un-associated with any known biological activity. But, the peptides derived from caseins potently bioactive which are utilized for development of many functional foods.

Whey Protein-based Bioactive Substances

Whey proteins comprise rest 20 per cent of the total milk proteins and consist of α-lactalbumin, β-lactoglobulin, bovine serum albumin, immunoglobulin, lactoferrin, transferrin, proteosepeptone fraction, growth factors, and hormones (Walzem, 2002). Many whey proteins are claimed to possess physiological activity in their native form and upon their degradation into bioactive peptides.

Table 15.1: Major Bioactive Protein Components of Bovine Colostrums and Milk

Protein	*Concentration (g/L)*		*Biological Activity*
	Colostrums	*Milk*	
Casein (αs1, αs2, β, and κ)	26.0	28.0	Ion carrier (Ca, PO_4, Fe, Zn, Cu), precursor for bioactive peptides immunomodulatory and anticarcinogenic
Lactoglobulin	8.0	3.3	Vitamin carrier, potential antioxidant, precursor for bioactive peptides, fatty acid binding
Lactalbumin	3.0	1.2	Effector of lactose synthesis in mammary gland, calcium carrier, immunomodulatory, precursor for bioactive peptides potentially anticarcinogenic
Immunoglobulins	20-150	0.5-1.0	Specific immune protection, precursor for bioactive peptides
Glycomacro-peptide	2.5	1.2	Antibiotic, antithrombotic, prebiotic, gastric hormone regulator
Lactoferrin	1.5	0.1	Antimicrobial, antioxidative, anticarcinogenic, anti-inflammatory, iron transport, cell growth regulation, precursor for bioactive peptides immunomodulatory, stimulation of osteoblast proliferation
Lactoperoxidase	Traces	Traces	Antimicrobial, synergistic effects with immunoglobulins, lactoferrin and lysozyme
Lysozyme	Traces	Traces	Antimicrobial, synergistic effects with immunoglobulins, Lactoferrin and lactoperioxidase
Serum albumin	1.3	0.3	Precursor for bioactive peptides
Growth factors	50μg-40mg/L	<1μg-2mg/L	Stimulation of cell growth, intestinal cell protection and repair, regulation of immune system

Source: Korhonen and Pihlanto, 2007

a) α-Lactalbumin

α-Lactalbumin, (α-La) which accounts for about 20 per cent of the whey proteins, is beneficial either as the intact whole molecule or peptides of the partly hydrolyzed protein (Chatterton *et al.*, 2006). High amounts of essential amino acids tryptophan and cystein present in α-La, which are the precursors of antihypertensive agents serotonin and glutathione, respectively have been linked to stress lowering effect of α - La (Markus *et al.*, 2000) as well as improved cognitive functions (Markus *et al.*, 2002). Ushida *et al.*, 2003 demonstrated protective effect of α-La against induced gastric mucosal injury. Bioactive peptides derived from the protein are associated with antihypertensive, antimicrobial, anticarcinogenic, immunomodulatory, opioid, and prebiotic effects (Chatterton *et al.*, 2006). α - La and its hydrolysates are well suited as an ingredient for infant formulae.

b) β-Lactoglobulin

β - Lactoglobulin (β - Lg) accounts for about 50 per cent of whey proteins of cow and buffalo milks. It possesses a variety of functional and nutritional characteristics that have made this protein a multifunctional ingredient in many food and biochemical applications. The derived bioactive peptides possess antihypertensive, antimicrobial, antioxidative, anticarcinogenic, immunomodulatory, opioid and hypocholesterolemic effects (Pihlanto and Korhonen, 2003).

c) Immunoglobulins

Milk and colostrum of all lactating species contain three major classes of immunoglobulins (Ig's- IgG, IgM and IgA) which provide passive immunity against invading pathogens. Ig's make upto 70-80 per cent of the total protein content in colostrum, whereas in mature milk, immunoglobulins account for only 1-2 per cent of the protein (Larson, 1992)

The major class of immunoglobulin in cow milk is IgG1 with concentrations of about 48 g/l in colostrum and 0.6 g/l in normal milk (Farrell *et al.*, 2004). Cow's milk immunoglobulins are effective means of preventing diarrhoea (Reddy *et al.*, 1988). Ig's link various parts of the cellular and humoral immune response system. They are able to prevent the adhesion of microbes, inhibit bacterial metabolism, agglutinate bacteria, augment phagocytosis of bacteria, kill bacteria through activation of complement - mediated bacteriolytic reactions, and neutralize toxins and viruses. Studies have revealed protective action of Ig's against various microbes including *Streptococcus mutans, Candida albicans, Helicobacter pylori*, rotavirus, *Cryptosporidium parvum, Escherichia coli* (Bostwick *et al.*, 2000). The various fraction comprising of these immunoglobulins can be enriched for manufacture of immune milk and related preparations (Mehra *et al.*, 2006). Immunoglobulins based products have found a growing worldwide market as dietary supplements for humans (Wheeler *et al.*, 2007).

d) Lactoferrin

Lactoferrin (LF) is an iron-binding glycoprotein present in milk. The concentration of LF in bovine colostrum and milk is about 1.5-5 mg/ml and 0.1 mg/l, respectively. Owing to its iron-binding properties, LF has been proposed to play a role as a bacteriostatic and antimicrobial agent against a range of microorganisms. Bioactive peptides with strong antimicrobial activity have also been derived from LF. LF has been found to inhibit the growth of *Escherichia coli, Salmonella typhimurium, Shigella dysenteriae, Listeria monocytogenes, Bacillus stearothermophilus* and *Bacillus subtilis* (Batish *et al.*, 1988). LF is also involved in phagocytic killing, immune responses, and acts as an antioxidant, anti-inflammatory, anticancer agent, and growth factor (Zimecki and Kruzel, 2007). Current commercial applications of bovine lactoferrin include infant formulas, iron supplements and drinks, fermented milks, chewing gums, immune-enhancing nutraceuticals, cosmetic formulas, and feed and pet care supplements (Tamura, 2004).

e) Lactoperoxidase

Lactoperoxidase (LP) is a prominent antimicrobial agent found in bovine milk and colostrum. The concentration of LP in colostrum and milk is approximately 11-45 mg/l and 13-30 mg/l, respectively. LP is a very effective bactericidal agent in the presence of SCN^- and H_2O_2. The enzyme LP, in the presence of H_2O_2, catalyses the oxidation of thiocyanate (SCN^-) and produces hypochlorous acid and derivatives which possess antimicrobial action (Visalsok *et al.*, 2004). LP system has been shown to be active against a wide range of microorganisms, including bacteria, viruses, fungi, molds, and protozoa in *in-vitro* studies (Seifu *et al.*, 2005). Indigenous LP in milk may be exploited for the cold-sterilization of milk and also an useful additive for infant formulae (Fox, 2001). Furthermore LP can be effective for preservation of different products *i.e.* meat, fish, vegetables, fruits, and flowers (Boots and Floris, 2006).

f) Lysozyme

Lysozyme (LZ) is an antimicrobial enzyme found in milk and colostrum. The concentration of LZ in colostrums and normal milk is about 0.14-0.7 and 0.07-0.6 mg/l, respectively. The enzyme LZ hydrolyses β-1→4 linkages between N-acetylmuramic acid and 2-acetylamino-2-deoxy-D-glucose residues in bacterial cell walls, resulting in cell lysis. The enzyme is toxic to Gram-positive and Gram-negative bacteria, such as *Pseudomonas aeruginosa, Salmonella typhimurium,* and *Listeria monocytogenes*. Combinations of LZ and LF are more bacteriostatic than either of the proteins alone (Suzuki *et al.*, 1989).

g) Growth Factors

Growth factors consist of heterogeneous group of proteins and peptides that initiate cellular growth and expression of differentiated function. Growth factors identified in bovine mammary secretions include: BTC (beta cellulin), EGF (epidermal growth factor), FGF1 and FGF2 (fibroblast growth factor), IGF-I and IGF-II (insulin like growth factor), TGF-β 1 and TGF-β 2 (transforming growth factor) and PDGF (platelet-derived growth factor) (Shing and Klagsbrun, 1984). The most abundant growth factors in bovine milk are EGF (2-155 ng/mL), IGF-I (2-101 ng/mL), IGF-II (2-107 ng/mL), and TGF-β 2 (13-71 ng/mL), whereas the concentrations of the other known growth factors remain below 4 ng/mL (Pouliot and Gauthier, 2006).

The EGF and BTC stimulate the proliferation of epidermal, epithelial, and embryonic cells. They also promote wound healing and bone resorption. The TGF - β family plays an important role in the development of the embryo, tissue repair, formation of bone and cartilage, and regulation of the immune system. IGF stimulate proliferation of many cell types and regulate some metabolic functions, *e.g.*, glucose uptake and synthesis of glycogen (Pouliot and Gauthier, 2006). Applications of growth factors based on bovine milk or colostrum have been proposed and the main health targets of those product concepts have been skin disorders, gut health, and bone health (Donnet - Hughes *et al.*, 2000).

h) Hormones

A large number of hormones are found in bovine colostrum and milk in miniscule concentrations. These hormones belong to the following main categories like, gonadal hormones (estrogens, progesterone, androgens), adrenal (glucocorticoids), pituitary (prolactin, growth hormone), and hypothalamic hormones (gonadotropin - releasing hormone, luteinizing-hormone-releasing hormone, thyrotropin-releasing hormone), and somatostatin (Jouan *et al.*, 2006) and various other like bombesin, calcitonin, insulin, melatonin, and parathyroid hormone. Hormones play important role in the regulation of specific functions of the mammary gland and in the growth of the newborn, including development and maturation of its gastrointestinal and immune systems.

i) Bioactive Peptides

Bioactive peptides are specific protein fragments influencing body in a positive way. Most of these are encrypted within the primary sequence of the native protein and are released generally through following ways: 1) hydrolysis by digestive enzymes; 2) enzymatic cleavage by proteases derived from microorganisms and/or plants; 3) food processing or manufacturing (acids, alkali, heating etc. Apart from these, recombinant DNA techniques have been experimented for the production of specific peptides or their precursors in microorganisms.

The functionality of bioactive peptides is based on their inherent amino acid composition and sequence. The size of active sequences may vary from two to twenty amino acid residues, and many peptides are known to reveal multifunctional properties (Meisel and FitzGerald, 2003). Biologically active peptides are produced from several dietary proteins during gastrointestinal digestion and fermentation, but milk is considered as main source of biopeptides, with specific nutritional, sensorial and functional properties. A brief outline of bioactive peptides derived from bovine milk, their precursors, and possible bioactive role is shown in Table 15.2.

Table 15.2: Bioactivity of Peptides from Milk Proteins

Bioactivity	*Precursor Protein*	*Names*
ACE-inhibitory	α-, β-casein	Casokinins
	α-lactalbumin, β-lactoglobulin	Lactokinins
Antimicrobial	Lactoferrin	Lactoferricin B
	α_{s1}-casein	Isracidin
Casoxins	α_{s2}-casein	Casocidin
Antithrombotic	κ-casein	Casoplatelins
Immunomodulatory	α-, β-casein, β-lactoglobulin	Immunopeptides
Mineral binding	α-, β-casein	Phosphopeptides
Opioid agonist	α-, β-casein	Casomorphins
	α-lactalbumin	α-lactorphins
	β-lactoglobulin	β-lactorphins
Opioid antagonist	κ-casein	casoxins
	Lactoferrin	Lactoferroxins

Source: Meisel and Bockelmann, 1999.

j) Glycomacropeptide

Glycomacropeptide (GMP) is released from the κ-casein molecule by the action of chymosin at the amino acid sequence of 105 (Phenylalanine) - 106 (Methionine). *In-vitro* studies have shown the role of GMP in inactivation of microbial toxins from *E. coli* and *Vibrio cholera.* It also inhibits adhesion of cariogenic *Streptococcus mutans* and *Streptococcus sobrinus,* bacteria and influenza virus, modulates immune system responses, and promotes growth of bifidobacteria (Mans and López-Fandiño, 2004). Wang *et al.* (2001) suggested that the high sialic acid content of GMP may deliver beneficial effects for brain development and improvement of learning ability. The physiological benefits make GMP a potential ingredient for many food applications (Thomä-Worringer *et al.*, 2006).

Bioactive Lipids from Milk

Conjugated Linoleic Acid

Milk fat consists of more than 400 fatty acids which occurs as triacylglycerols. Milk fat is the richest source of conjugated linoleic acid (CLA). CLA content ranges from 2 to 53.7 mg/g milk fat. The major CLA isomer in milk fat is 9- cis, 11-trans, also called rumenic acid (Collomb *et al.*, 2006).

Several workers have reported the positive effects of CLA namely antiatherogenic, antidiabetic, antiobesity, immune modulation and protection against various cancer, *e.g.*, breast, prostate, and colon cancer (Collomb *et al.*, 2006).

Polar Lipids

Polar lipids are mainly located in the milk fat globule membrane (MFGM). MFGM is a highly complex biological structure that surrounds the fat globule stabilizing it in the continuous aqueous phase of milk and preventing it from enzymatic degradation by lipases. The polar lipid content of raw milk is reported to range between 9.4 and 35.5 mg per 100 g of milk. Among polar lipids, phospholipids, sphingolipids and butyric acid are gaining popularity. The polar lipids are reported to posses a wide variety of physiological functionalities. These compounds are secondary messengers involved in transmembrane signal transduction and regulation, growth, proliferation, differentiation, and apoptosis of cells. They also play a role in neuronal signalling and are linked to age - related diseases, blood coagulation, immunity, and inflammatory responses (Pettus *et al.*, 2004). Sphingolipids and their derivatives possess anticancer, cholesterol - lowering, and antibacterial activities (Rombaut and Dewettinck, 2006). Butyric acid and butyrate have been shown to inhibit development of colon and mammary tumors (Parodi, 2003).

Bioactive Carbohydrates from Milk

Lactose Derived Bioactives

Lactose is the major carbohydrate present in milk and occurs in milk of all mammals with only a few minor exceptions. The approximate concentration of lactose in mammalian milk is between 2.0 per cent and 10 per cent. Lactose acts as a substrate for growth of lactic acid bacteria which play beneficial role in many physiological activities. Lactose is the precursor for many bioactive molecules which have received a greater attention in the functional food market.

a) Lactulose

Lactulose is formed in small quantities during heating of milk. It can also be produced enzymatically with β-galactosidases and fructose as galactosyl acceptor (Lee *et al.*, 2004). Lactulose is the first lactose derivative that was commercialized (Timmermans, 1997). It is a non-digestible carbohydrate which is fermented in the human colon (Nilsson and Nyman, 2005) and possess bifidogenic effects which are less pronounced when compared with fructo-oligosaccharides or galacto-oligosaccharides (Bouhnik *et al.*, 2004). Lactulose is widely applied as laxative and administration of lactulose or lactitol is a standard treatment for chronic hepatic encephalopathy (Als-Nielsen *et al.*, 2004).

b) Lactitol

Lactitol is produced by chemical hydrogenation of lactose. It is a strong laxative and its metabolism by humans is insulin-independent. Lactitol hydrolyses into galactose and sorbitol in the human intestine and most of the lactitol is metabolized to short chain fatty acids (SCFA) by the colonic microflora (Dills, 1989). SCFA reduces the colonic pH thereby inhibiting the growth of unwanted microflora in the colon. Moreover, lactitol is used as an alternative to lactulose in the treatment of hepatic encephalopathy (Als-Nielsen *et al.*, 2004).

c) Lactobionic Acid

Lactobionic acid produced by chemical oxidation of lactose (Gerling, 1997), is a strong chelator of calcium and is used in calcium supplements in pharmaceuticals.

d) Lactosucrose

The trisaccharide lactosucrose is derived from transfructosylation of lactose. Lactosucrose is not resorbed in the upper intestine and is, thus, available for hydrolysis and metabolism by the colonic microflora. Lactosucrose has a bifidogenic effect and its consumption was reported to decrease fecal pH and to inhibit growth of colonic clostridia (Ogata *et al.*, 1993).

e) Galacto-oligosaccharides

Galacto-oligosaccharides (GOS) are non-digestible carbohydrates, which are resistant to gastrointestinal digestive enzymes, but are fermented by specific colonic bacteria (Sako *et al.*, 1999). GOS are produced via transgalactosylation of lactose by β-galactosidases (Matthews, 2005). Ito *et al.* (1993) demonstrated that consumption of GOS resulted in a significant degree of bifidogenesis. Teuri *et al.* (1998) also reported improvement in gastrointestinal performance upon GOS consumption. Chonan and Watanuki (1995) showed that the apparent absorption of calcium in rats was stimulated by feeding GOS. GOS also recovered the absorption of magnesium and thus suppressed the calcification of the heart and the kidney under the conditions of high phosphorus and calcium dietary concentrations (Chonan *et al.*, 1996).

Minerals and Vitamins

Milk is a good source of some minerals esp. calcium and vitamins. Calcium is important for bone formation. Milk is the only known vegetarian source of vitamin $B_{12.}$ Some epidemiological evidence suggested that higher intake of calcium, especially from dairy products, is associated with the maintenance of blood pressure. Calcium plays protective role in prevention of calcium cancer. Milk calcium phosphate binds bile salts and prevent their toxic effects (Van der Meer and Lapre, 1991).

Non-protein Nitrogen Fraction

Among non-protein nitrogen fractions (NPN) of milk, nucleotides, nucleosides, and nucleobases have gained popularity for their wide variety of biological functions. These components are suggested to be acting as pleiotrophic factors in the development of brain functions (Schlimme *et al.*, 2000). Nucleotides act as exogenous anticarcinogens in the control of intestinal tumor development (Michaelidou and Steijns, 2006). Because of the important role of nucleotides in infant nutrition, some infant and follow - up formulae have been supplemented with specific ribonucleotide salts.

Functional Dairy Foods

A wide variety of functional dairy foods has been developed in the recent past by incorporation of various dairy or non-dairy ingredients and their health claims validated through *in-vitro* and *in-vivo* trials. Functional ingredients have also been incorporated to make up the losses in functionality during processing of the dairy products. Some important dairy product categories added with specific functional ingredients and their health claims are discussed in the following sections.

Milk

Liquid milk itself acts as a good carrier for many bioactive ingredients. Owing to the beneficial effects of ω-3 fatty acids, numerous studies have been conducted to successfully incorporate these into liquid milk. Dairy Farmers (Australia) developed dairy product in which all the saturated milk fat was replaced with the healthier mono-unsaturated and ω-3 fats. Similarly products were developed by Dawn Dairy (Ireland) and Parmalat (Italy). Chocolate flavoured milk fortified with ù-3 has been introduced in market by Neilson Dairy (Dairy Oh, Canada) and Parmalat (Beatrice, Canada). Milk fortified with phytosterol was observed to be effective in reducing low density lipoprotein (LDL) (Hansel *et al.*, 2007).

Several probiotic milk preparations were developed by incorporating probiotic bacteria during the packaging of milk. But the viability of the probiotics in these products was less when compared to fermented probiotic milk products. *B. longum* ATCC 15708 added milk was found to be better tolerated by lactose maldigesters (Jiang *et al.*, 1996). BRA sweet milk was prepared by adding *B. infantis, Lb. reuteri* and *Lb. acidophilus* to pasteurized cold milk before packaging (Rothschild, 1995). Milk is one of the important carriers of herbal nutraceuticals. Phenolic compounds of herbs are a good alternative for the synthetic antimicrobial agents used in food industry. Phenolic compounds namely, ferulic acid, tea catechins, oleuropein, ellagic acid and p-coumaric acid have been reported to inhibit the growth of pathogenic bacteria (*S. enteritidis, S. aureus, Listeria monocytogenes*) and fungi in milk (Schaller *et al.*, 2000). Polyphenol-rich extracts of green tea (catechins) were added to fruit-flavoured milk drinks without any significant effect on their aesthetic appeal and flavour (Bender and Bender, 2005).

Melatonin a naturally occuring hormone found in animals and in some other living organisms, including algae is especially effective in overcoming sleeplessness. The melatonin concentration of cow's milk obtained at night is roughly four times higher than in milk collected during the day (Özer and Kirmaci, 2010). Ingman Dairy, Finland introduced the world's first ever high-melatonin premium milk (under the Night Time brand) in 1999.

Vitamin D is essential for the improvement of calcium absorption; fortification of semi-skimmed or non-fat milks with vitamin D is required. Daly *et al.* (2006) postulated that consumption of milk enriched with calcium and vitamin D-3 is a simple, nutritionally sound and cost-effective strategy to reduce age-related bone loss. Singh *et al.* (2007) manufactured fortified milk with addition of inorganic calcium salts. Iron supplementation through food fortification is a low-cost strategy to prevent and treat iron deficiency anaemia. Saini *et al.* (1993) demonstrated that enrichment of buffalo milk with iron (10 ppm) did not cause any flavour defect. Other minerals such as magnesium and selenium are also potentially useful in the manufacture of functional dairy beverages. Recent efforts to improve the selenium level in cow's and sheep's milk through natural measures instead of adding selenite to milk have been successful. Research supported by the AgResearch Research and Capability Fund and Pre- Seed Fund in New Zealand showed that supplementation of milking cows and sheeps with selenium, either by injection or swallowing a selenium capsule (Senrich) (AgResearch, 2007).

Fermented Dairy Foods

Fermented dairy products with many therapeutic properties are an excellent medium to generate an array of products that fit into the current consumer demand for health-based foods.

a) Yoghurt

Yoghurt is a fermented milk product containing a mixture of *Streptococcus salivarius ssp. thermophilus* and *Lactobacillus delbrueckii ssp. bulgaricus* (Tamime and Marshall, 1997). Several functional ingredients like, probiotics, prebiotics, synbiotics, ù-3-fatty acids, CLA, isoflavones and herbs have been added to prepare functional yoghurts with desired functionality and acceptable sensory quality. Yoghurts are regarded as the best carriers for many probiotic bacteria. Herbs not only impart physiological benefit but also prevent the growth of unwanted bacteria and promote the growth of beneficial lactic acid bacteria in yoghurts. Addition of catechin to bifidobacteria-containing yoghurt has been reported to improve the survival of the bifidobacteria during storage (Akahoshi and Takahashi, 1996). Similar effect was noticed for Aloe vera supplemented yoghurt (Pszczola, 1998). Rosenthal *et al.* (1997) reported that tea catechins and ferulic acid inhibit the growth of pathogenic bacteria (coliforms and salmonella) with little effect on yoghurt lactic acid bacteria.

Yoghurts with added conjugated linoleic acid (CLA) are also available in the world market today. In these products, CLA is either added directly or produced within the yoghurt by the action of probiotic bacteria on free linoleic acid. Lin (2003) demonstrated that the addition of linoleic acid increased content of non-fat yoghurt significantly without affecting the sensory properties of the final product.

Low calorie yoghurts have been prepared by incorporating fat substitutes and artificial sweeteners to serve the needs of diabetic population. Aspartame and Acesulfame-K are most popular sweeteners for the formulation of numerous low-calorie and sugar free yoghurts, milk beverages; whey based beverages and cultured milk products (Lotz *et al.*, 1992). Sugar esters have also been tried for development of low-calorie yoghurt (Farooq and Haque, 1992).

b) Dahi and its Variants

Fermented milk products like *dahi, lassi* and *shrikhand* are prominent in people's diet and about 10 per cent of total milk produced in India is utilized for these products. Sensorily acceptable calcium enriched *dahi* was manufactured by Ranjan *et al.*, 2006. Rajpal and Kansal (2008) reported the anti-carcinogenic effects of probiotic *dahi* added with *L. acidophilus* and *B. bifidum.* Several attempts have been made to incorporate different additives into *shrikhand* to address the growing interest in the diversification of food products to attract a wider range of consumers. The pulp of fruits such as apple, mango, papaya, banana, guava and sapota (Dadarwal *et al.*, 2005) and incorporation of probiotic organisms (Geetha *et al.*, 2003) have been tried in *shrikhand.*

Lassi, is a refreshing, delicious and easily digestible beverage can serve as an excellent medium to carry probiotics and functional ingredients. Kumar (2000) developed low calorie *lassi* using aspartame at different levels. Similarly, Khurana (2006) prepared low- calorie fruit *lassi* using mango, banana and pineapple and artificial sweeteners. Sudhakar (2004) developed probiotic *lassi* using a combination of various probiotic cultures.

c) Cheese

Probiotics and herbs are mostly used in the manufacture of functional cheeses till date. Proper strain selection is very much important for preparing probiotic cheeses (Gilliland, 2001) because of compatibility issues. Some cheeses may be particularly suitable for the delivery of probiotic bacteria relative to fermented milks such as yoghurt, because of lower acidity and presence of complex matrix of protein and fat that provides protection to probiotic microorganisms during their passage through the gastrointestinal tract (Donnelly, 2003). Various cheese *i.e.* Turkish white brined, Feta-type, Cheddar, Philippine white soft, Edam, Emmental, Domiati, Ras, Soft, Herrgård cheeses, Quarg, and cheese-based dips have been tried as delivery systems for viable probiotic micro-organisms.

To maintain and improve the viability of beneficial bacteria in probiotic cheeses, several technologies i.e spray-drying (probiotic (Lb. paracasei NFBC 338 milk powder as adjunct culture) (Gardiner *et al.*, 2002) and microencapsulation has also been employed to protect probiotic organisms and improve viability. Encapsulated *B. bifidum, B. infantis* and *Bifidobacterium longum* were used in the manufacture of

Crescenza cheese. Another method of introduction of probiotics, particularly into semi-hard and hard cheeses, is through the addition of a dried culture during salting of curd. This method minimises the losses of bacterial cells to whey and eliminates the effects of competition with lactic acid bacteria during milk ripening (Dinakar and Mistry, 1994). Alternately the probiotic bacteria can be grown in milk hydrolysate (*i.e.* increase of the biomass of the cells) before using them in cheese making (Gomes *et al.*, 1998).

Apart from beneficial effects of probiotics, their addition into cheese has the additional advantages like release of bioactive peptides into cheese matrix and inhibition of pathogens by release of certain antimicrobial compounds. These organisms apparently produced bioactive peptides with anti-hypertensive properties during maturation of the cheese. *Lb. rhamnosus* GG, in cheese has been reported to prevent dental caries also (Abou-Dawood, 2002). Some strains of bifidobacteria reduced the levels of pseudomonas in Cottage cheese (O'Riordan and Fitzgerald, 1998). Commercial probiotic cultures of *Lb. rhamnosus* and *P. freudenreichii subsp. shermanii* with anti-clostridial effects and activity against contaminating yeasts and moulds are also available (Hansen, 1997).

Herbs have also found their applications in cheese making. Ripened cheese varieties containing herbs are traditional in Turkey and Syria. The most popular of these cheeses are *Otlu, Otlu Cacik* and *Otlu L*or which are consumed as a part of almost every meal (Hayaloglu and Fox, 2008). Besides offering functionality, herbs also impart characteristic appearance, flavour, aroma and shelf-life to the cheeses. Herbs and their essential oils show antimicrobial activity against many pathogenic bacteria without having any adverse effect on the added starter culture in the cheese manufacture (Bullerman and Gourma, 1987; Manju *et al.*, 2014). This property could be advantageous in mould-ripened cheeses where the growth of moulds is desirable while the production of mycotoxins may present a health risk. Along with increased proteolysis and lipolysis, herbs also provide a typical vitamin and mineral profile in herby cheeses leading to better digestibility and sensory quality when added at desired levels (Hayaloglu and Fox, 2008). About 25 kinds of herbs especially, *Allium* spp., *Thymus* spp., *Ferula* spp., *Anthriscus nemorosa, etc.*, in single or in combination were used to make herbal cheeses (Hussain *et al.*, 2011). Indian herbs such as coriander, curry leaf, spinach and amla, have been incorporated into *sandesh* (Bandyopadhyay *et al.*, 2007) to get their benefits.

Herbal extracts have also been used as coagulants in the preparation of cheeses. Lamasa *et al.* (2001) reported that proteases extracted from *Cynara cardunculus* L. and selectively hydrolyses whey proteins which may eventually lead to more digestible and functional peptides that can be incorporated into food formulae, to bring about favourable contributions to texture and taste, as well as reduction of allergenic effects (Schmidt and Poll, 1991). However, addition of certain kinds of herbs may facilitate histamine formation in herb cheese which may lead to poisoning.

d) Other Fermented Dairy Products

Kefir, viili, skyr, ymer, labneh, calpis sour milk and Greek-style yoghurt are some of the fermented milks which have gained popularity as healthy beverages. Recently, 'bio' kefir has been marketed in Poland containing *Bifidobacterium spp.* and/or *Lb. acidophilus*. Recently, Salem *et al.* (2013) prepared functional labneh with addition of herbs, dill and parsley. Probiotic cultured buttermilks prepared with the addition of different probiotic bacteria on the market under different brand names in several European countries (El-Kholy *et al.*, 2003).

Fat-rich Dairy Products

Ghee is the most widely used milk product in the Indian sub-continent and is considered to be the supreme cooking medium. Besides being a valuable source of fat soluble vitamins A, D, E and K, ghee improves the growth rate and digestibility upon consumption (Kansal and Bhatia, 2007). Kehar *et al.* (1956) observed improvement in digestibility of protein and biological value when cow milk ghee was added to a diet sub optimal in vitamin A. Mineral absorption from diet also increases with ghee consumption.

Milk fat, particularly ghee absorbs all the medicinal properties of the herbs with which it is fortified, without losing its own attributes. About 55-60 medicated ghee types are reported in Ayurvedic literature and they have also been used in the treatment of various diseases (Pandya and Kanwajia, 2002). At NDRI herbal ghee incorporating functional attributes of *Terminalia arjuna* has been developed for providing beneficial effects against cardiovascular diseases.

Addition of herbs to butter not only imparts health benefits but also prevents hydroperoxides formation (oxidation). Sage and rosemary extracts are widely used for this purpose (Rižnar *et al.*, 2006). Herbal extracts have antioxidant activity many times stronger than synthetic antioxidants like BHA or BHT (Lee *et al.*, 2003). Ozkan *et al.* (2007) reported that addition of *Satureja cilicica* essential oil in butter exhibited strong antioxidant activity in a concentration dependant manner.

Heat and Acid Coagulated Dairy Products

Khoa is an important traditional Indian dairy product prepared by partial desiccation of milk followed by heat coagulation. These products are rich in fat or sugar or both. Because of the high fat and sugar contents, they remain inedible for a section of population. Prabha (2006) developed low-fat *khoa* (fat replacement by WPC-70 and Simplesse®) and sugar-free *burfi* sucrose with sucralose. Chetana *et al.* (2004) developed a technology for the production of sugar-free *gulabjamun* using aspartame. Jayaprakash (2003) developed a technology for preparing low-calorie or diabetic *rasogolla* using a high-intensity sweetener (aspartame) and a bulking agent (sorbitol). Kantha and Kanawjia (2007) made successful attempts to develop soy fiber fortified low-fat *paneer*.

Infant Foods

The composition of milk of each species is unique and tailored to sustain growth and development of its own offspring. Human milk contains specific proteins, lipids designed to be easily digestible and certain unique biochemical and immunological factors which provide protection to the newborn against infective agents in environment (Koletzko *et al.*, 2000). Human infants should ideally be nursed on mother's milk, which constitutes nature's best food. However, in the event of lactation failure, insufficient milk secretion, and where mothers are suffering from transmittable diseases, human milk substitutes serve as savers of precious life during vulnerable stages of infancy. Bovine milk as such or with certain modifications has been widely used for infant feeding.

Various methods have been proposed for the introduction of probiotic organisms into the infant gut and infant formulas. Incorporation of probiotics in dried preparations, which will enhance the proliferation of probiotic organisms in the gut was proposed by Saavedra *et al.* (2004). Neslac, a commercialised milk powder contains *B. animalis subsp. lactis* BB-12 for infants (Playne *et al.*, 2003). Incorporation of prebiotics such as oligosaccharides (Kunz and Rudolff, 2002), and lactulose (Strohmaier, 1997) into infant formulae has also been reported by many workers. Several infant formula preparations were made by incorporating functional ingredients like lactoferrin, lysozyme, vitamins, amino acids, and minerals.

Frozen Dairy Products

Ice-cream and other frozen desserts have potential as carriers of many functional ingredients, but the effect of freeze stress upon the activity of functional ingredients must be considered during manufacture and extended storage. In recent years, many reports have appeared on addition of probiotics to frozen desserts. The addition of probiotics into ice-cream may be direct (*i.e.* blending of ice-cream mix and probiotic cells immediately prior to freezing), or it may involve fermentation of the milk for proliferation of probiotic cells prior to blending with the ice-cream mix. Encapsulation, freeze-drying, and co-encapsulation of different micro-organisms for probiotic ice-cream preparation have also been evaluated (Kailasapathy and Sultana, 2003) and the results showed that free cells and freshly encapsulated cells without freeze-drying demonstrated the best survival rates. Strain having good resistance to freeze-stress with minimum flavour problems should be chosen for the manufacture of good quality probiotic frozen desserts with beneficial effects (Vasiljevic and Shah, 2008). Recently, Kumar (2009) developed a synbiotic ice-cream using FOS and *L. acidophillus* (NCDC 13).

Herbs have also found their applications into ice-cream manufacture. Some herbs (*Aloe vera*, etc.) have been found to stimulate the growth of probiotic bacteria *in-vivo* and *in-vitro* (Hussain *et al.*, 2013).

Whey Based Beverages

Whey is a by-product obtained during the process of cheese production (Jelicic *et al.*, 2008). It is a valuable source for lactoferrin, lactoperoxidase, immunoglobulins, and growth factors. So far, several whey based beverages are available in the world food market. Thirst quenching whey beverages namely Rivella, and Rivella-green were developed using herbal extracts from green tea (Jelen, 2009). Acid whey was used in combination with fruit juices for whey beverage preparations as it is more compatible with the acidic flavour of fruits. In recent past special attention has been paid to the production of fermented whey based beverages/products with probiotic bacteria (Pavunc *et al.*, 2009). Almeida *et al.* (2008) produced a fermented probiotic beverage using Minas frescal cheese whey. Addition of whey protein concentrates to cheese whey stimulated the growth of *Lb. acidophilus* (Matijevic *et al.*, 2008).

Milk and Cereal Based Products

Even though milk is regarded as a complete food, it lacks in iron, dietary fibre and contains lesser amounts of essential amino acids. Cereals (wheat, rice, sorghum, pearl millet, maize and oats) and millets (sorghum, barley and other minor millets) are important sources of dietary fibre, protein, carbohydrates, vitamins B complex and E, iron, trace minerals, and fiber. Cereal based milk products not only offer the advantage of consuming both the milk and cereal nutrients in a palatable form but also complements its nutrition (Hussain *et al.*, 2012).

Raabadi a cereal based fermented milk beverage popular in Northern India is made from cereal flour (germinated/non-germinated). This provides efficient means for utilizing underutilized cereals. Technology has been developed at NDRI for preparing *raabadi* using wheat, pearl millet and sorghum. Helland *et al.* (2004) prepared probiotic milk based puddings using *Lb. rhamnosus* GG, *Lb. acidophilus* LA-5 and 1748, and *B. animalis subsp. lactis* BB-12. Several milk solid and cereal based composite dairy foods including weaning foods, biscuits and extruded snack have also been prepared at NDRI.

Conclusion

Since time immemorial milk and milk products have been valued for their nutritional benefits. Keeping in view of the present functional food regime, researchers have focused much attention on the physiological benefits of milk derived bio-actives. Milk and milk products being very nutritious foods, if incorporated with its bioactives or any functional ingredients derived from other sources will validate the definition of functional foods. However, knowledge of interactions of functional ingredients with food constituents which certainly changes sensory and textural aspects of food requires considerable attention. Processing conditions which may alter the functionality of the bioactives need to be carefully designed. Insufficient knowledge about any of elements may lead to development of foods with unacceptable quality and/or with loss of functionality. In many countries regulations regarding functional foods are not yet well established. Concerning their key role in human health and well being, regulatory bodies should be set up for functional foods. Statements on food labels that characterize the relationship of any food or food component to a disease or health-related condition must be preapproved by these regulatory authorities before their use.

References

Abou-Dawood SAI 2002. Survival of non-encapsulated and encapsulated *Bifidobacterium bifidum* in probiotic *Karish* cheese. Egypt J Dairy Sci, 30: 43-52.

AgResearch 2007 Selenium enriched milk straight from the cow. (URL: http: //www. agresearch.co.nz/publications/intouch/AgResearch_News_Dec2007.pdf.)

Akahoshi R, Takahashi Y 1996. Yoghurt containing *Bifidobacterium* and process for producing the same. PCT-International Patent Application, WO96/37113A1 (cited from FSTA 1997-08-P0149).

Almeida KE, Tamime AY, Oliveria, MN 2008. Acidification rates of probiotic bacteria in *Minas* frescal cheese whey. LWT-Food Sci Technol, 41: 311-316.

Als-Nielsen B, Gluud LL, Gluud C 2004. Non-absorbable disaccharides for hepatic encephalopathy: Systematic review of randomized trials. Brit Med J, 328: 1046-1050.

Bandyopadhyay M, Chakraborty R, Raychaudhuri U 2007. Incorporation of herbs into sandesh, an Indian sweet dairy product, as a source of natural antioxidants. Int J Dairy Technol, 60(3): 228-233.

Batish VK, Chander H, Zumdegeni KC, Bhatia KL, Singh RS 1988. Antibacterial activity of lactoferrin against some common food-borne pathogenic organisms. Aust J Dairy Technol, 5: 16-18.

Bender AE, Bender DA 2005. A Dictionary of Food and Nutrition. Oxford University Press, New York.

Boots JW, Floris R 2006. Lactoperoxidase: From catalytic mechanism to practical applications. Int Dairy J, 16: 1272-1276.

Bostwick EF, Steijns J, Braun S 2000. Lactoglobulins. In: Natural food antimicrobial systems. Naidu AS (Ed.) CRC Press, Boca Raton. pp. 133-158.

Bouhnik Y, Raskine L, Simoneau G, Vicaut E, Neut C, Flourie B 2004. The capacity of nondigestible carbohydrates to stimulate fecal bifidobacteria in healthy humans: A double-blind, randomized, placebo-controlled, parallelgroup, dose-response relation study. Am J Clin Nut, 80: 1658-1664.

Bullerman H, Gourma LB 1987. Effects of oleuropein on the growth and aflatoxin production by *Aspergillus parasiticus*. Z Lebensm Unters For, 20: 226-228.

Chatterton DEW, Smithers G, Roupas P, Brodkrob A 2006. Bioactivity of β-lactoglobulin and α-Lactalbumin-Technological implications for processing. Int Dairy J, 16 (11): 1229-1240.

Chetana R, Manohar B, Reddy SRY 2004. Process optimization of *gulab jamun*, an Indian traditional sweet, using sugar substitutes. Eur Food Res Technol, 219: 386-392.

Chonan O, Takahashi R, Yasui H, Watanuki M. 1996. Effects of β-1–4 linked galactooligosaccharides on use of magnesium and calcification of the kidney and heart in rats fed excess dietary phosphorus and calcium. Biosci Biotech Bioch, 60: 1735-1737.

Chonan O, Watanuki M 1995. Effect of galactooligosaccharides on calcium absorption in rats. J Nutr Sci Vitaminol, 41: 95-104.

Collomb M, Schmid A, Siebe R, Wechsler D, Ryhanen EL 2006. Conjugated linoleic acids in milk fat: Variation and physiological effects. Int Dairy J, 16: 1347-1361.

Dadarwal R, Beniwal BS, Singh R 2005. Process standardization for preparation of fruit flavoured *shrikhand*. J Food Sci Technol, 42(1): 22-26.

Daly RM, Brown M, Bass S, Kukuljan S, Nowson C 2006. Calcium and vitamin D-3 fortified milk reduces bone loss at clinically relevant skeletal sites in older men: a 2-year randomized controlled trial. J Bone Miner Res, 21: 397–405.

Dills WL 1989. Sugar alcohols as bulk sweeteners. Annu Rev Nutr, 9: 161-186.

Dinakar P, Mistry VV 1994. Growth and viability of *Bifidobacterium bifidum* in Cheddar cheese. J Dairy Sci, 77: 2854-2864.

Donnelly L 2003. Gut feeling. Dairy Ind Int, 68(5): 19-20.

Donnet - Hughes A, Duc N, Serrant P, Vidal K, Schiffrin EJ 2000. Bioactive molecules in milk and their role in health and disease: The role of transforming growth factor - beta. Immunol Cell Biol, 78: 74-79.

El-Kholy WI, El-Shafei K, Sadek ZI 2003. The use of immobilized *Lactobacillus* metabolites as a preservative in soft cheese. Arab Universities Journal of Agricultural Sciences, 11: 607-622.

Farooq K, Haque ZU 1992. Effect of sugar esters on the textural properties of nonfat low calorie yoghurt. J Dairy Sci, 75: 2676-2680.

Farrell HM, Jimenez-Flores R, Bleck GT, Brown EM, Butler JE, Creamer LK 2004. Nomenclature of the proteins of cows' milk-sixth revision. J Dairy Sci, 87: 1641-1674.

Fox PF 2001. Milk proteins as food ingredients. Int J Dairy Technol, 54: 41-55.

Gardiner GE, Bouchier P, O'Sullivan E, Kelly J, Collins JK, Fitzgerald G, Ross RP, Stanton C 2002. A spray-dried culture for probiotic cheese manufacture. Int Dairy J, 12: 749-756.

Geetha VV, Sarma KS, Reddy VP, Reddy YK, Moorthy PRS, Kumar S 2003. Physico-chemical properties and sensory attributes of probiotic shrikhand. Ind J Dairy Biosci, 14(1): 58-60.

Gerling KG 1997. Large scale production of lactobionic acid-use and new applications. In Whey International Dairy Federation, Brussels, Belgium pp. 251-261.

Gilliland SE 2001. Probiotics and prebiotics. Applied Dairy Microbiology In Marth EH, Steele, JL (eds), 2nd edn. Marcel Dekker, New York pp. 327-344.

Gomes AMP, Malcata FX, Klaver FAM, Grande HJ 1998. Incorporation and survival of *Bifidobacterium* spp. strain Bo and *Lactobacillus acidophilus* strain Ki in a cheese product. Neth Milk Dairy J, 49: 71-95.

Hansel B, Nicole C, Lalanne F 2007. Effect of low fat fermented milk enriched with plant sterols on serum lipid profile and oxidative stress in moderate hypercholesterolemia. Am J Clin Nutri86: 790-796.

Hansen SL 1997. Protective cultures for cheese. Meieriposten, 86: 261-262.

Hayaloglu AA, Fox PF 2008. Cheeses of Turkey: 3. Varieties containing herbs or spices. Dairy Sci Tech 88: 245-256.

Helland MH, Wicklund T, Narvhus JA 2004. Growth and metabolism of selected strains of probiotic bacteria in milk and water-based cereal puddings. Int Dairy J, 14: 957-965.

http: //www.leatherheadfood.com/, Accessed on 5th April, 2013.

Hussain SA, Garg FC, Pal D 2012. Effect of different preservative treatments on the shelf-life of sorghum malt based fermented milk beverage. J Food Sci Technol DOI: 10.1007/s13197-012-0657-4.

Hussain SA, Raju PN, Singh RRB, Patil GR 2013. Potential herbs and herbal nutraceuticals: Food applications and interactions with food components. Crit Rev Food Sci Nutr, Accepted for publication, DOI: 10.1080/10408398.2011.649148, manuscript ID: BFSN-2011-0407.

Hussain SA, Rani R, Singh RRB 2011. 'Potential herbs for incorporation into dairy products' In "Functional Dairy Foods: Concepts and Applications". Tomar SK, Singh R, Singh AK, Arora S, Singh RRB (eds) S.S. Publishing House, Delhi, India. pp: 471-522.

Ito M, Deguchi Y, Matsumoto K, Kimura M, Onodera N, Yajima, T 1993. Influence of galactooligosaccharides on the human fecal microfl ora. J Nutr Sci Vitaminol, 39, 635–640. Japanscan (2002) Legal affairs, 19(12) February, 32 (www.japanscan.co.uk).

Jayaprakash KT 2003. Technological studies on the manufacture of rasogolla using sweeteners. M Tech Thesis, National Dairy Research Institute (Deemed University), Karnal.

Jelen P 2009. Whey-based functional beverages. In Functional and Speciality Beverage Technology, Press.Paquin P (ed). CRC, New York. pp 259–296.

Jelicic I, Botanic R, Tratnik R 2008. Whey based beverages-new generation of dairy products. Mljekarstvo, 58: 257–274.

Jiang T, Mustapha A, Saviano DA 1996. Improvement of lactose digestion in humans by ingestion of unfermented milk containing *Bifidobacterium longum*. J Dairy Sci, 79: 750-757.

Jouan PN, Pouliot Y, Gauthier SF, Laforest JP 2006. Hormones in milk and milk products: A survey. Int Dairy J, 16: 1408-1414.

Kailasapathy K, Sultana K 2003. Survival and β-d-galactosidase activity of encapsulated and free *Lactobacillus acidophilus* and *Bifidobacterium lactis* in ice-cream. Aust J Dairy Technol, 58(3): 223–227.

Kansal VK, Bhatia E 2007. Nutraceutical properties of dairy ghee In: Souvenir on International Conference of Traditional Dairy Foods, Nov, 14-17, Karnal, India.

Kantha KL, Kanawjia SK 2007. Response surface analysis of sensory attributes and yield of low fat paneer enriched with soy fiber. Ind J Dairy Sci, 60(4): 230-238.

Kehar ND, Krishnan TS, Chanda R 1956. Studies on fats, oils and vanaspati. Manager of publications, Vet. Res. Inst., Izatnagar, India.85-90.

Khurana HK 2006. Development of technology for extended shelf life fruit lassi. PhD Thesis National Dairy Research Institute, Deemed University, Karnal.

Koletzko B, Hernell O, Michaelsen KF 2000. Short and long-term effects of breast feeding on child health. In: Advances in experimental medicine and biology, Vol. 248. Kluwer Academic/Plenum Publishers, New York, pp: 1-447.

Korhonen H, Pihlanto A 2007. Bioactive peptides from food proteins. In: Handbook of Food Products Manufacturing: Health, Meat, Milk, Poultry, Seafood, and Vegetables, Hui YH (ed) John Wiley and Sons, Inc. Hoboken, NJ. pp: 5-38.

Kumar M 2000. Physico-chemical characteristics of low-calorie lassi and flavoured dairy drink using fat replacer and artificial sweetener. MSc thesis National Dairy Research Institute, Deemed University, Karnal.

Kumar O 2009. Physico-chemical and sensory attributes of synbiotic ice-cream. MSc Thesis National Dairy Research Institute, Deemed University, Karnal.

Kunz C, Rudolff S 2002. Health benefits of milk-derived carbohydrates. Fresh Perspectives on Bioactive Dairy Products, Document No. 375, International Dairy Federation, Brussels pp. 72–79.

Lamasa EM, Barrosa RM, Balcao VM, Malcata FX 2001. Hydrolysis of whey proteins by proteases extracted from *Cynara cardunculus* and immobilized onto highly activated supports. Enzyme Microb Technol. 28: 642-652.

Larson BL 1992. Immunoglobulins of the mammary secretions. In: Advanced Dairy Chemistry-1 Proteins, Fox PF (ed.) Elsevier Applied Science, London, pp: 231-254.

Lee SE, Hyun JH, Ha JS, Jeong HS, Kim JH 2003. Screening of medicinal plant extracts for antioxidant activity. Life Sci 73: 167-179.

Lee YL, Kim CS, Oh DK, 2004. Lactulose production by b-galactosidase in permeabilized cells of *Kluyveromyces lactis*. Appl Microbiol Biot, 64: 787-793.

Lin TY 2003. Influence of lactic cultures, linolenic acid and fructo-oligosaccharides on conjugated linoleic acid concentration in non-fat yoghurt. Aust J Dairy Technol, 58: 12-14.

Lotz A, Klug C, Kreuder K 1992. Sweetener stability. Dairy Ind Int, 57: 27-28.

Manju G, Hussain SA, Mishra SK, Chand Ram 2014. "Natural Antimicrobials for Preservation of Food" In: Dairy and Food Processing Industry – Recent trends: Part-I. Ed. Mishra BK, (ed.) Biotech Books, New Delhi, India, pp: 204-230.

Mans MA, López-Fandiño R 2004. κ-casein macropeptides from cheese whey: Physicochemical, biological, nutritional, and technological features for possible uses. Food Rev Int, 20: 329-355.

Markus CR, Olivier B, de Haan EH 2002. Whey protein rich in α-lactalbumin increases the ratio of plasma tryptophan to the sum of the large neutral amino acids and improves cognitive performance in stress-vulnerable subjects. Am J Clin Nutr, 75: 1051-1056.

Markus CR, Olivier B, Pamhuysen GE, Van der Gugten J, Alles MS, Tuiten A 2000. The bovine protein α - lactalbumin increases the plasma ratio of tryptophan to the other large neutral amino acids, and in vulnerable subjects raises brain serotonin activity, reduces cortisol concentration and improves mood under stress. Am J Clin Nutr 75: 1051-1056.

Matijevic B, Botanic R, Tratnik L, Jelicic I 2008. The influence of whey protein concentrate on growth and survival of probiotic bacteria in whey. Mljekarstvo, 58: 243–255.

Matthews BW 2005. The structure of *E. coli* b-galactosidase. C R Biol, 328: 549-556.

Mehra R, Marnila P, Korhonen H 2006. Milk immunoglobulins for health promotion. Int Dairy J, 16: 1262-1271.

Meisel H, Bockelmann W 1999. Bioactive peptides encrypted in milk proteins: proteolytic activation and thropho-functional properties. A Van Leeuw 76(1-4): 207-215.

Meisel H, FitzGerald RJ 2003. Biofunctional peptides from milk proteins: mineral binding and cytomodulatory effects. Curr Pharm Design, 9: 1289-1295.

Michaelidou A, Steijns J 2006. Nutritional and technological aspects of minor bioactive components in milk and whey: Growth factors, vitamins and nucleotides. Int Dairy J, 16: 1421-1426.

Mulvihill DM, Ennis MP 2003. Functional milk proteins: production and utilization. In Advances in Dairy Chemistry, Vol 1: Proteins. Fox PF, McSweeney PLH. Kluwer Academic/Plenum Publishers, New York pp. 1175-1228.

Nilsson U, Nyman M 2005. Short-chain fatty acid formation in the hindgut of rats fed oligosaccharides varying in monomeric composition, degree of polymerisation and solubility. Brit J Nutr, 94: 705-713.

O'Riordan K, Fitzgerald GF 1998. Evaluation of bifidobacteria for the production of antimicrobial compounds and assessment of performance in Cottage cheese at refrigerated temperature. J Appl Microbiol, 85: 103-114.

Ogata Y, Fujita K, Ishigami H, Hara K, Terada A, Hara H, Fujimori I, Misuoka T 1993. Effect of a small amount of 4G-beta-D-galactosylsucrose (lactosucrose) on fecal flora and fecal properties. J Jpn Soc Nutr Food Sci, 46: 317-323.

Özer BH, Kirmaci HA 2010. Functional milks and dairy beverages. Int J Dairy Technol, 63(1): 1-15.

Ozkan G, Simsek B, Kuleasan H 2007. Antioxidant activities of *Satureja cilicica* essential oil in butter and in vitro. J. Food Eng. 79: 1391-1396.

Pandya NC, Kanawjia SK 2002. Ghee: A Traditional Nutraceutical. Indian Dairyman, 54(10): 67-75.

Parodi PW 2003. Anti-cancer agents in milk fat. Aust J Dairy Technol, 58: 114-118.

Pavunc AL, Turk J, Kos B, Beganovic J, Frece J, Mahnet S, Kirin S, Suskovic J 2009. Production of fermented probiotic beverage from milk permeate enriched with whey retentate and identification of present lactic acid bacteria. Mljekarstvo, 59: 11–19.

Pettus BJ, Chalfant CE, Hannun YA 2004. Sphing inflammation: Roles and implications. Curr Mol Med, 4: 405-418.

Pihlanto A Korhonen H 2003. Bioactive peptides and proteins. In Advances in Food and Nutrition Research. Taylor SL (ed) Elsevier Inc., San Diego, CA. pp: 175-276.

Playne MJ, Bennet LE, Smithers GW 2003. Functional dairy foods and ingredients. Aust J Dairy Technol, 58: 242–264.

Pouliot Y, Gauthier SF 2006. Milk growth factors as health products: Some technological aspects. Int Dairy J, 16: 1415-1420.

Prabha S 2006. Development of a technology for the manufacture of dietetic *burfi*. Ph.D. Thesis. National Dairy Research Institute (Deemed University), Karnal.

Pszczola DE 1998. Production and Potential food application of cyclodextrins, Food Technol, 42: 96-100.

Rajpal S, Kansal VK 2008. Buffalo milk probiotic dahi containing *Lactobacillus acidophilus*, *Bifidobacterium bifidum* and *Lactococcus lactis* reduces gastrointestinal cancer induced by dimethylhydrazine dihydrochloride in rats. Milchwissenschaft, 63(2): 122-125.

Ranjan P, Arora S, Sharma GS, Sindhu JS, Singh G 2006. Sensory and textural profile of curd (*dahi*) from calcium enriched buffalo milk. J Food Sci Technol, 43(1): 38-40.

Rižnar K, Èelan Š, Knez Ž, Škerget M, Bauman D, Glaser R 2006. Antioxidant and antimicrobial activity of rosemary extract in chicken frankfurters. Food Chem. Toxicol. 71(7): C425-C429.

Rombaut R, Van Camp J, Dewettinck K 2006. Phospho-and sphingolipid distribution during processing of milk, butter and whey. Int J Food Sci Technol, 41: 435-443.

Rosenthal I, Rosen B, Bernstein S 1997. Phenols in milk Evaluation of ferulic acid and other phenols as antifungal agents. Milchwissenschaft 52: 134-138.

Rothschild P 1995. Internal defenses. Dairy Ind Int, 60(2): 24-25.

Saavedra JM, Abi-Hanna A, Moore N, Yolken RH 2004. Long-term consumption of infant formulas containing live probiotic bacteria: tolerance and safety. Am J Clin Nutr, 79(2): 261-267.

Saini SPS, Bains GS, Jain SC 1993. Flavour of iron fortified buffalo milk. Res Ind, 38: 146-149.

Sako T, Matsumoto K, Tanaka R 1999. Recent progress on research and applications of non - digestible galacto - oligosaccharides. Int Dairy J, 9: 69-80.

Salem AS, Salama, WM, Hassanein AM, El-ghandour HM 2013. Enhancement of Nutritional and Biological Values of *Labneh* by Adding Dry Leaves of *Moringa oleifera* as Innovative Dairy Products. World Appl Sci J 22(11): 1594-1602.

Schaller F, Rahalison L, Islam N, Potterat O, Hostettmann K, Stoeckli-Evans H, Mavi S 2000. A new potent antifungal 'quinone methide' diterpene with a cassane skeleton from *Bobgunnia madagascariensis*. Helvetica Chimica Acta 83: 407-413.

Schlimme E, Martin D, Meisel H 2000. Nucleosides and nucleotides: natural bioactive substances in milk and colostrum. Brit J Nutr, 84 (Suppl.1): S59-S68.

Schmidt DG, Poll JK 1991. Enzymatic hydrolysis of whey proteins. Hydrolysis of a-lactalbumin and b-lactoglobulin in buffer solutions by proteolytic enzymes. Neth Milk Dairy J, 45: 225-40.

Seifu E, Buys EM, Donkin EF 2005. Significance of the lactoperoxidase system in dairy industry and its potential applications: a review. Trends Food Sci Tech 16: 137-154.

Shing YW, Klagsbrun M 1984. Human and bovine milk contain different sets of growth factors. Endocrinology, 115(1): 273-282.

Singh G, Arora S, Sharma GS, Sindhu JS, Kansal VK, Sangwan RB 2007. Heat stability and calcium bioavailability of calcium-fortified milk. LWT-Food Sci Technol, 40: 625-631.

Sloan AE 2004. The top ten functional food trends. Food Technol, 58: 28-51.

Strohmaier W 1997. Lactulose, an innovative food ingredient - physiological aspects. Proceedings of Food Ingredients Europe Conference, Porte de Versailles, Paris, pp. 69–72

Sudhakar M 2004. Studies on bioprotective properties of probiotic lassi. M.Sc. Thesis National Dairy Research Institute (Deemed University), Karnal.

Suzuki T, Yamauchi K, Kawase K, Tomita M, Kiyasawa I, Okongi S 1989. Collaborative bacteriostatic activity of bovine lactoferrin with lysozyme against *E. coli* 0111. Agr Biol Chem, 53: 1705-1706.

Tamime AY, Marshall VME 1997. Microbiology and technology of fermented milks. In Microbiology and Biochemistry of Cheese and Fermented Milk, Law BA (ed), 2nd edn., Blackie Academic and Professional, London, pp. 57–152.

Tamura Y 2004. Production and application of bovine lactoferrin. Bulletin of the International Dairy Federation, 389: 64-68.

Teuri U, Korpela R, Saxelin M, Montonen L, Salminen S 1998. Increased fecal frequency and gastrointestinal symptoms following ingestion of galactooligosaccharide containing yoghurt. J Nutr Sci Vitaminol, 44: 465-471.

Thomä-Worringer C, Sörensen J, López-Fandiño R 2006. Health effects and technological features of caseinomacropeptide. Int Dairy J, 16: 1324-1333.

Timmermans E 1997. Lactose derivatives: Functions and applications. In Whey, International Dairy Federation. Brussels, Belgium, pp. 233-250).

Ushida Y, Shimokawa Y, Matsumoto H, Toida T, Hayasawa H 2003. Effects of bovine α-lactalbumin on gastric defense mechanisms in naive rats. Biosci Biotech Biochem 67: 577-583.

Van der Meer R, Lapre JA 1991. Calcium and colon cancer. Bulletin of the International Dairy Federation, 255: 55-59.

Vasiljevic T, Shah NP 2008. Probiotics-rom Metchnikoff to bioactives. Int Dairy J, 18: 714-728.

Visalsok T, Shigeru H, Satoshi Y, Souichi K 2004. Effects of a lactoperoxidase-thiocyanate-hydrogen peroxide system on *Salmonella enteritidis* in animal or vegetable foods. Int J Food Microbiol, 93: 175-183.

Walzem RL, Dillard CJ, German JB 2002. Whey components: millennia of evolution create functionalities for mammalian nutrition: what we know and what we may be overlooking. Cr Rev Food Sci 42: 353-375.

Wang B, Brand-Miller J, McVeagh P, Petocz P 2001. Concentration and distribution of sialic acid in human milk and infant formulas. Am J Clin Nutr, 74: 510-515.

Wheeler TT, Hodgkinson AJ, Prosser CG, Davis SR 2007. Immune components of colostrums and milk-a historical perspective. J Mammary Gland Biol, 12: 237-247.

Wu JP, Ding XL (2002). Characterization of inhibitory and stability of soy protein-derived Angiotensin-I-Converting Enzyme inhibitory peptides. Food Res Int, 35: 367-375.

Zimecki M, Kruzel ML (2007). Milk-derived proteins and peptides of potential therapeutic and nutritive value. Journal of Experimental Therapy and Oncology, 6: 89-106.

2016, Dairy and Food Product Technology Pages 245–252
Editors: Birendra Kumar Mishra and Subrota Hati
Published by: BIOTECH BOOKS, NEW DELHI

Chpater 16

"Nakamsua": A Traditional Fermented Fish Product of Garo Hills, Meghalaya

N. Balamurgan[1], B.K. Mishra[1] and B. Paul[2]

[1]Department of RDAP, North-Eastern Hill University, Tura Campus, Tura – 794 002, Meghalaya
[2]Department of Zoology, Don Bosco College, Tura, Meghalaya

Introduction

The Garo Hills district of Meghalaya is inhabited by tribal dwellers, the majority of whom are Garo tribes who are very fond of fish and dry-fish. Fish is a daily constituent of their diet. They have antiquated an interesting method of preservation of dry-fish through fermentation in hollow stump of bamboo. The fermented fish preparation is referred to as *Nakamsua*. The methodology of preparation for this product involves the use of a special kind of potash locally known as *Kalchi* which is obtained by burning dry pieces of plantain stems, dowsing the ashes in water and straining in conical shaped bamboo strainer. The dry fish and bamboo stumps are washed with *kalchi* followed by drying the fish again and cramming the semi dry fish into the bamboo stumps. They are then tightly mobbed with dried citrus leaves, capped with plantain leaves and stored away for a month or two during which fermentation of the fish takes place. The fully fermented fish is found to be "aromatic" and relished by the Garos. *Nakamsua* is used as a condiment in their everyday food preparations. It is referred to as a soul-stirring food item and is included in almost every food preparations along with vegetables and/or meat. It is believed to have medicinal properties and used by many during certain crises periods. Some rural villages where *Nakamsua* is prepared were visited and the process was observed and documented. This article reflects the expertise of the ethnic Garos' who are unwittingly exploiting the natural microbial consortium in fermentation of this fish product. At present, this product is prepared for local consumption at house-hold level without much consideration to good manufacturing practise. Scientific studies on *Nakamsua* would enhance its value addition and create public awareness with respect to its hygiene, nutritional value, health benefits and help to produce high-quality stable product with increased shelf life. Such studies would encourage commercialization of the product which would be a fillip to the rural economy.

The custom of fermenting fish was a significant preservation method before refrigeration; canning and other modern preservation techniques came to light especially in developing countries. Fermented foods in varied forms are widely consumed by different races of the world and have their own models of fermentation. Fermented fish processing is an artisanal activity and the process differ from one country to another (Sarojnalini and suchitra, 2009).

Fermented fish have, for many years, have been popular in Africa and among the Southeast Asian countriese specially in Thailand, Kampuchea, Malaysia, Philippines, and Indonesia where the use of fermentation as a preservation method for fish has been of great value since earliest times. There are a number of fish fermentation products such as fish sauce which are mainly prepared for flavour enhancers and condiments. However there are also fermented fish prepared as a staple food for years in almost every country in Europe, Africa, Middle East, Asia and South East Asia generally as a condiment for rice dishes (Petrus *et al.*, 2013). It has also become very popular in the developed countries due to their high nutritive value and organoleptic characteristics.

The earliest reported fermented fish sauce is *garum*, which is known to have been popular in the Roman era. It is made from the viscera and blood of mackerel. Other fish sauces, for example, *botargue* and *ootarides*, were produced in Italy and Greece in the 19th century. Another sauce reported to be produced in ancient Greece was *aimeteon*, which was made from Tunny viscera and blood. *Nuoc-mam* is a fish sauce prepared from small fish in Southeast Asia. The fish are fermented in earthenware containers in a high concentration of salt for several months. The clear amberliquid that rises is separated and consumed. *Shoittsuru* is the fermented fish of Japan sometimes referred to as fish soy; its origin may predate soy sauce. *Burongdalag* is a blend of rice and the fish prepared by fermentation in the Philippines. In southern India and Sri Lanka, pickled or Colombo curd fish have been known for many years. In this food, fish and salt in a 3:1 ratio are mixed in concrete tanks, dried tamarind fruit is added, and the mixture pickled (Jashbhai, *et al.*, 2008).

In Africa the "*lanhouin*" is a popular salted, fermented and dried fish usually consumed in southern Benin (Anihouvi *et al.*, 2006). In Indonesia Peda is a wet fermented mackerel (*Rastrilliger* sp.), where fresh mackerel put in layers in basket and topping with 30 per cent (w/w) salt and fermented at room temperature for 3 months or more. This traditional fermented fish product without drying process could be grouped as intermediate moisture food. This process can be classified as spontaneous fermentation where only natural microorganisms involved in this process and therefore wide variation of final products quality are found. Other popular traditional fermented fish product in South Kalimantan, Indonesia is fermented fresh water fish or locally known as *Wadi Betok*, which are widely consumed in South and Central Kalimantan for decades. The whole fresh water fish after added with 30 100 per cent (w/w) salt were layered in plastic jars and left for spontaneous fermentation for 7 days to 4 months at room temperature (Petrus *et al.*, 2013). Fermented fish such as Pla-ra in Thailand, Pa-dag in Lao PDR and Pra-hoc in Cambodia are prepared by mixing fish with salt or roasted rice or roasted rice bran and then put into a jar with a complete lid. These products contained fascinated odour due to roasted rice and roasted rice bran mixed with the fish and were also high in nutrients of protein, calcium and phosphorus (Udomthawee, *et al*, 2012). In the Philippines, fermented fishery products consist of *bagoong* (fish paste) and *patis* (fish sauce). These products aregenerally used as condiments. Also *bw"ongisda* (fermented rice fish mixture) and *bw-onghipon*, also known as *bataobatao* (fermented shrimp ricemixture). These fermented products are acidic with a cheese-like aroma (*Minerva, 1992*).Likewise in Japan, *Narezushi*, is fish fermented together with rice or another starch, has a characteristic taste that develops from the auto-digestion of meat (Fukuda *et al.*, 2014). Traditional fish sauces in Japan, include "*Shotturu*" is made from sandfish, "*Ishiru*" is made from squid, "*Ikanago-Shoyu*" is made from sand lance fish, and so on (Fukuda, *et al.*, 2014). Similarly Yu-lu is a traditional fermented fish sauce widely consumed as a condiment in the southern and eastern parts of china (Jaing *et al.*, 2007).

In India fermented fish products can be encountered throughout the North-Eastern states and the process of fermentation is unparalleled from the rest of the world. For example the 'Ngari' is an indigenous fermented fish product of Manipur. It is prepared from small and less-priced sun dried fishes such as *Puntius sophore* (Ham) and *Puntius ticto* (Ham) subjecting to fermentation in the absence of salt for 5 to 6 months or more at room temperature. The process of 'Ngari' preparation involves a brief washing of the sun dried fishes, followed by draining and drying for24-48 h. The fishes are then pressed hard using stoneroller to breakdown head and bones. Before filling the fishes, a thin layer of mustard oil was applied to the inner wall of the earthen pot *'Kharung'* to check porosity. Fishes are pressed hard mechanically inside the pots using wooden stick. The pots were then sealed airtight and incubated at room temperature for 6 months (Sarojnalini, 2009). *Shidal* is also a similar salt free, semi fermented and solid fish based product especially of Assam. It is prepared exclusively from small carps locally known as *puthi* (*Puntius* spp.). *Shidal*-chutney is a highly relished cuisine made from this product amongst many people of Northeast India. It has several local names like *seedal, seepa, hidal*and *shidal*in Assam, Tripura, Arunachal Pradesh and Nagaland (Muzaddadi *et al.*, 2013; Ahmed, *et al.*, 2013). *Sindol* is an indigenous fermented fish product of Assam, India. It is prepared from small sun-dried fish *Amblypharyn godonmola*, in the presence of salt for 2 to 3months or more at room temperature. *Tungtap*is a popular fermented fish (*Puntius* spp. and/ or *Danio* spp.) product, commonly prepared and consumed by the *Khasi* and *Jaintia* tribes of Meghalaya in North-Eastern state of India. Sun-dried pre-salted fish (*Puntius* spp. or *Danio* spp.) is mixed with fish fats to create semi-anaerobic condition and packed in earthen pots at room temperature and left for traditional fermentation for a period of six months under slightly acidic condition and the end product is consumed along with regular meals.

Microbiology of Fermented Fish

A fermented fish can be described as any fishery product that has undergone degradative changes through enzymatic or microbiological activities either in the presence or absence of salt. Microorganisms play a crucial role in the stages of fish fermentation leading to the degradation of tissue proteins. Also fermentation snipes the ability of spoilage microbials to despoil fish by making the fish muscle more acid and hence undesirable bacteria usually ceases to multiply when the pH drops below 4.5thereby preventing putrefaction. Proteolysis and liquefaction that occur during fish fermentation is largely due to autolytic breakdown of the fish tissues.

Lactic acid bacteria (LAB) and yeasts are found to be the dominant microorganisms during fish fermentation. Several species belonging to the genera *Leuconostoc, Weissella, Pediococcus, Lactococcus, Enterococcus* and *Streptococcus* have been isolated from fermented fish, but, *Lactobacillus* strains are found to be the most abundant. The characteristic smell of fermented fish is the result of enzymatic and microbiological activity in the fish muscle. It is suggested that the organic acids produced during the fermentation of fish in Asia are mainly lactic acid (Das, 2014).

Fermented fish products in Japan are often produced by filamentous fungi like *KOJI* because the use of KOJI on fish fermentation improves fermented products. *KOJI*, which is filamentous fungal fermented cereal, is used widely to seed or culture various fermented food and beverages such as miso, soy sauce, sake, and so on. *KOJI* is important enzyme producer, including gamylase, protease, lipase, and so on. *KOJI* include not only filamentous fungi but also bacteria and yeasts. Effects of mixed culture in filamentous fungi fermentation are an improvement of fermented food nutrient and functionality and high efficiency of material conversion especially in fish fermentation products. *KOJI* is important enzyme producer, including amylase, protease, lipase, and so on. The fermented fish has a characteristic taste that develops from the auto-digestion of fish meat (Fukuda, *et al.*, 2014).

Thus Fermentation not only extends the shelf life but also imparts a unique taste, flavor and enhances the nutritional quality of end products. Thus it is not just a dietary source of food but also a good source of nutrition.

Fish Fermentation by the Garo Tribes

The Garo tribes of the state of Meghalaya follow an intriguing method of fermenting fish which is unique from the rest of India and the world. The procedure involves two stages. The first step is the preparation of a kind of local potash from burnt banana leaves called *kalchi* and the next step involves the preparation of the fish for fermentation. The end product of fermentation is referred to as *Nakamsua*.

Figure 16.1: Step-wise Preparation of *Kalchi*.

A. Preparation of *Kalchi*

Kalchi is prepared from plantain stems. The plantain stems are collected, sun dried and burnt to obtain ash. The ash so obtained are dowsed in water and strained in conical shaped bamboo strainer in bottles. The clear filtrate collected is called *Kalchi.*

Figure 16.2: Step-wise Preparation of *Nakamsua.*

B. Preparation of *Nakamsua*

The small varieties of dry fishes namely the *Nawari* or the *chenda* are preferred. Hollow bamboo stumps are used for fermenting fish. The fish is washed properly with *kalchi* and water. Likewise the hollow bamboo stumps used are also washed with *kalchi*. The fish is then semidried again in sun and then crammed tightly into bamboo stumps with help of slender bamboo sticks. They are then tightly mobbed with dried citrus leaves and further capped with mud and/or plantain leaves and stored away for a month or two during which fermentation takes place.

Conclusion

Scientific research on the interesting indigenous fermented fish product *Nakamsua* of the Garo Hills is wanting. At present, these products are prepared for local consumption at house-hold level under non-sterile and marginally controlled conditions. Scientific evaluation with respect to the physico-chemical, sensory, microbial association and probiotic efficacy of this fermented fish product is paramount to systemically and scientifically upgrade, commercialize this product and to establish its beneficial effects. Improved and optimised production methods will further help to economically and feasibly produce this product with increased shelf life in future at reduced production costs and help to stimulate to the rural economy of Garo hills.

References

Anihouvi V.B, Ayernor G.S, Hounhouigan J.D.L and Sakyi-Dawson E., 2006, Quality characteristics of *lanhouin*: A traditionally processed fermented fish product in the republic of Benin, African Journal of Food Agriculture and Nutrition, vol. 6, No.1: 1-15.

Armaan Ullah Muzaddadi and Prasanta Mahanta, 2013, Effects of salt, sugar and starter culture on fermentation and sensory properties in *Shidal* (a fermented fish product), *African Journal of Microbiology* Research Vol. 7(13), pp. 1086-1097.

C.H. Sarojnalini and T. Suchitra, 2009, Microbial profile of starter culture fermented fish product 'Ngari' of Manipur. *Indian J. Fish*, 56(2): 123-127.

Jashbhai B. Prajapati and Baboo M. Nair, 2008, The History of Fermented Foods, *Handbook of Fermented Functional Foods*, Second Edition, edited by Dr. Edward R. Farnworth, CRC Press, Taylor and Francis Group, Boca Raton, pp: 1-24.

Jin-Jan Jaing, Qing-Xiao Zeng, Zhi-Wei Zhu, Li-Yan Zhang, 2007, Chemical and Sensory changes associated Yu-lu fermentation process – A traditional Chinese fish sauce, *Food Chemistry*, 104: 1629-1634.

Kotchanipha Udomthawee, Kasem Chunkao, Achara Phanurat and Khunnaphat Nakhonchom, 2012, Protein, Calcium and Phosphorus Composition of Fermented Fish in the Lower Mekong Basin, *Chiang Mai J. Sci.*, 39(2) : 327-335.

Manoj Kumar Das, (2014), Starter culture development of Sindol-A fermented fish product of Assam, *International Journal of Agricultural Science and Research* (IJASR), 4(2): 107-114.

Minerva SD. Olympia, 1992, Fermented fish products in the Philippines, *Applications of Biotechnology to traditional fermented foods*, National Research Council, National Academy Press, Washington, D.C. pp: 131-139

Petrus, Hari Purnomo, Eddy Suprayitno and Hardoko, 2013, Physicochemical characteristics, sensory acceptability and microbial quality of *Wadi Betok*a traditional fermented fish from South Kalimantan, *Indonesia International Food Research Journal 20(2): 933-939*

Sarifuddin Ahmed, Krushna Chandra Dora, Sreekanta Sarkar, Supratim Chowdhury and Subha Ganguly, 2013, Quality analysis of shidal - a traditional fermented fish product of Assam, North-East India, *Indian J. Fish.*, 60(1): 117-123

Tsubasa Fukuda, Manabu Furushita, Tsuneo Shiba and Kazuki Harada, 2014, Fish Fermented Technology by Filamentous Fungi, *Journal of National Fisheries University*, 62: 0163168.

2016, Dairy and Food Product Technology
Editors: **Birendra Kumar Mishra and Subrota Hati**
Published by: **BIOTECH BOOKS, NEW DELHI**

Pages 245–264

Chapter 17

Value Addition in Minor Fruits: Jamun and Jackfruit

A.K. Chaurasiya[1] and A.K. Singh[2]

[1]Department of Horticulture, NEHU, Tura Campus, Meghalaya
[2]M.S. Swaminathan School of Agriculture, CUTM, Gajapati – 761 211, Odisha

Introduction

When 70 per cent of Indian populations are engaged only in production activities of Agriculture, we need more and more entities and systems to add value. In doing so, we need to look at food wastages and their prevention, improvement in value addition of horticultural produces through adoptable processes, harnessing untapped food resources, utilizing by-products and assuring food quality and safety. All these have to be interlinked with extension of shelf life, which is also value addition.

India is the second largest producer of fruits and vegetables in the world. The total production of fruits and vegetables is 81.285 and 162.186 million tones, respectively during 2012 – 2013 in India. Though India is producing various kinds of major and minor fruits but hardly 2.2 per cent of the production is commercially processed whereas more than 50 per cent of the produce is processed in developed countries (Rasul, 2001). However, the level of processing in the major fruit producing countries are Brazil- 70 per cent, USA –60-70 per cent, Malaysia –83 per cent and Israel –50 per cent. International trade in processed fruit products is around US $ 9200 million.

The production of fruits in India has recorded a growth rate of 3.9 per cent, whereas the fruit-processing sector has grown at about 20 per cent per annum. However, the growth rates have been extensively higher for frozen fruits and vegetables

(121 per cent) and dehydrated fruits and vegetables (20 per cent). There exist over 4000 fruit processing units in India with an aggregate capacity of more than 12 lakh MT. (less than 4 per cent of total fruits produced). It is estimated that around 20 per cent of the production of processed fruits is meant for exports, the rest caters to the defence, and institutional sectors and house hold consumption. India has exported processed fruits and vegetables in 1999-2000 valued 86.0 US $ million (1.5 per cent) and during 2000- 2001 total valued 122 US $ million and international is 2 per cent.

People's growing interest in health is reflected in concern about cholesterol and low calorie foods, organic foods and antioxidant rich foods etc; today's market is flooded with blended fruit based beverages. We have a habit of concentrating only on major fruit based products and ignoring the potential of under exploited fruit products in the context of value addition. Thus to give the variety to the taste of the consumer innovative items are always asked for.

Fruits are said to be minor/under exploited mainly due to lack of planting materials; lack of intervention on production system and fitness with existing farming system; lack of information on post harvest management, processing and value added product; lack of information on access to market; lack of national policy though they have high nutritional and medicinal values. In order to improve the social and nutritional status of the people and also to make a dent in export promotion, it is worthwhile to increase the manufacture of high quality, delicious and variable processed products from under exploited fruits in the present day value addition programme.

North East India is endowed with diverse climatic conditions, which are capable of producing different minor fruit crops as a orphan way as well as organic. In spite of enormous nutritive and medicinal values, the cultivation of minor/underutilized fruit crop is restricted, as a result of this they are not available in plenty and a large number of them are unknown in the world market. There is always demand from consumers all over the world for new food products, nutritious and also delicately flavoured and attractively coloured.

Under these circumstances, the processed product has been formulated to develop value added products from some of the underutilized crops like Jamun and Jackfruit which are highly perishable and impossible to keep them for more than 24 hours under ambient condition *e.g.,* Jamun; some of them are not easy to eat out of hand The seasonal nature with short storage life even under low temperature condition, necessitate processing of the jackfruit.

Importance of Minor Fruits

Jamun (*Syzygium cuminii*) belongs to the family Myrtaceae. It is indigenous to India. The refreshing and curative properties of jamun make it one of the useful medicinal plants of India. Fruits are a good source of iron, an effective medicine against diabetes, heart and liver trouble. The ripe fruit can be processed as soft beverages and also fermented one. A good quality jam, jelly and pickle can also prepare from this fruit. A method of extraction of the jamun juice has been standardized (Sukla *et al.,* 91). Jamun juice mixed with mango juice is very effective for quenching thirst for diabetic patients. Its juice being highly acidic in nature is not consumed as such but juice is stoma chic, carminative and diuretic apart from having cooling and digestive properties.

A method of extraction of the jamun juice has been standardized (Shukla *et al.*, 1991) for the preparation of ready-to-serve beverage (nectar). It has been found that the maximum yield of jamun juice with a high level of the anthocyanin and other soluble constituents was obtained by grating the fruit, heating up to 70°C and passing the heated mass through a basket press (Ramanjaneya, 1985). Roy (1999) suggested that a maximum yield of jamun juice with a high level of anthocyanin and other soluble constituents can be obtained by grating fruits, heating it to 60°c and passing the grated mass through a basket press.

The attractive colour due to anthocyanin pigments is a major quality attributes in jamun beverages. The anthocyanin pigments are destroyed at high storage temperature. Refrigerated storage has stabilizing effect on the anthocyanins of the fruit juices (Ponting *et al.*, 1960). Khurdiya and Roy (1985) have been studied the quality of jamun juice and nectar during storage at different temperature including the cool chamber developed by Roy and Khurdiya (1982) with special reference to stability of colour. They observed that juice was acceptable up to 8 months storage at room temperature and 12 months at cool temperature, whereas the nectar can be stored for 10 months and 12 months at room and at cool temperature respectively. A ready to serve beverages (nectar) of jamun was prepared with 25 per cent juice, 18° Brix and 0.6 per cent acidity, it was found to be highly acceptable by Khurdiya and Roy (1985).

Jackfruit (*Artocarpus heterophyllus* lamk.) is known as the "poor man's food". It is generally used for culinary purposes in northern India as well as for table purpose in other parts of country. Ripe fruit can be processed for preparing the jam, squash, ready to serve beverages, chutney and papad. Fruits at early stage of the ripening may be utilized for making the preserve and candy.

A puree was prepared by adding water to the fruit in the ratio of 3:1 followed by passing through a pulper. The resulting puree was then mechanically pressed to obtain the juice (Seow and Shanmugam, 1992).

Blended R.T.S. was prepared as per F.P.O. specifications (Giridharilal *et al.*, 1995) by mixing fruit juice (100 ml) with the soya milk whey at 50:50 ratio, sugar and citric acid. The processed product was pasteurized at 80° C (Saravan Kumar and Manimegalai, 2002).Squash was prepared by mixing syrup with 17 per cent juices to maintain the T.S.S of 52° Brix and 1 per cent acidity (Bhatia *et al.*, 1995).

Extraction of Juice

Commonly juice is extracted from fresh healthy fruits by crushing and pressing technology. But, there are different types of machine has been developed such as Screw type juice extractor and fruit pulper. These are mostly used for juice extraction but extraction of juice varies from type of fruit to fruit, because of differences in their structure and composition.

A method of extraction of the jamun juice has been standardized (Shukla *et al.*, 1991). It has been found that the maximum yield of jamun juice with a high level of the anthocyanin and other soluble constituents was obtained by grating the fruit, heating up to 70°C and passing the heated mass through a basket press (Ramanjaneya, 1985). The jackfruit puree was prepared by adding water to the fruit in ratio of 3:1 following by passing through a pulper. The resulting puree was then mechanically pressed to obtain the juice (Seow and Shanmugam, 1992).

Clarification of Juice

Prasad and Mali (2000) reported that after the extraction, juice kept for 3-4 hours to down the coarse tissue particles. The supernatant solution was siphoned of leaving the coarse particles.

Bottling and Storage

The jamun products (RTS, squash and syrup) stored in colourless glass bottles were acceptable even after six months at room temperature (Kannan and Thirumaran, 2001; Gavande *et al.*, 1995).

Krishnaveni *et al.* (2001) reported that the jackfruit RTS packed in green glass bottle and stored at ambient temperature have better retention of ascorbic acid and beta carotene contents. Sensory quality found highly acceptable even after storing for 6 months at room temperature. R.T.S. of jackfruit filled in sterilize glass bottle (capacity 200 ml) leaving 1 inch headspace and capped air tightly (Saravan Kumar and Manimegalai, 2002).

FLOW-SHEET FOR EXTRACTION OF JUICE

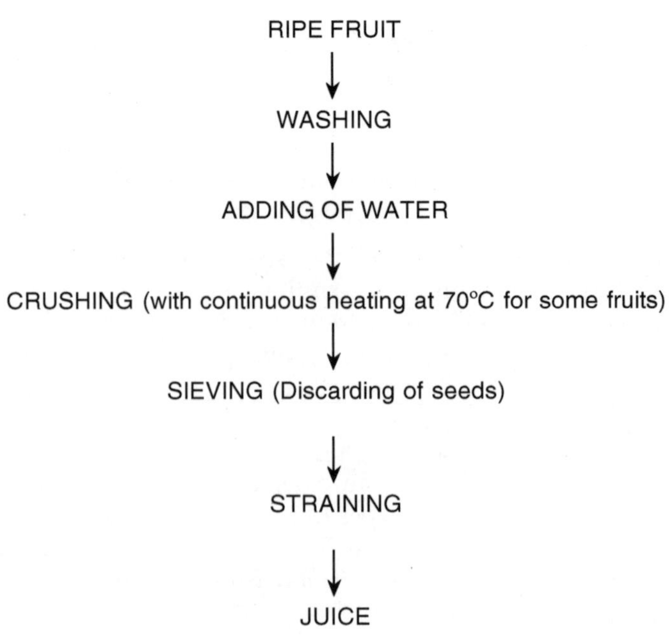

FLOW-SHEET FOR R.T.S. AND NECTAR

FRUIT (Pulp/juice)

↓

MIXING WITH STRAINED SYRUP SOLUTION
(sugar +water+ acid, heated just to dissolve) According to recipe

↓

HOMOGENIZATION

↓

BOTTLING

↓

CROWN CORKING
PASTEURIZATION (at about 90° C) FOR 25 MINUTES

↓

COOLING

↓

STORAGE

FLOW-SHEET FOR PROCESSING OF SQUASH

FRUIT (Pulp/juice)

↓

PREPARATION OF SYRUP SOLUTION
(sugar +water+ acid, heated just to dissolve)

↓

STRAINING

↓

MIXING WITH JUICE

↓

ADDITION OF PRESERVATIVE
(0.6gKMS or 1.0g sodium benzoate/liter squash)

↓

BOTTLING AND CAPPING

↓

STORAGE

Sourc: Srivastava and Kumar, 2003.

Name of Products	FPO specification, 1955 (Minimum Ingredients)		Other Ingredients	
	Juice (per cent)	TSS (°Brix)	Juice (per cent)	TSS (°Brix)
Ready to serve	10	10	10-12	10-14
Squash	25	40	25-30	40-50
Nectar	20	15	20-25	15-20

Jamun

The trees are evergreen, attain medium to huge size. Fruits are attractive black or purple in color, sweet in taste. The stone, bark, leaves etc., are used as an effective medicine against diabetes (passing of sugar in the urine) by the followers of Ayurvedic system.

Jamun Juice

Ingredients

Jamun fruits 1.0 kg

Water ½ litre

Sodium Benzoate 150 ppm./litre, final product

Procedure

Fresh ripe fruit wash properly and discord the spoiled fruits. Fruits add with water keep on heater for heating till just for soften. Remove from heater. Pulping by pulper machine. Straining with the help of muslin clothes and collects the dark purple color juice. Filled in sterilized bottles with adding of preservative, capping and sealing. Pasteurized at 80°C for 20 minutes. Stored at refrigeration or room temperature.

Jamun RTS (Ready to Serve)

Ingredient for one litre RTS

Juice 120 ml

Sugar 120 gm

Citric acid 2.5 gm

Water 850 ml

Preservative as Sodium Benzoate - 150 ppm./litre, final product

Preparation of Syrup

Boil the sugar with water in stainless steel utensil, and add the citric acid to the boiling syrup for clarification. Strain the syrup and cool it.

Add the syrup little by little to the mixed with pulp and stirring, and get final RTS. Filled in sterilized bottles with adding of preservative, capping and sealing. Pasteurized at 80°C for 20 minutes. Stored at refrigeration or room temperature.

Jamun Squash

Ingredient for one litre Juice

Juice	1000 ml
Sugar	1860 gm
Citric acid	42.0 gm
Water	2000 ml

Preservative as Sodium Benzoate - 1.0g./litre, final product

Procedure

Same as RTS. Filled in sterilized bottles with adding of preservative, capping and sealing. Stored at refrigeration or room temperature

Jackfruit

It is indigenous to India, giving the largest single fruit (40 kg.) is heavy yielder producing more than many of the fruit trees. It is classed as commercial fruit but is seldom cultivated in regular plantations. It is a popular fruit in South, North and North East India. The ripe fruit is tastier. In ripe fruits is more wastage.

Nutritional Value

Water (73.1 per cent), Protein (0.6 per cent), Fat (0.6 per cent), Carbohydrate (23.4 per cent), Fibre (1.8 per cent) and Ash (0.5 per cent).

TSS.- 18.2 ° Brix, Acid – 0.14 per cent, Total sugar – 10.26 per cent, Reducing sugar – 5.26 per cent and TSS/Acid ratio – 130.

Procedure

Extraction of Pulp

Fresh ripe fruit cuts into in four sections from vertical to horizontal section. Separate the cubes and remove the seeds and collect the fresh flesh. Pulping by pulper machine and collect the pulp. Filled in sterilized bottles with adding of preservative, capping and sealing. Pasteurized at 80°C for 20 minutes. Stored at refrigeration or room temperature. In room temperature can be stored up to 10 month, but in refrigeration can be stored up to 2 years.

Jackfruit RTS (Ready to Serve)

Ingredient for one litre RTS

Juice	110 ml
Sugar	110 gm
Citric acid	3.0 gm

Water 850 ml

Preservative as Potassium metabisulphite 0.2. gm./litre final product

Procedure

Preparation of Syrup

Boil the sugar with water in stainless steel utensil, and add the citric acid to the boil syrup for clarification. Strain the syrup and cool the syrup.

Add the syrup little by little to the mixed with pulp and stirring, and get final RTS. Filled in sterilized bottles with adding of preservative, capping and sealing. Pasteurized at 80°C for 20 minutes. Stored at refrigeration or room temperature.

Jackfruit Jam

Ingredient for 1.0 Kg pulp

Pulp	1000 gm
Sugar	750 gm
Citric acid	8.0 gm
Pectin	1 gm. (not necessory)

TECHNICAL FLOW-CHART FOR PROCESSING OF JAM

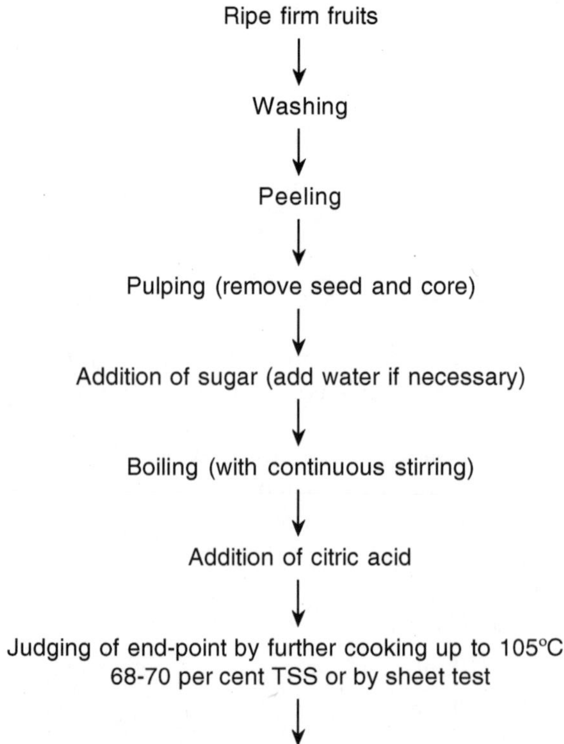

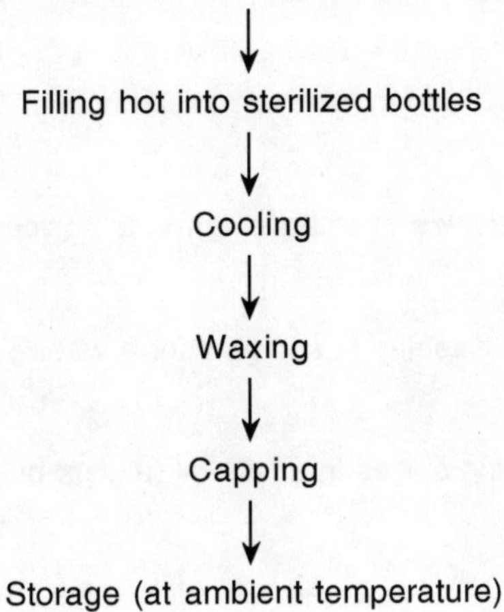

Procedure

Mix the pulp with sugar. Cook in low flme with continuous stirring. Add the citric acid and pectin to continue stirring till the mixture acquires the consistency of jam. Remove from fire and allow cooling. A well-set jam should fall out the spoon when poured. Pack to sterile bottles and seal. Store at room temperature up to one year acceptable.

Jackfruit Pickle

The preservation of fruit or vegetables in common salt or vinegar is called pickling. Spices and oil may also be added in pickle. Common salt @ 15 per cent prevents its spoilage. Vinegar too @ 2 per cent acts as a preservative.

TECHNICAL FLOW-CHART FOR PREPARATION OF PICKLE

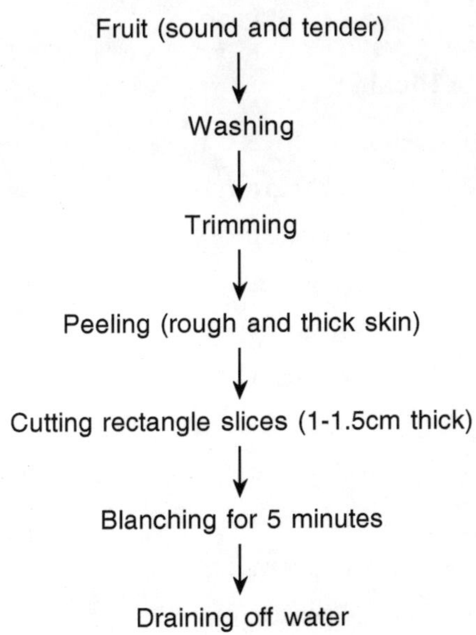

Ingredient for Jackfruit Pickle

Ingredients

Jackfruit pieces	2000 gram
Salt	350 gram
Turmeric powder	15 gram
Nigela seed	40 gram
Red chilli powder	15 gram
Fenugreek seed	60 gram
Clove	10 pieces
Cumin powder	10 gram
Cinnamon powder	10 gram
Large cardamom	4 pieces

Small cardamom	10 pieces
Black pepper powder	10 gram
Coriander powder	10 gram
Asafetida	5 gram
Onion	50 gram
Ginger	15 gram
Garlic	15 gram
Oils	800 ml

References

Bhatiya, B.S., Siddapa, G. and Lal, G.(1995). Development of products from jackfruit : part III- jackfruit preserve, candy, chutney, and dried bulb. *Indian Food Packer* **9**: 7.

Gavande, U.K., Joshi G.D. and Wasker, D.P. (1995). Storage of jamun (*Syzygium cuminii*) fruit product. *ASEAN Food Journal* **10** (2): 54-56.

Girdhari Lal, Siddapa, G.S. and Tandon, G.L. (1995). Preserves candied and crystallized fruits. In: *Presevation of Fruit and Vegetables.* ICAR, New Delhi.

Kannan, S. and Thirumaran, A.S. (2001). Storage life of jamun products. *Processed Food Industry* **5**(1): 18-19.

Khurdiya, D.S. and S.K. Roy (1985). Processing of jamun fruit into ready-to-serve beverages. *J. Food Sci. Technol.*, **22:** 27.

Krishaveni, A., Manimegalai, G and Saravan Kumar, R. (2001). Storage stability of jackfruit R.T.S. beverages. *Journal of Food Science and Technology*, **36** (6): 601-02.

Ponting, J.D., Samshuck, D.W. and Brekke, J.E. (1960). Colour and deterioration in grape and berry juices and concentrates. *Food Res.*, **25**: 471.

Prasad, R. N. and Mali, P. C. (2000). Changes in physico- chemical characteristics of pomegranate squash during storage. *Indian J. Hort.*, **5**(1): 18-21.

Ramanjaneya, K.H.(1985). Studies on some aspects of jamun and its processing. Ph.D. thesis. I.A.R.I., New Delhi.

Rasul, N. (2001). Value addition due to food processing and income distribution amongst the poor. *Indian Food Industry*, **20**(6): 17-20.

Roy, S. K. and Khurdiya, D.S. (1982). Keep vegetables fresh in summer. *Indian Hort.*, **27**: 5.

Roy, S. K., Pal, R. K. and Ramanjaneya, K. H. (1999) Jamun: a potential fruit for processing. *Indian Horticulture* **44** (2): 9-11.

Saravan Kumar, R. and Manimegalai, G.(2002). Storage stability of soya milk whey based jackfruit juice blended R.T.S. beverages. *Processed Food Industry* (Dec.): 42-47.

Seow, C.C. and Shanmugam, G. (1992). Storage stability of canned jackfruit juice at tropical temperature. *J. Food. Sci. Technol.*, **29** (6): 371-74.

Shukla, K. G., Joshi, M.C., S. Yadav and N.S. Bisht (1991). Jamun wine making: standardization of a methodology and screening culture. *J. Food Science Tech.*, **28** (3): 142.

Srivastava, R.P. and Kumar, S. (2003). Unfermented and fermented fruit beverages.In: *Fruit and Vegetable Preservation: Principles and Practices.* International Book Distributing Company, Chaman studio building, 2nd floor, Charbag, Lucknow-226004. pp.175-193.

2016, Dairy and Food Product Technology *Pages 265–280*
Editors: **Birendra Kumar Mishra and Subrota Hati**
Published by: **BIOTECH BOOKS, NEW DELHI**

Chapter 18
Post Harvest Technology in Ber (*Ziziphus* spp.)

A.K. Singh[1] and A.K. Chaurasiya[2]

[1]M.S. Swaminathan School of Agriculture,
CUTM, Gajapati – 761 211, Odisha
[2]Department of Horticulture,
NEHU, Tura Campus, Meghalaya

Introduction

Ber is a tropical and subtropical fruit native to the northern hemisphere [Lyrene, 1979]. It belongs to the genus *Ziziphus* of the family Rhamnaceae and order Rhamnales. The family has 50 genera and more than 600 species [Bailey, 1947] of which the species *Z. jujube* Mill (Chinese jujube or Chinese date), *Z. mauritiana* Lamk (Indian jujube or ber) and *Z. spinachristi* (L.) wild (Christ's thorn) are the most important in terms of distribution and economic significance. *Z. mauritiana* is commonly cultivated throughout the northwest of India and in the drier parts of South India [Brandis, 1906; Sebastian and Bhandari, 1990]. Ber is also found in disturbed areas near settlements and along roads in African countries [Johnston, 1972] where fruits are harvested from naturally seeded plantations and sold in local markets. *Z. jujube* Mill is cultivated in China, Korea and in parts of Southeast Africa [Lin and Cheng, 1995]. Ber is being cultivated on an estimated area of 22,000 hectares. The yield potential varies from one to two quintals per tree per annum (Yamadagni, 1985). It is considered an underutilised fruit crop in semi-arid regions of the world and can be successfully cultivated in the marginal ecosystem of the subtropics and tropics [Pareek, 2001].

Ber fruit is generally eaten fresh and is a rich source of ascorbic acid, essential minerals and carbohydrates [Jawanda *et al.*, 1980; Pareek *et al.*, 2002]. It is richer than apple and mango in vitamin C, protein and minerals and contains higher phosphorus

Table 18.1

Characteristics	*Major Cultivars of Ber*							
	Katha	*Bagwari*	*Umran*	*Chhuhara*	*Ilaichi*	*Karaka*	*Mundia Murhra*	*Narama*
Appearance	Golden yellow	Yellow to reddish brown	Golden yellow	Greenish yellow	Yellow to reddish brown	Yellow to reddish brown	Greenish yellow	Light green
Avg. Fruit wt. (g)	18.5	16	21	12.5	3.6	23	22	17.5
Pulp:stone rati	25	13.3	19.6	11.3	24.8	–	16.7	12.9
Moisture (per cent)	74.33	77.92	77.81	76.4	73.97	86.6	88.13	79.3
TSS °Brix	21.4	18	22.7	17.2	24.7	11.8	13	16.8
Acidity (per cent)	0.1	0.11	0.29	0.35	0.22	0.31	0.22	0.25
Reducing sugar (per cent)	4.54	5.94	4.38	3.72	3.91	5.88	4	3.7
Total sugar (per cent)	19.65	16.17	14.84	16.23	16.98	9.97	11.1	14.9
Ascorbic acid (mg/100g)	97.76	126.46	150.99	146.69	129.41	103.5	174.6	146.5

and iron than orange [Khera and Singh, 1976]. In general, the fruits contain 85.9 per cent moisture, 12.8 per cent carbohydrates, 0.8 per cent protein, 0.1 per cent fats, 0.8 per cent iron, 0.03 per cent each of calcium and phosphorus, and 70 I.U. vitamin A/ 100 g with an energy value of 55 calories/100 g [Yamdagni, 1985].

Freshly harvested fruits can be stored in containers such as gunny, net or polythene bags, cloth packs and boxes for 4-15 days at room temperature (25-35 °C) without loss of organoleptic quality.

- ✰ Fruits can be stored in a cool chamber for 6-10 days. However, the high humidity which develops in the cool chamber is conducive to spoilage.
- ✰ Fruits can retain acceptable organoleptic quality for 3 weeks kept in polythene bags and baskets at 13 °C in an incubator.
- ✰ Kept in polythene bags in a cold storage at 10 °C, fruits of some cultivars can be stored for 28-42 days.
- ✰ Fruits can be stored frozen at -18 °C for 6 weeks.

The shelf-life depends mainly on the ber cultivar, storage temperature, packaging method and stage of harvest. During storage, the fruits loose weight and shrivel, change colour and become red, loose acidity and ascorbic acid, but gain in sweetness.

The storage life of ber fruits is extremely short and the rapid perishability of the fruits is a problem. At ambient temperature a shelf-life of 2–4 days is common. Due to the surplus of fruits in the local markets during peak season, a substantial quantity goes to waste, resulting in heavy postharvest losses. A cost and returns analysis performed by [Gupta *et al.,]* showed that ber production is highly remunerative but requires proper handling with respect to preharvest, harvesting and postharvest treatments, packaging, transportation, storage, postharvest pathology, processing, etc. Profits could be enhanced if efforts to increase production are supplemented with efforts to minimize postharvest losses and enhance shelf life.

Ber fruits are usually stored at ambient/room temperature (25–35 °C) from harvest until their consumption. *et al.,* observed that Umran and Sanaur-2 fruits could be stored for up to 10–12 days at room temperature. Panwar, 1981 reported that ber fruits remained in marketable condition for about one week. Ripe fruits of ber when stored at room temperature without any treatment remained for up to 7 days [Bal, 1982]. Gupta *et al.,* observed that the shelf-life at room temperature was the longest in Sanaur-5, followed by Ponda, Reshmi and Umran cultivars. Pareek and Gupta, 1988 observed the shelf-life of Gola and Kaithli cultivars at ambient temperature for up to 7 and 10 days, respectively. *Z. spinachristi* fruits could be stored for 6 days at room temperature [Abbas *et al.,* 1990; Al-Niami and Abbas, 1988]. Golden yellow colour ripe fruits of Umran could be stored for about one week at 30 °C [Anonymous, 1990]. The fruits of cultivar Gola were suitable for eating for up to 8 days of storage [Anonymous, 1990]. Siddiqui and Gupta, 1995 observed 38 per cent PLW, 35.3 per cent over ripening and 13.4 per cent decay loss in cultivar Umran at room temperature after 6 days of storage. In contrast to this Gupta and Kadam, 1995 reported that ber fruits stored at ambi- ent temperature had a short life of 3 days only. Under ambi- ent conditions, ber fruits showed a high degree of pathologi- cal infection and loss in colour, and could be stored for only 9 days [Vishal Nath *et al.*, 1999]. Pareek *et al.*, 1983observed that the storage environment did not affect the levels of total sugars in the fruits. It was observed that Virosil Agro (2.5 and 5 per cent) and Bavistin (1 per cent) maintained the original levels of reducing sug- ars during the storage period.

Storage in Zero Energy Cool Chamber

High temperature and moderate humidity at the time of fruit maturity (February to March) leads to the attack of different micro-flora that caused decay, increased PLW and reduced shelf-life and quality. These factors lead to heavy losses which can be minimised by storing fruits in a zero energy cool chamber (ZECC). ZECC have been designed to enhance the shelf-life of fruits and vegetables by lowering the temperature and increasing the relative humidity inside the chambers via passive evaporative cooling [Roy and Khurdiya, 1986]. Ber fruits of Kaithli, Umran and Gola cultivars could be stored in these chambers for 14, 15 and 18 days, respectively [Anonymous, 1988-1989]. Siddiqui and Gupta, 1990 stored fruits in these chambers up to 6–10 days. The fruits of cultivar Gola were found to be in acceptable condition up to 12 days of storage in ZECC. There was a decrease in fruit firmness, specific gravity and or- ganoleptic score with a corresponding increase in acidity of fruits under ZECC [Ramkrishan and Godara, 1993]. Mean PLW (10.36 per cent) was recorded for Umran cultivar stored in ZECC after 12 days of storage which were organoleptically acceptable, while PLW was more than double (24.13 per cent) under ambient temperature [Fageria, 1999; Dhaka, 2000]. The significant reduction in PLW under cool chamber was due to prevailing higher humidity and lower temperature, which lowered the transpiration rate as well as ethylene production at the lower temperatures.

Cold Storage

Several studies have examined the effect of low temperature on postharvest changes in the chemical constituents of ber fruits and on their storage behaviour [Jawanda *et al.*, 1980; Al-Niami and Abbas, 1988; Al-Niami *et al*, 1989]. Jawanda *et al.*, observed that Umran and Sanaur-2 fruits could be stored for up to 30 and 40 days, respectively, in commercial cold storage (0–3.3°C). Jain *et al.*, 1981 reported that fruits stored in perforated polythene bags and baskets at 13°C in biochemical oxygen demand (BOD) incubators remained at acceptable organoleptic quality for up to 3 weeks. Panwar, 1981 stored Umran and Kaithli ber up to 42 days at 10 °C. The shelf-life of Gola and Kaithli cultivars of ber at 1.7 °C was found to be 42 and 28 days, respectively [Pareek and Gupta, 1988]. In cold storage (10°C, 79 per cent relative humidity), fruits of cultivars Gola, Kaithii and Umran remained acceptable for up to 42, 28 and 35 days, respectively [Anonymous, 1989]. The golden yellow coloured ripe fruits of Umran could be stored for about 3 weeks at low temperatures ranging from 0–4°C [Anonymous, 1990]. According to Monthira, 1982 fruits could be stored in perforated polythene bags for 8, 16 and 24 days at 15°C, 10°C and 5°C, respectively. Fruits stored at 5 °C lost only 48 per cent of their weight during the entire 12-week storage duration, while fruits stored at 22 and 15°C lost 70 and 75 per cent of their weight, respectively. At 3 weeks of storage more than 40 per cent of fruits had shrivelled under the 22°C and 15°C storage temperatures compared with only 3 per cent under the 5°C storage temperature [Tembo *et al*, 2008]. The difference in storage life seems due to variation in year of production, regions, orchard management practices, irrigation geometry, maturity stages, locations of the fruits and the time of harvesting fruits from tree, and the storage environment, etc.

Post-harvest Treatments

To extend the shelf-life and to reduce decay losses in storage ber fruits can be treated as follows (optional):

They can be dipped in cold water for 2 hours or exposed to cold air for 4 hours immediately after harvest to remove field heat.

They can be dipped in calcium chloride or ascorbic acid solution.

They can be treated with growth regulators (*e.g.* cycocel), waxed or fumigated.

They can be sprayed with fungicides, *e.g.* thiobendazole (at 500 ppm) or 0.2 per cent $ZnSO_4$. Fungicides should only be used provided cer- tain health and safety regulations are followed (see box below).

Handling of Ber Fruit after Harvest

1. Grading
2. Remove under-ripe, over-ripe, damaged and misshapen fruits. Grade the remaining fruits manually or by passing them through sieves into 2 or 3 levels on the basis of size and colour according to the grading standard below. Wash graded fruits using chlorinated water (100 ppm) and drain them for further processing or pack them for storing.

Grade	*Standard*
A	Shining yellow, large (>35mm) to medium size (25-35mm) fruits of uniform shape with no blemishes.
B	Uneven yellow or yellow red, large (>35mm) to medium size (25-35mm) fruits, of uniform shape with some blemishes.
C	Red, large (>35mm) to small (<25mm) fruits. Uneven yellow, small (<25mm) fruits.

Pack ber fruits either for proper storage or for safe transport to local or distant markets:

- ☆ In small packages of 1-2 kg in perforated 150 gauge polythene or nylon-net bags or cardboard cartons.
- ☆ In large packages of 10-20 kg in gunny bags, net bags, cloth packages or wooden or plywood boxes with holes or slits.
- ☆ For transportation, corrugated cartons of about 10 kg are the most suitable packaging material. For short distances, cheaper materials can be used if cushioning and ventilation is provided.
- ☆ Ber fruits should be packed in non- organic materials to avoid spoilage caused by microbes during storage and transport.

Physico-chemical characteristics of major ber cultivars are provided in Table 18.2. Extensive studies have been carried out using ber fruits to prepare various processed products, such as candy [Pareek, 1983; Gupta *et al.*, 1980; Gupta and Kadam, 1995], dehydrated products [Gupta and Kadam, 1995, Agrawal *et al.*, 1997], juice and wine [Gupta *et al.*, 1995; Khurdia, 1980], jam and jelly [Gupta *et al.*, 1981], and shreds and powder [Patil *et al.*, 1999]. Ber fruits are highly nutritious, rich in ascorbic acid and contain fairly good amount of vitamin A and B, minerals like calcium, phosphorus and iron. Ber fruits are also higher in calorific value and ascorbic acid content than the orange (Jawanda and Bal, 1978). Most fruits are seasonal and highly perishable and it is estimated that the total loss of fruits in India, for want of adequate post harvest care, transportation and storage is around 20-30 per cent (Madan and Ullasa, 1993).

It is a fruit of Indian origin, which finds a place in the ancient Indian scriptures (Parek and Vashistha, 1983). It is a rich source of ascorbic acid. Further, cost of its production is low and therefore, it is considered poor man's apple (Kudachikar *et al.*, 2000). Ber is consumed mainly as a fresh fruit and small quantities are processed. It can be processed into delicious products such as juice (Wasker and Garande 1999), RTS beverage (Khurdia 1980), chutney and pickles (Bal and Singh 1978), jam, jellies, dried fruit and powdered pulp (Sood *et al.*, 1980) and candy (Unde *et al.*, 1998). Dried bers can be used in the preparation of various Indian delicious like halwas, pinnies, laddoos, ice cream, pudding, fruit chats and fruit salads (Aggarwal *et al.*, 1997).

Composition of Fresh Ber Fruit

Table 18.2: Food Value/100 g Edible Portions

Constituent	Range
Moisture (per cent)	79-82
Starch (per cent)	0.72-1.15
TSS	12-21
Total sugars (per cent)	3.1-4.5
Reducing sugars (per cent)	1.4-9.7
Non-reducing sugars (per cent)	1.3-9.7
pH	4.2-4.9
Acidity (per cent)	0.13-1.42
Protein (per cent)	0.96-1.75
Total Ash (per cent)	0.34-0.45
Ascorbic acid (mg/100gm)	39-166

Uses

1. The fruit is eaten raw or pickled or used in beverages. It is quite nutritious and rich in Vitamin C.
2. It is second only to guava and much higher than citrus or apples. In India, the ripe fruits are mostly consumed raw, but are sometimes stewed. Slightly under ripe fruits are candied by a process of pricking, immersing in a salt solution.

3. Ripe fruits are preserved by sun-drying and a powder is prepared for out-of-season purposes. It contains 20 to 30 per cent sugar, up to 2.5 per cent protein and 12.8 per cent carbohydrates.
4. Fruits are also eaten in other forms, such as dried, candied, pickled, as juice, or as ber butter.
5. The dried ripe fruit is a mild laxative.
6. The seeds are sedative and are taken, sometimes with buttermilk, to halt nausea, vomiting, and abdominal pains in pregnancy. They check diarrhea, and are poultice on wounds.

Value Added Products and Processing

Ber fruits may be processed into several products, such as jam, jelly, pulp, juice, powder, dried ber, candy, slices, tutti-frutti and wine. The by-products of the processing industries may be utilized for the extraction of pectin.

Juice and Pulp

Juicy varieties of the ber are better suited for preparing the juice (Khurdya and Singh 1975). A process for the extraction of juice from the ber fruits has been standardized (Kadam *et al.*, 1991). The juice is extracted, allowed to stand at 4°C for two hours. After removal of the scum, juice is passed through four layers of the muslin cloth. Then, it is filtered through an ordinary filter paper. The recovery of the juice is about 40 per cent of the fresh weight.

FLOW-CHART FOR EXTRACTION OF PULP AND JUICE

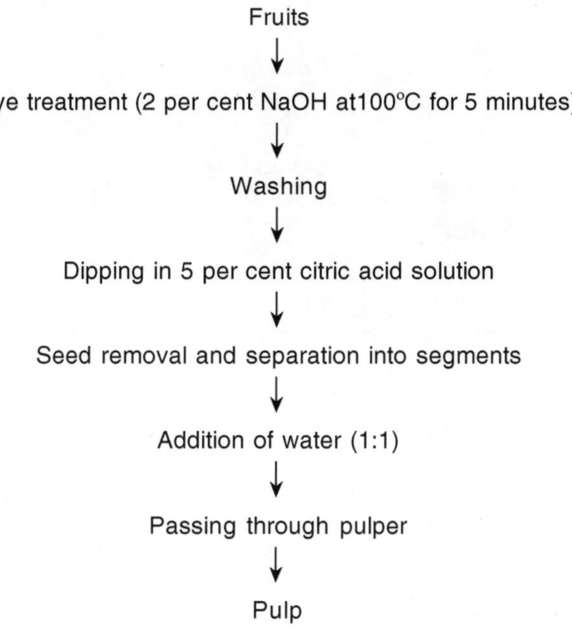

Ready to Serve

A good quality RTS may be prepared from ber juice at relatively low cost ber juice (10 per cent) can be used for ready to serve beverage (Kadam *et al.*, 1991).Carbonated beverage of ber was highly acceptable and has excellent keeping quality for three months (Khurdiya 1989 and Khurdiya 1990).

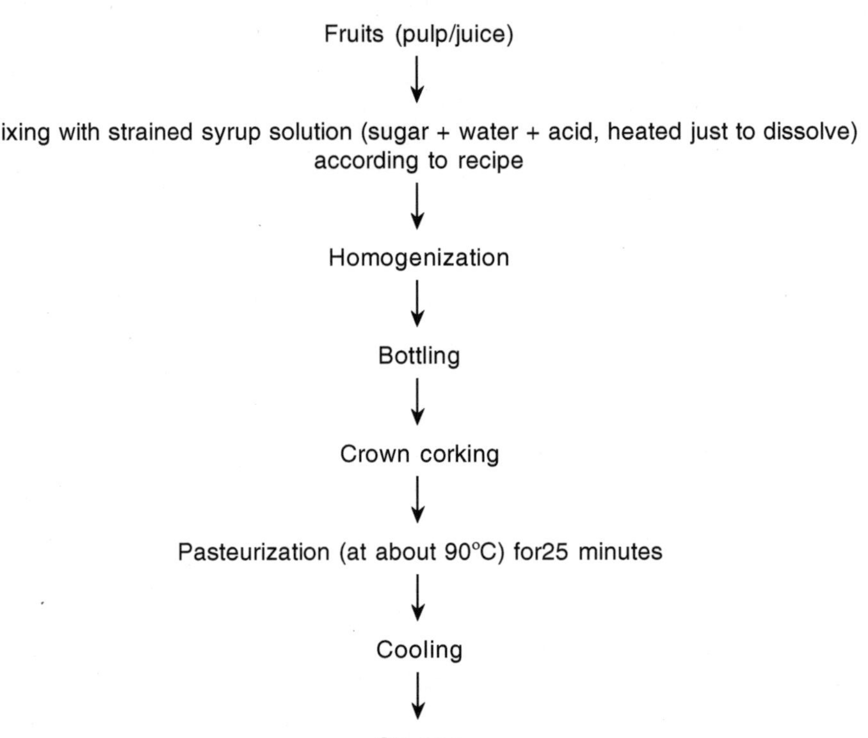

Dried Ber and Powder

Excellent quality ber was prepared by treating fruits with sulphur dioxide at 3.5-10gm/kg for 3 hours fallowed by drying in sun for 7-10 days or in a cabinet drier at 60°C for 20-35 hours until the moisture content reduces to 15 per cent (Khurdya and Singh 1975). The product may be consumed as such or it may be reconstituted in10 per cent sugar solution to be consumed as a liquid beverage. A process for ber fruits drying has been developed (Chavan *et al.*, 1992) from which powder has successfully been prepared.

TECHNICAL FLOW-CHART FOR PROCESSING OF DEHYDRATED FRUIT (BAEL)

Fruits (mature green)

↓

Washing

↓

Peeling (Remove the hard shell)

↓

Cut into1-1.5cm thick slices

↓

Fumigate the slices of fruit with sulphur dioxide fumes

↓

Dehydrated at 55-60°C in oven up to a constant weight

↓

Packed in polyethylene bags

↓

Storage at ambient temperature

TECHNICAL FLOW-CHART FOR PROCESSING OF FRUIT POWDER

Fruits (mature green)

↓

Washing

↓

Peeling/Remove the hard shell

↓

Cut into1-1.5cm thick slices

↓

Fumigate the slices of fruit with sulphur dioxide fumes

↓

Dehydrated at 55-60°C in oven up to a constant weight

↓

Grinding fruit slices

↓

Packed pulp powder in polyethylene bags

↓

Storage

Candy and Tutti-fruity

Earlier attempts to prepare the candy resulted in having dark brown coloured product (Gharate, 1984; Gupta *et al.*, 1980; Gupta *et al.*, 1981). Recently a method of preparing ber candy has been standardized to obtain a product with an attractive golden yellow colour and better storage life (Kadam *et al.*, 1991).

Wine

Ber juice can also be used for the preparation of wine. Kainsa and Gupta (1979) reported the preparation of wine from the ber pulp. Adsule *et al.* (1993) also reported that good quality wine prepared from ber juice.

TECHNICAL FLOW-SHEET FOR PROCESSING OF FRUIT WINE

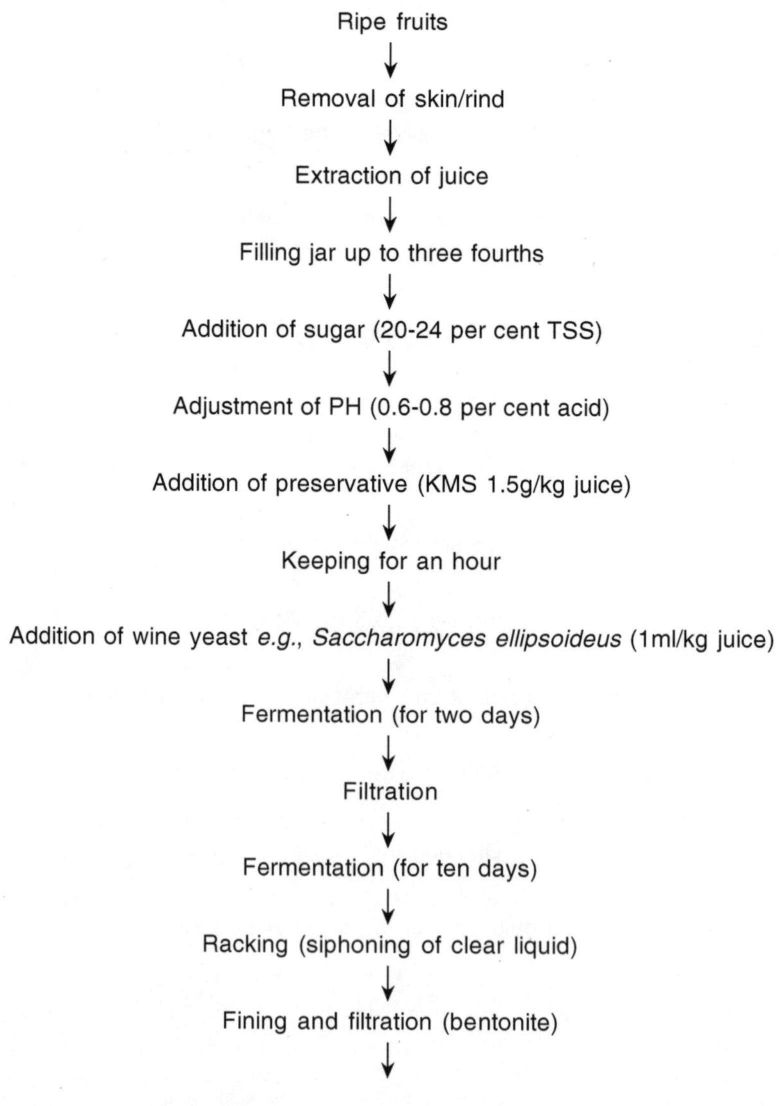

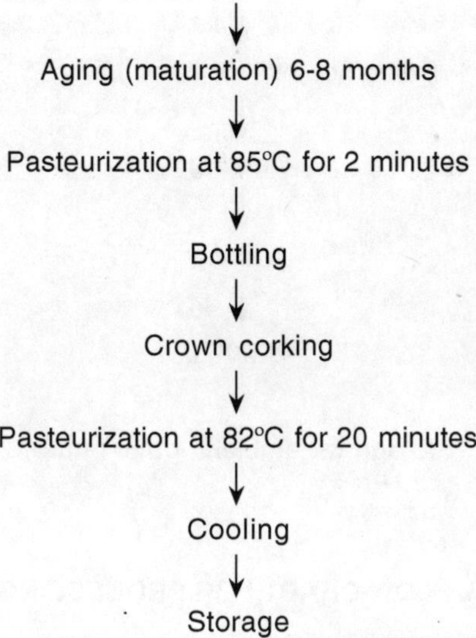

Jam and Jelly

An excellent jelly can be prepared from unripe ber fruits by adding proper amount of pectin and acid to this fruit or by mixing it with other fruits (Morton k. and J. Morton 1955).The jam prepared from ber fruits was leathery due to the presence of mucilaginous compound in the fruits.

TECHNICAL FLOW-CHART FOR PROCESSING OF JAM

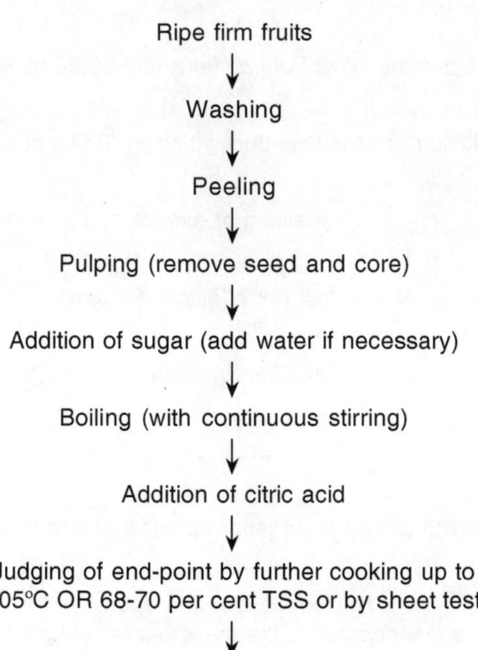

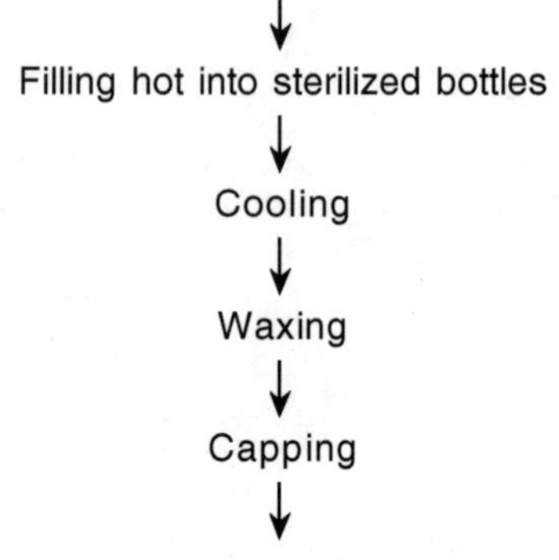

TECHNICAL FLOW-CHART FOR PROCESSING OF JELLY

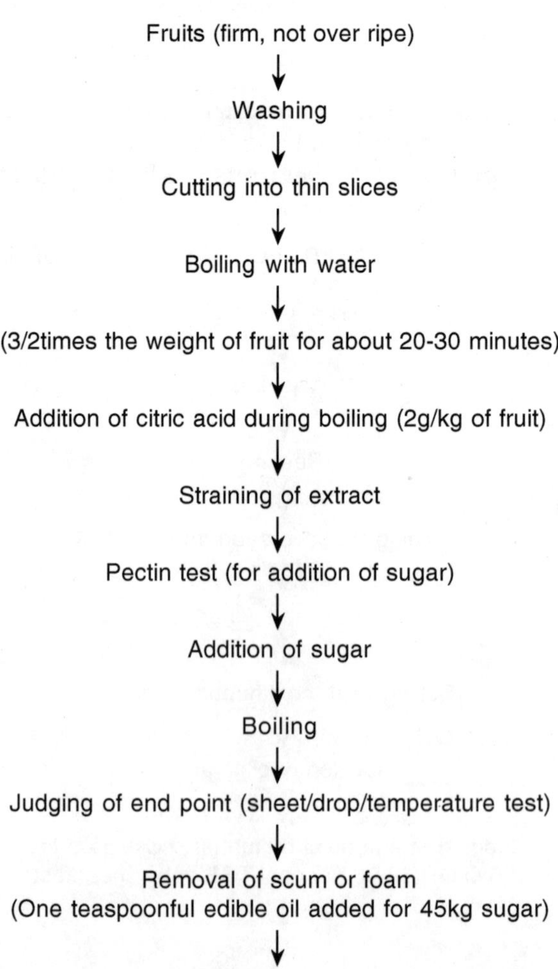

↓

Colour and remaining citric acid added

↓

Filling hot into clean sterilized bottle

↓

Waxing

↓

Capping

↓

Storage (at ambient temperature)

References

Abbas MF, Al-Niami, J. H. and Asker, M. A. (1990). The effect of temperature on certain chemical constituents and storage behaviour of jujube fruits (*Ziziphus spina christi* L. Willd.) cv. Zaytani. Haryana Journal of Horticultural Science. 19: 263–267.

Adsule, R.N., Chougule, B.A., Kotecha, P.M. and Kadam, S.S. (1993). Processing of ber II. Preparation of wine. *Beverage and Food World.*, **19**: 16.

Agarawal, P., Kaur, B. and Bal, J.S. (1997). Studies on dehydration of different cultivars for making ber chuharas. *J. Food Sci. Technol.*, **34**: 534-536.

Al-Niami, J. H. and Abbas, M. F. (1988). The effect of temperature on certain chemi- cal changes and the storage behaviour of jujube fruits (*Ziziphus spina- christi* L. Willd.). Journal of Horticultural Science. 63: 723–724.

Al-Niami, J. H., Abbas, M. F. and Asker, M. A. (1989). The effect of temperature on some chemical constituents and storage behaviour of jujube fruit cv. Zayoui. Basrah Journal of Agricultural Science. 2: 31–36.

Anonymous. (1990). Annual Report, Central Food Technological Research Insti- tute, Mysore, India.

Anonymous. Annual Progress Report. 1988–89. All India Coordinated Research Project on Postharvest Technology of Horticultural Crops. Indian Council of Agricultural Research, New Delhi, India. 1989.

Bailey, L. H. (1947). The standard cyclopaedia of Horticulture. Macmillan and Company, New York.

Bal, J. S. (1982). A study on biochemical changes during room and refrigerator storage of ber. Progressive Horticulture. 14: 158–161.

Bal, J. S. and Singh, P. (1978). Developmental physiology of ber (*Ziziphus mau- ritiana* Lamk.) var. Umran. I. Physical changes. Indian Food Packer. 32: 59–61.

Brandis, D. (1906). Indian trees. An account of trees, shrubs, woody climbers, bamboos and palms indigenous or commonly cultivated in British India. Bishen Singh Mahendra Pal Singh, Dehra Dun, India. : 169–172.

Chavan, U.D., Adsule, R.N. and Kadam, S.S. (1992). Processing of ber III. Preparation of ber powder and tutti-frutti. *Beverage and Food World.*, **20**: 28-29.

DGCIS -Directorate General of Commercial Intelligence and Statistics, Kolkata: Export Summary Data.

Dhaka, R. S., Lal, G., Fageria, M. S. and Agrawal, M. (2000). Studies on zero energy cool chamber for storage of ber (*Ziziphus mauritiana* Lamk.) fruits under semi-arid conditions. Annals of Arid Zone. 39: 439–441.

Fageria, M. S., Dhaka, R. S., Sharma, B. M. and Gujar, K. D. (1999). Effect of time of harvest on postharvest behaviour of ber fruits cv. Mundia under semi-arid conditions. In: Faroda AS, Joshi NL, Kathiya S and Kar A. (Eds.) Man- agement of Arid Ecosystem. Arid Zone Research Association of India and Scientific Publisher, Jodhpur, India. 365–368.

Gharte, A.N. (1984).Studies on preparation of candy from ber. M.Sc. (Ag.) Thesis submitted to MPAU, Rahuri, India.

Gupta, A. K., Panwar, H. S. and Vashishtha, B. B. (1983). Studies on physico-chemical changes during development and maturity in ber fruit cv. Gola. Punjab Horticulture Journal. 23: 186–190.

Gupta, O. P, Kainsa, R. L., Chauhan, K. S. and Dhawan, S. S. (1981). Post harvest studies in ber fruits (*Ziziphus mauritiana* Lamk.) IV. Comparison of sugar and gur for the preparation of candies. Haryana Agricultural University Jour- nal of Research.1: 369–392.

Gupta, O. P. and Kadam, S. S. (1995). Ber. In: Salunkhe DK and Kadam SS (Eds) Handbook of Fruit Science and Technology. Marcel Dekker Inc New York.: 387.

Gupta, O. P., Kainsa, R. L. and Chauhan, K. S. (1980). Postharvest studies on ber fruits (*Ziziphus mauritiana* Lamk.). 1. Preparation of candy. Haryana Agricultural University Journal of Research. 10: 163.

Jain, R. K., Chitkara, S. D. and Chauhan, K. S. (1981). Studies on physico-chemical characters of ber (*Ziziphus mauritiana* Lamk.) fruits cv. Umran in cold storage. Haryana Journal of Horticultural Science. 106: 141–146.

Jawanda, J. S., Bal, J. S., Josan, J. S. and Mann, S. S. (1980). Studies on the storage of ber fruits II. Cool temperature. Punjab Horticultural Journal. 20: 171– 178.

Jawanda, J.S. and Bal, J.S. (1978). The ber highly paying and rich in food value. *Indian Hort.* **23**: 19-21.

Johnston, M. C.(1972). Rhamnaceae. In: Milne RE and Polhill RM (Ed), Flora of Tropical East Africa. Crown Agents, London.

Kadam, S.S., Chavan, U.D. and Dhotre, V.A. (1991). Processing of Ber 1. Preparation of ready-to-serve beverage and candy. *Beverage and Food World.*, **18**: 13-15.

Kainsa, R.L. and Gupta, O.P. (1979). Post harvest studies on ber fruits (*Zyziphus mauritiana* Lamk.) II. Preparation of wine. *Haryana Agril. Univ. Res. J.*, **9**: 260.

Khera, A. P. and Singh, J. P. (1976). Chemical composition of some ber cultivars (Ziziphus *mauritiana* L.). Haryana Journal of Horticultural Science. 5: 1: 21–24.

Khurdia, D.S. (1980). Studies on dehydration of ber (*Zyziphus mauritiana* Lam) fruit. *J. Food Sci. Technol.* **17**: 127-130.

Khurdiya, D. S. and Singh, R. M. (1975). Ber and its products. Indian Horticulture. 20: 5, 25.

Khurdiya, D.S. (1989). Carbonation in fruit beverages 1. *Beverage and Food World.* **16:** 9.

Khurdiya, D.S. (1990). A study on fruit juice based carbonated drink. *Indian Food Pack.*, **44:** 45.

Kudachikar, V. B., Ramana, K. V. R. and Eipeson, W. E. (2000). Pre and post-harvest factors influencing the shelf life on ber (*Ziziphus mauritiana* Lamk.): A review. Indian Food Packer. Jan.–Feb.: 81–90.

Lin, M. J. and Cheng, C. Y. (1995). A taxonomic study of the genus Ziziphus. Acta Horticulturae. 390: 161–165.

Madan, M. and Ullasa, B.A. (1993). Post harvest losses in fruits. *Advances in Horticulture* (eds. Chadha, K. L. and Pareek, O.P.), Malhotra Publishing House, New Delhi, **4:** 1795-1810.

Monthira, S. (1982). Effect of low temperature and packaging material on storage life of Bombay jujube (*Ziziphus mauritiana* Lamk.). Bangkok, Thailand.

Morton, K. and Morton, J. (1955). Fifty tropical fruits of Nassau. Text House, Coral Gables, New Delhi, India.

Panwar, J. K. (1981). Postharvest physiology and storage behaviour of ber fruit (*Ziziphus mauritiana* Lamk.) in relation to temperature and various treat- ments. Thesis Abstracts, Haryana Agricultural University, Hisar. 7: 64–65.

Pareek, O. P. (1983). The Ber. Indian Council of Agricultural Research. New Delhi, India.

Pareek, O. P. (2001). Ber. International Centre for Underutilized Crops, Southampton, UK : 13.

Pareek, O. P. and Gupta, O. P. (1988). Packaging of ber, datepalm and phalsa. In: A souvenir on packaging of fruits and vegetables in India. Agri-Horti Soci- ety, Hyderabad, India. 91–103.

Pareek, O. P., Bhargava, R., Vishal Nath and Vashishtha, B. B. (1999). Changes in metabolite composition in postharvest treated ber fruits under ambient and cool storage environment. 4th Agricultural Science Congress, Jaipur, India. February 21–24, 212.

Pareek, S., Fageria, M. S. and Dhaka, R. S. (2002). Performance of ber genotypes under arid condition. Current Agriculture. 26: 63–65.

Parek, O.P. and Vashistha, B.B. (1983). Delicious ber varieties of Rajasthan. *Indian Hort.*, **28**: 13-16.

Patel, R.K., Singh, A., Yadav, D.S. and Des, L.C. (2008). Underutilized fruits of North Eastern Region, India. Underutilized and Underexploited Horticultural Crops, (ed. K.V. Peter), **4**: 223-238.

Patil, D. M., Katecha, P. M. and Kadam, S. S. (1999). Drying of ber preparation of shreds and powder. Processed Food Industry August. : 14–15.

Ramkrishan, N. R. and Godara, R. K. (1993). Physical and chemical parameters as affected by various storage conditions during storage of Gola ber (*Ziziphus mauritiana* Lamk.) fruits. Progressive Horticulture. 25: 60–65.

Roy, S. K. and Khurdiya, D. S. (1986). Studies on evaporative zero energy input cooled chambers for storage of horticultural produce. Indian Food Packer, 40: 26–31.

Sebastian, M. K and Bhandari, M.M. (1990). Edible wild plants of the forest areas of Rajasthan, India. Journal of Economic and Taxonomic Botany, 14: 689–694.

Sood, D.R., Wagle, D.S., Nainawatee, H.S. and Srivastav, H.C. (1980). Quality attributes of some bers (*Ziziphus jujuba*) strains. *Indian Nutr. Dietet.*, **17**: 447-451.

Tembo, L., Chiteka, Z. A., Kadzere, I., Akinnifesi, F. K. and Tagwira, F. (2008). Storage temperature affects fruit quality attributes of ber (Ziziphus *mauritiana* Lamk.) in Zimbabwe. African Journal of Biotechnology. 7: 8: 3092– 3099.

Unde, P. A., Kanwade, V. L. and Jadhav, S.B. (1998). Effect of syruping and drying methods of quality methods of ber candy. *J. Food Sci. Technol.*, **35:** 259-261.

Vishal Nath, Bhargava, R. and Pareek, O. P. (1999).Improvement in shelf life of ber (*Ziziphus mauritiana* Lamk.) by postharvest treatments. 4th Agricultural Science Congress, Jaipur, India. February 21–24, 210.

Wasker, D.P. and Grande, V.K. (1999). Standardization of a method of juice extraction from ber fruit. *J Food Sci. Technol.*, **36**: 540-541.

Yamadagni, R. (1985). Ber. In: *Fruits of India, Tropical and Subtropical* (ed. By Bose, T.K.), Naya Prakash, Calcutta, India.

Index

N

O

P

W

Y

Z